Environmental Geography

Environmental Geography

Third Revised Edition

H.M. SAXENA

RAWAT PUBLICATIONS

Jaipur • New Delhi • Bangalore • Guwahati • Kolkata

ISBN 978-81-316-0848-7 (HB)

First Edition, 1999
Second Edition, 2004
Third Revised Edition, 2017

Reprinted, 2023

Published by
Prem Rawat for **Rawat Publications**
Satyam Apts, Sector 3, Jawahar Nagar, Jaipur 302 004 (India)
Phones: 0141 265 1748 / 265 7006
E-mail: info@rawatbooks.com
Website: www.rawatbooks.com

New Delhi Office
4858/24, Ansari Road, Daryaganj, New Delhi 110 002
Phone: 011 2326 3290

Also at *Bangalore, Guwahati* and *Kolkata*

Typeset by Rawat Computers, Jaipur
Printed at Nice Printing Press, New Delhi

Contents

List of Figures

Preface

Geography, right from its beginning, deals with the man-environment relationship in spatial context, and today it has become an interdisciplinary science concerned with the surface of earth and its natural environment as well as the human intervention, which has changed the physical and cultural landscape over the years. *Environmental Geography* is, therefore, a close analytical scrutiny of the various facets of environment, its degradation as well as its management.

The first two editions of the book have been received very well by students and scholars for which I am grateful to them. This is the third edition, which has been thoroughly revised and updated. For the updated edition several new topics have been added in order to keep the text contemporary. Apart from new additions almost in all the chapters, four new chapters have been incorporated on Biomes, Biodiversity, Towards Sustainability and Climate Change.

First four chapters are related to the concepts of environment, ecology, ecosystems and biomes. Human impact on the natural environment and degradation of environment are the themes of chapter five and six, while chapters seven to eleven are related to various types of pollution, environmental hazards/disasters and population and ecological crisis. Conservation of natural resources, management of environment and environmental impact assessment have been discussed in chapters 12, 13 and 14 respectively. The two major environmental issues, i.e. biodiversity and climate change have been analysed in detail in chapters 15 and 16 followed by two chapters on sustainable development. Governance and geo-politics of the environment have been discussed in chapter 19 while global environmental issues and environmental problems and planning in India have been examined in chapters 20 and 21 respectively. The last chapter is devoted to environmental ethics and legislations in India.

Environmental geography is a fast growing branch of the subject, so its contents are also varied and changing according to time. In this book, the author has tried to incorporate as much information as possible. The book will be useful not only to geography students but also to the students of environmental science and management. Topics such as ecology, biodiversity, climate change, sustainable development, etc. are also part of the syllabus of civil services and other competitive examinations.

Many scholars and friends have had a role, either direct or indirect, in writing this book. Acknowledgements have been made throughout the book as a mark of gratitude towards the scholars whose works I have consulted to clarify much of what I wanted to say.

My wife, Nirmal, as ever, has been my greatest strength and inspiration. Without her active cooperation as well as the support of the other members of my family, this book would have been much harder to write. To all these and others unnamed, I sincerely say, 'Thank you'.

H.M. Saxena

1

Introduction: Environment, Ecology and Geography

Environment is the source of life on earth and it not only directs but also determines the existence, growth and development of mankind and all its activities. Primitive man ate wild fruits – hunted and fished – and relied heavily on nature for his very existence. Even these basic activities damaged nature to some extent. At that time, man did not concern himself with conservation methods, in other words, there was no need for it because nature was able to cope with whatever damage was inflicted by quickly restoring its resources.

As society developed, man's impact on environment grew in scope and strength. Until very recently, we adhered to the following dictum: 'we cannot expect favours from nature; we must take them'. For quite a long time, no thought was given to the possible consequences of such 'taking' from nature. But, the consequences were significant and pervasive. Mankind has long trusted in nature's potential and restorative powers, though there was no reason to believe that these powers were inexhaustible. Only recently, man has come to realize the necessity for conservation. Undoubtedly, this is the result of revolution in science and technology which has dramatically increased man's ability to use natural environment and its resources. Nature has been increasingly damaged, restorative capabilities have been progressively weakened, and human environment is deteriorating day by day to the point of affecting not only the quality of life but even the very existence of life. At this point, man realized the need to preserve and improve his environment, which is now shown as the ecological approach.

The human environment is the earth we live on. It includes all the physical parts of the earth, such as air, soil, minerals, rocks and water, and all its living organisms, such as animals and plants. Environmental science provides an approach towards understanding the environment of our planet and the impact of human life upon that environment. It is also a search for solutions to the environmental problems that confront us. In fact, environmental science is a vast subject, including all those branches of knowledge which deal with ecology as an interdisciplinary science drawing upon subjects like zoology, botany, geography, social and natural sciences. The study of environment basically deals with the study of 'environment' and 'ecology'. Therefore, it is necessary to discuss the various aspects of environment and ecology. Among other branches of knowledge, the status of geography is unique because of its interdisciplinary nature. Basically, geography is the study of the interrelationship between man and environment. This aspect is of prime importance in present times, because, while other disciplines deal with specific aspects of environment, such as forest, soil, air, water or even aspects like social, economic, demographic, etc., geography deals with the environment in its totality. It also analyses the 'time' and 'space' relationship between man and environment.

Geography, right from its beginning, deals with various components of the environment, i.e. relief, climate, natural vegetation, soil, water bodies, etc. not only in isolation but in their relation to man and his activities. The concept of man-environment relationship has been developed during the initial stages of the development of geography and still forms the basis of geographical knowledge and research. It was during 1950 to 1970, that there was a shift in geographical studies when more emphasis began to be given to quantitative and locational analyses. But soon, the situation changed and it was felt throughout the world that geographers can contribute a great deal towards environmental studies. With the result, the branches like 'Geography of Environment', 'Ecological Geography' and 'Geography of Environment and Ecology' have been developed. The present book is an introduction to various facets of the geographical study of environment and ecology. Prior to other discussions, it is necessary to understand the nature of environment and ecology.

Environment

The term 'environment' etymologically means surroundings. Literally, it is an English word formed by two words, i.e. 'environ' and 'ment' which means 'encircle' or 'all round'. Thus, environment is a complex of many variables which surrounds man as well as all living organisms. Any external force, substance or condition, which surrounds and affects the life of the organism in any way becomes a factor of its environment. These factors have been variously called as environmental factors, ecological factors or simply, factors and may be living biotic, as well as non-living or abiotic. The sum of all these living and non-living factors makes the environment of an organism. The place where an organism lives is called 'habitat'. It is also known as 'milieu' which means 'total set of surroundings'. In fact, habitat is a part of the earth's surface where life originates and develops according to the geographical conditions. Moore (1967), in his *A Dictionary of Geography*, has defined habitat as 'the natural environment of a plant or animal. It is often expressed as one of the main natural regions or one of their sub-divisions, which are recognized in geography as equatorial forests, the tundra'. According to Goudie (1994), 'environment is the representative of physical components of the earth'. The encyclopaedic definitions of environment are:

The sum total of all conditions, agencies and influences which affect the development, growth, life and death of an organism, species or race. (*The Universal Encyclopaedia*)

The entire range of external influences acting on an organism, both the physical and biological (in other organisms) forces of nature surrounding an individual. (*Encyclopaedia Britannica*)

Components of Environment

Environment is a comprehensive term which refers to our surroundings, thus includes all the abiotic and biotic components. The four basic components of the environment are – atmosphere, hydrosphere, lithosphere and biosphere.

1. *Atmosphere:* The atmosphere forms a protective shell over the earth and is a composition of gases, dust particles and water vapours. It sustains life on the earth by providing oxygen for human respiration, wild fauna in natural ecosystems and domestic animals. It also protects us from hostile environment of outer space as well as from ultraviolet rays. The atmosphere is composed of nitrogen, oxygen, argon, carbon dioxide and trace gases.
2. *Hydrosphere:* The hydrosphere covers three-quarters of the earth's surface. A major part of the hydrosphere is the marine ecosystem in the oceans, while only a small part is freshwater. The freshwater in rivers, lakes and glaciers is perpetually being renewed by a process of evaporation and rainfall whereas some of this freshwater is stored in underground aquifers. Only about one per cent of available freshwater is fit to be used for human consumption and other use.
3. *Lithosphere:* Lithosphere is the surface of the earth. It consists of soil, minerals, organic matter, etc. The soil cover is the basis for agriculture, while a large number of minerals provide raw materials to industries. On the other hand, oil, coal and gas are extracted from underground sources.
4. *Biosphere:* This is relatively a thin layer on the earth in which life can exist. Biosphere comprises living organisms and faunal environments and their interaction with environment – atmosphere, hydrosphere and lithosphere as depicted in Figure 1.1.

Figure 1.1

Physical Environment and its Components

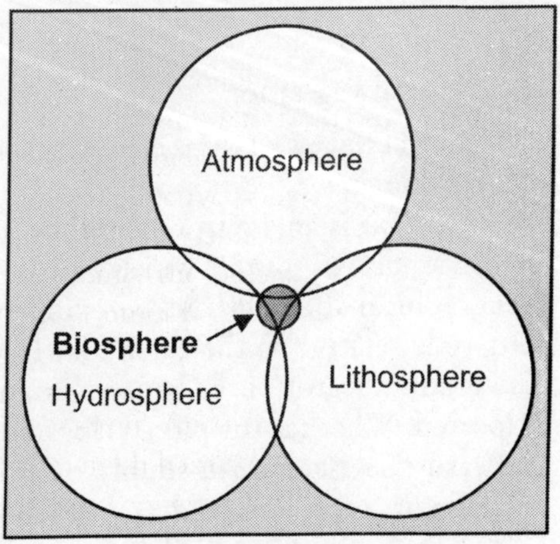

Environmental Factors

As described, environment is a complex phenomenon of several factors which are the products of the forces and processes of nature, and, according to White and Renners, are universal and inherent all over the earth. These factors are classified into two groups as under:

1. *Direct factors*: These are light, temperature, soil, air, water, soil nutrients, etc.
2. *Indirect factors*: These are soil structure, soil organisms, altitude, wind, slope, etc.

According to Oosting (1948), the environment is a complex of variable factors or causes, which include: (a) substances (soil, water), (b) conditions (temperature, light), (c) forces (wind, gravity), (d) organisms (plants, animals), and (e) time. Another ecologist Daubenmire (1959) classified these factors into seven coordinate headings as: (a) soil, (b) water, (c) temperature, (d) light, (e) atmosphere, (f) fire, and (g) biotic factors. Others have grouped these elements as follows:

1. climatic or aerial factors,
2. topographic or physiographic factors,
3. edaphic factors, and
4. biotic factors

While studying environment, geographers have also considered the above mentioned factors. These factors are termed as 'physical factors'. The natural environment of any part of the earth's surface can be classified into following fourteen elements:

Abstract Elements

1. Location
2. Situation
3. Geometrical position
4. Size or aerial space
5. Form

Physical Elements

1. Physiography/landforms
2. Climate
3. Rocks and minerals
4. Soils
5. Surface waters of the land
6. Underground water
7. The ocean and its coast

Biotic Elements

1. Natural vegetation
2. Animal life

A brief review of these elements is necessary for the proper understanding of environment.

Location is an environmental factor which indicates relative occurrence in space or, in other words, it is an indicator of space relationship. Geographical location can be expressed in absolute terms of latitude and longitude relative to the earth's surface as a whole. The latitudinal location also is an indicator of the broad climatic zone in which a specified area/place is located. Whether the place is located in tropical, temperate or cold region, can easily be ascertained simply by its latitudinal location. Apart from this, location in terms of land and sea is also a factor which affects the local environment. This is what is known as *situation* and it means the occurrence of a place with regard to the fixed features of nature. The basic types of situation are: (i) *continental situation*: (a) intermontane situation, (b) midland situation, (c) piedmont situation, and (d) river plain situation; and (ii) *maritime situation*: (a) littoral situation, (b) peninsular situation, (c) insular situation, (d) pelagic situation, and (e) isthmian situation. All these situations have had their impact on the environment in one way or the other. Similarly, *size*, i.e. aerial space, is also an environmental factor which controls the climate as well as regional environment.

Relief or landform is a main physical factor which affects the environment variously. The impact of relief can be seen on climate, natural vegetation, soils as well as on settlement of man and economic development. Because of favourable relief, the river valley civilizations have developed in China, India, Mesopotamia, and even today relief is a controlling environmental factor for all types of economic development. The various forms of landforms developed in areas like hilly, plain, plateau, arid, glacial, coastal, limestone, etc. regions have had specific natural environment and ecological settings. Mountains are least hospitable to human cultures. The topographical effect on environment can be seen on precipitation, development of snow fields and glaciers, rate of erosion, variations in vegetation types, soils, etc.

Climate of a region is determined by such methodological factors like temperature, pressure, winds, relative humidity, rainfall, evaporation rate, etc. These factors are concerned mainly with the aerial environment of an organism. Climate is not only directed but determines the entire ecological balance of a region. The other environmental factors like natural vegetation, soil, water bodies as well as man's activities are in one way or the other related with climate. The various forms of ecosystem, such as desert, coastal, and glacial are the products of climate. The climate is highly affected by man's activities in the form of pollution, which has now become a global problem. The direct effects of temperature and rainfall can be seen on vegetation and animals.

Soil is also an important environmental factor. Plants depend for their nutrients, water supply and anchorage upon the soil. Soil system is indeed very complex and dynamic, undergoing continuous change, and the rate of such change being influenced by a number of other factors of the environment. Soil formation depends upon parent material, climate, vegetation, relief and time. A properly managed soil is an asset. If it is misused, the soil can be permanently damaged and will create environmental degradation.

Natural vegetation is an index of other environmental factors such as climate, soil, landforms, while by itself, it is a factor of environment. It controls the temperature and helps in rainfall by increasing humidity. It is also a controller of soil erosion. Forests are assets to mankind and are the natural resources of national importance. Great damage has been done to forests after the Industrial Revolution and now the world is facing several problems due to deforestation which has become the main cause of environmental degradation both in developed as well as developing countries.

Among other environmental factors, water bodies and animals are also important. All these factors of the environment are not independent but are linked with each other in such a way that none can be separated from the other. Therefore, the environment should be taken into consideration in its entirety. The total impact of environment is reflected in all man's developmental activities – economic, social, political or cultural.

Ecology

Ecology is the science of the relations of all organisms to their environment. The two components of nature, organisms and their environment are not only complex and dynamic but also interdependent. Ecology is a relatively new science, dealing with the various principles which govern such relationships between organisms and their environment.

In 1859, St. Hilaire Isidore Geoffory, a French zoologist, proposed the term 'ethology' for 'the study of the relations of organisms within the family and society in the aggregate and in the community'. Similarly, a British naturalist, St. George Jackson Mivart, used the term 'hexicology', which he defined in 1894 as 'devoted to the study of the relations between the organisms and their environment as regards the nature of the locality ...' The term 'ecology' was coined by combining two Greek words, viz. 'oikos' (meaning 'house' or 'dwelling place') and 'logos' (meaning 'the study of') to denote the relationships between the organisms and their environment. Although there is some controversy about the original coining of the term but there is a consensus that German biologist, Ernst Haeckel first gave substance to this term. He used this term first in 1886 but defined for the first time in 1870 as follows: 'By ecology we mean the body of knowledge concerning the economy of nature – the investigation of the total relations of the animal both to its inorganic and to its organic environment, including above all, its friendly and inimical relations with those animals and plants with which it comes directly and indirectly into contact.'

Now, ecology is a systematic and well-developed science. Some of the definitions of ecology are as follows:

> Ecology is defined as the study of the relations of organisms, or groups of organisms to their environment, or the science of the interrelations between living organisms and their environments. (Odum)

> Ecology is the science of all the relations of all organisms to their environments. (Taylor)

> Ecology is the science which treats the organisms in relation to their environment. (Moore)

> Ecology is the science dealing with the relationship of organisms to one another and to other factors that comprise their environment. (*Encyclopaedia Britannica*)

> Ecology is the study of plants and animals in relation to their environment. (Haggett)

Ecology is the science of the mutual relationship of organisms to their environment. (Monkhouse and Small)

It becomes clear from the above definitions that ecology is the study of the interrelationship between all living organisms and environment. All its components, i.e. energy, air, water, soil, vegetation, animals as well as man have mutual relations, responsible for the development of a particular ecological system. The interrelationship of ecological elements has been depicted in Figure 1.2.

Figure 1.2
Ecological Relationship

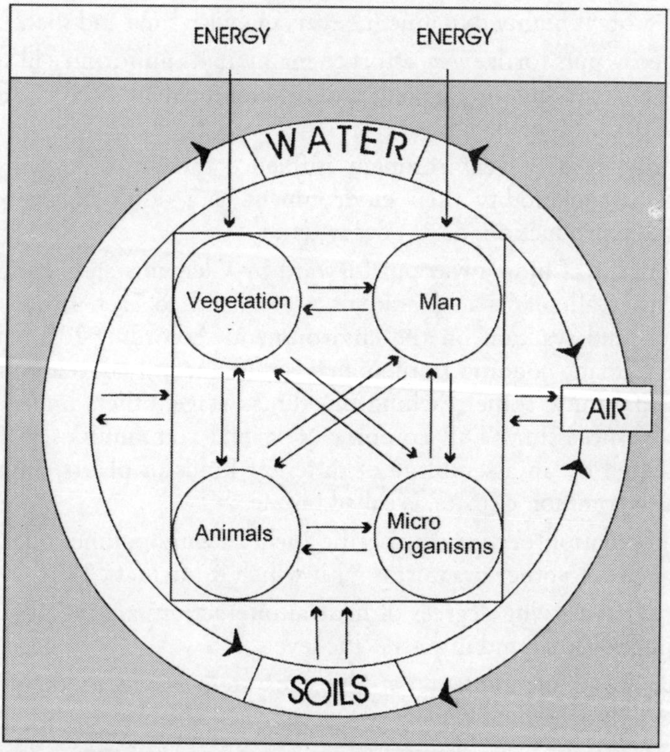

All the elements – biotic and abiotic – are not only interrelated but both these components function in an orderly manner as a definite system. Thus, 'structure and function' provide a base of study of ecosystem. The spectrum of modern ecology can also be defined in terms of levels of organization. It begins with the groups of organisms, up to ecosystem level. In ecology, the term 'population' is used to include groups of individuals of any one kind of organism. Likewise, community or biotic community in the ecological sense includes all the populations occupying a given area. The community and the non-living environment function together as an ecological system or ecosystem. The present status occupied by ecology is quite evident by its role in the solution of socio-economic and environmental problems of pollution. Its role in managing water, soil, forest and land resources as well as in agricultural development and conservation of resources has become important. The international problems of environmental pollution also needs an ecological approach. In fact, sustainable development is nothing but development within the framework of an ecosystem.

Basic Concepts of Ecology

The basic concepts of ecology are as follows:

1. All living organisms and their environment are mutually reactive, affecting each other in various ways.
2. The environment, which is a complex phenomenon of several interrelated factors is highly dynamic (i.e. varying with time and space).
3. The species puts forth every effort to maintain its uniformity in structure, function, reproduction, growth and development by preservation of its genetic pool.
4. Not only does the environment influence the life of organisms, but organisms too modify their environment as a result of their growth, disposal, reproduction, death, decay, etc.
5. The concept of biome was put forward by Clements and Shelford, who stated that all plants and animals are related to each other by their coaction and reaction on the environment. According to them, under similar climatic conditions, there may simultaneously develop more than one community, some reaching the climax stage, others under different stages of succession. This complex of several communities in any area, represented by an assemblage of different kinds of plants, animals, etc. sharing a common climate, is called biome.
6. Each population occupies a specific niche, a unique functional position with respect to other organisms with which it interacts.
7. There exists varying degrees of neutral interactions among organisms, at both inter-specific and intra-specific levels.
8. The chemical components of the ecosystem move in defined cycles known as biochemical cycles.
9. The successful growth of an organism is governed by limiting factors. The minimal and maximal levels of tolerance for all ecological factors of a species vary seasonally, geographically and according to the age of the population.
10. The exploitation of ecosphere by man's activities is the main cause of the disruption of the ecosphere. It leads to a setback of the state of development and reduction in the stability of the ecosystem.

Main Sub-divisions and Specialized Fields of Ecology

On the basis of the approach of study, the main sub-divisions of ecology can be summarized as follows:

Main Sub-divisions and Specialized Fields of Ecology

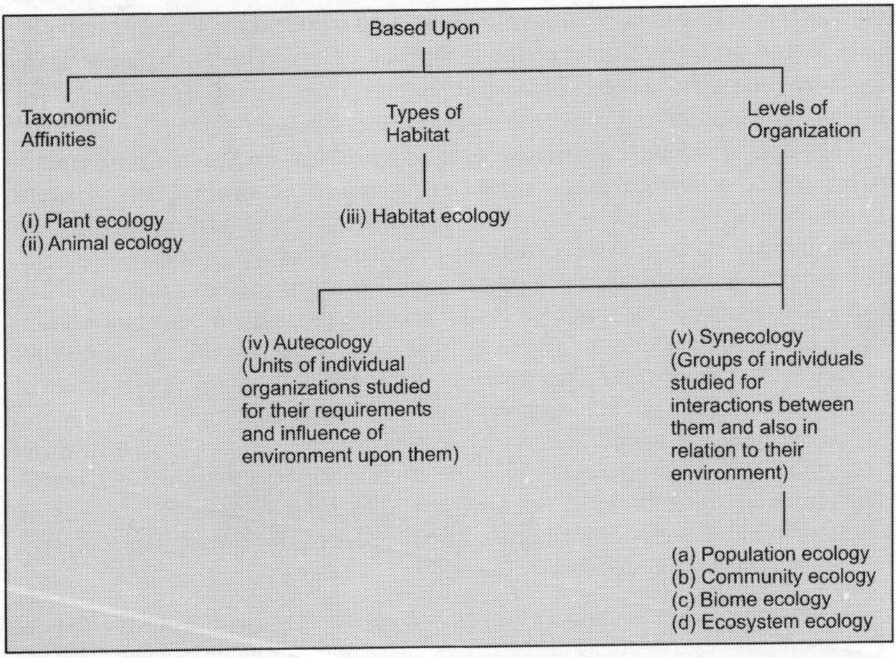

The scope of ecology has widened and with the introduction of human welfare approach (applied ecology), several specialized fields have been developed. The recognized specialized areas of ecology are as follows:

1. Fresh water ecology
2. Marine ecology
3. Estuarine ecology
4. Terrestrial ecology
5. Paleocology
6. Zoogeography
7. Phytogeography
8. Statistical ecology
9. Systems ecology
10. Space ecology
11. Generecology
12. Production ecology
13. Radiation ecology
14. Ecological energetics
15. Applied ecology
16. Pollution ecology
17. Conservation ecology
18. Human ecology

Geography of Environment and Ecology

Geography and ecology are complementary to each other. Today, ecology is an interdisciplinary science, drawing upon various disciplines like zoology, botany, geography and other natural and social sciences. Geography, in its broad spectrum, is a group of sciences concerned with the surface of the earth, including the distribution of life upon it, its physical features, the evolution of landforms and life, the present and the past climate and the human intervention which has changed the physical and cultural landscape over the years. Similarly, the study of environment is a primary concern of geography.

Geographers not only analyse the environment and its various components in isolation but deal with environmental impact on man and his activities and vice versa. Man is an integral part of the biosphere today. With a myriad of activities, born out of the intellect of the human race, man has radically changed the natural environment and the ecosystem existing therein.

The spatial aspect is associated with geographical studies of environment. In fact, every ecosystem has its own geographical conditions, which needs complete analysis and this aspect provides a base for geography of environment. Today, terms such as 'human ecology', 'social ecology', 'ecological technology' and 'ecological equilibrium in the environment' have sprung into existence to connote man's relation to environment and conservation of nature, which is also a primary objective of the geography of environment. Moore (1967) has observed that 'ecology lies on the frontiers of so many other subjects, including various branches of geography'.

Geography is a science, analysing spatial distribution based on 'cause and effect'. The geographical data are very useful for effective analysis of the environmental and ecological aspects (now GIS or Geographical Information System provides detailed information for any part of the world). Gerasimov, a Russian geographer, has observed:

> Geography has accumulated a huge volume of information on the natural conditions and resources of the entire world and on the degree and forms of their development and economic use. It is no less important that in handling these new formidable tasks geography can rely on the complex system of its branches. (*Geography and Ecology*, 1983)

Strahler and Strahler have discussed the geographical aspects and its relationship with environment in their book *Geography and Man's Environment* (1977) and have observed:

> The phrase 'man and the land' is often used to convey the essence of geography. While the relationship of humans to their earthly environment is a core concept in geography, an equally important idea is that the relationship has a certain distribution in space. Geographers recognize that the quality of life layer varies from place to place in terms of richness or poverty of life forms capable of being supported.

Ackerman (1965) has stated that 'its goal is nothing less than an understanding of the vast, interacting system comprising all humanity and its natural environment on the surface of the earth'. Savindra Singh has stated that 'environmental geography may be defined as the study of spatial attributes or interrelationships between living organisms and natural environment in general and between technologically advanced "economic man" and his natural environment in particular'. At the time of the 27th IGU in 1992, preliminary *Bulletin of Geography* in *Discovery* states: 'Geography is the science of space and place. Its subject matter is the natural and human phenomena that make up the world's environment and places.'

In fact, geography's integrating concepts and processes concern the whole world ecosystem of which man is the dominant part. In *Geography and Ecology* (1983), Gerasimov says: 'The concept of close mutual links and interactions of all environmental components serves as a scientific basis for constructive geographic research.' Gerasimov has observed some new themes in geo-ecological field, such as:

1. increasing the effectiveness of territorial organization of social production;
2. further discovery of natural resources needed for society's development;
3. goal-oriented transformation of the environment;
4. planned progressive improvement of the general quality of the environment;
5. development of new natural-technical structural complexes;
6. environmental monitoring and its rational organization with systematic observation of and control over the environment and its man-induced changes;
7. global anthropogenic and biospherical monitoring for control over global changes of the entire environment;
8. the development of scientific (geographic) forecasts of anthropogenic changes in the environment; and
9. optimizing the role of the environment in the mixed natural-technological systems.

An effective protection and improvement of the environment requires a planned as well as rational utilization of natural resources. Therefore, geographical studies of environment and ecology are necessary not only for the knowledge and utilization of resources but also for their planning and conservation or in other words for the sustainable development. The kaleidoscopic changes in the ecological patterns of the earth which are going on today in the era of science and technology need a careful system analysis. The progress of mankind has to be examined in the context of ecological preservation.

Scope and Importance of the Geography of Environment

Environmental geography has emerged as a specialized branch of geography, which is multi disciplinary in nature. On one hand, it is related with other subjects like – biology, chemistry, physics, sociology, economics, public administration, management, etc. while on the other hand, it is closely linked with all other branches of geography, both physical and human. The geographical studies of environment are concerned with ecosystems of the world and their habitat, economy and society, degradation of environment, environmental management as well as planning. Environmental geography is basically the study of spatial attributes of interrelationships between biotic and abiotic components and between man and natural environment. The physical components, i.e. land, air, water and energy are closely related to geography, both physical (geomorphology, climatology, soil geography, etc.) and human (cultural, economic, social, etc.). In fact, geography is the only discipline that pursues the study of environment in its totality.

The scope of environmental geography is widespread because it deals with all the aspects of environment and its relation with man. The scope and importance of environmental geography is evident from the following points:

1. The core of environmental geography is the man-environment relationship. It explains the role of physical factors in shaping the human activities as well as the human impact on environment. This relationship varies in different geo-economic and ecological regions and is responsible for the present stage of development. Thus, environmental geography helps in understanding the role of physical factors in development and planning.
2. Like other geographical studies, environmental geography is concerned with spatial attributes of all the phenomena related to environment. The spatial aspects of relief, climate, soils, natural vegetation, water bodies, pollution, natural hazards, biodiversity, resources, etc. are the subject matter of the geography of environment, which not only explains the present pattern of their distribution but also helps in their conservation and planning.
3. The study of ecosystems has been done both in biology and environmental geography. The interactions and mutual interdependence of abiotic and biotic components of the biosphere are studied in both the subjects. But higher order system often termed as 'geobiosystem' as spatial unit of various dimensions (global, continental and regional) falls within the realm of environmental geography.
4. Basically, environmental geography is concerned with the life supporting layer of the earth having atmospheric, lithospheric and hydropheric components, which is responsible for the support of all types of life. In other words, the study of biosphere is the main concern of environmental geography.
5. The environmental geography deals with the spatio-temporal relationship of man-environment. The changing nature of man-environment relation not only explains the growth of civilization but also the role of scientific and technological development in this relationship. The concepts like determinism, possibilism, neo-determinism, probablism and voluntarism have been developed in geography, which shows the temporal aspect of this relationship. Similarly, man-environment relationship in various economies, i.e. developed and developing shows the level of development as well as human impact of natural environment.
6. Environmental geography deals with all the three aspects of man, viz. physical, social and economic. The 'physical man' is one of the organism's populations or biological communities. He is in need of basic elements of the environment like habitat, air, water and food like other biological populations. While 'social man' performs the social functions through social organizations, the 'economic man' derives and utilizes resources from the physical and biotic environment with his skills and technologies. All these aspects in totality are discussed in environmental geography.

7. The study of 'biomes' is another aspect of environmental geography. Biomes are the regional-scale assemblages of ecosystems. The terrestrial biomes like grassland, desert, forest, etc. have distinct climatic and vegetation characteristics and accordingly, pattern of life and economy. Human influences are becoming an increasingly important elements affecting the nature of biomes.
8. Environmental geography also deals with biodiversity. Its pattern at global, national and local level is a significant element of environment. The earth has enormous variety of plants and animals, both domesticated and wild, as also a wide array of habitats and ecosystems. But now, loss of biodiversity has become a threat to the environment. The geography of environment not only explains the spatial pattern of biodiversity, but also the causes of their loss as well as measures of their conservation, both, *In-situ* and *Ex-situ.*
9. Inter-disciplinary nature is the main characteristic feature of environmental geography. Its scope covers both physical and social science, because on the one hand, physical environment is the prime concern, while on the other hand, impact of environment on human beings or society is analysed. Another aspect is the human impact on environment, which is responsible for degradation.
10. Degradation of environment, both physical and by human activity and their spatial pattern is discussed in environmental geography. Physical degradation of environment due to soil erosion, volcanic activities, flood, drought and famines, desertification, thunder-storms, cyclones, etc. is common in various parts of the world. Similarly, the environmental degradation by human action is seen in the form of deforestation, pollution, acid rain, harmful effects of mining and use of pesticides, etc. The environmental geography not only depicts the areas of degradation but also analyses the causes, effects and measures to be taken for their control.
11. Environmental pollution in the form of air, water, noise and soil is widespread in the world. The detailed study of pollution is the main content of environmental geography. It not only discusses the global and regional pattern of pollution problems but also their causes, effects and measures of control.
12. The study of causes, effects and management of environmental hazards is also an aspect of the geography of environment. Natural hazards originate as highly dynamic processes which are unpredictable and uncontrollable. Those natural hazards that cause loss of human lives and great material damage to economies are called disasters. There are hazards caused by geophysical forces such as cyclone, volcanic eruption, tsunami, tornado, avalanche, flood, drought and also induced by man as forest fire, explosion, oil spill, radioactive fallout, water pollution, etc. The environmental geography examines the aerial extent of damage zone, intensity and duration at a point, and predictability of the event.

13. The climatic change, green house effect, ozone layer depletion, global warming, rising of sea level are the international problems of environment and are within the scope of environmental geography.
14. The most important aspect of the environmental geography now-a-days is the *environmental management*. The environmental management is a wide term used for all the efforts done towards the conservation and planning in the direction of *sustainable development*. There is a need for integrated environmental management to control pollution, desertification, soil erosion, deforestation, climatic change, ozone layer depletion, loss of biodiversity, etc. The environmental legislation and education, monitoring and mapping are also the parts of environmental management. The sustainable development is the need of the present time and geographical studies are always helpful in this objective.

References

Ackerman, E.A. (1965), *The Science of Geography,* National Academy of Sciences, Washington, DC.

Gerasimov, I.P. (1983), *Geography and Ecology,* Progress Publishers, Moscow.

Goudie, A. (ed.) (1994), *The Encyclopedic Dictionary of Physical Geography*, Blackwell, Oxford .

Haggett, P. (1972), *Geography: A Modern Synthesis*, Harper & Row, New York.

Moore, W.G. (1967), *A Dictionary of Geography*, Praeger, New York.

Odum, E.P. (1971), *Fundamentals of Ecology*, W.B. Saunders Co., London.

Strahler, A.N. and A.H. Strahler (1977), *Geography and Man's Environment*, John Wiley, New York.

2

Man-Environment Relationship in Geography

The whole fabric of geographical study is based on man-environment relationship or, in other words, geography is the study of the relationship between man and environment. While most of the other disciplines have started to study this relationship only after alarm over environmental problems, geographers have developed certain concepts like determinism or environmentalism, possibilism, neo-determinism, etc. in their early writings and have also explained the impact of nature on man and vice versa. The study of these concepts is still relevant for better understanding of the changing pattern of man-environment relationship.

Early Thoughts

The question as to how far the physical factors like relief, climate, water bodies, soils, etc. affect man is an old one and ancient philosophers have explained it in their writings. Hippocrates (420 BC) writes in his discussion on *Airs, Waters and Places* about the easy going life of the Asiatics endowed by nature with very favourable physical environments and compares these with those of European countries, with a hard hand of nature upon them. He also contrasts the tall, gentle, brave folk of the most windy mountain lands with the lean, sinewy, inhabitants of the dry lowlands. Aristotle has also recorded similar observations in his *Politics*: 'The inhabitants of the colder countries of Europe are brave, but deficient in thought and technical skill, and as a consequence of this they remain free longer than others, but are wanting in political organization and unable to rule their neighbours. The peoples of Asia, on the contrary, are thoughtful and skilful but without spirit, whereas their permanent condition is one of subjection and slavery. Strabo, in his geographical writings, has explained how shape, climate and space relations of Italy affected the rise and the strength of Rome.

Emphasizing the impact of environmental factors, Bodin, in his writing in the second half of the 16th century, describes the peoples of northern lands as brutal, cruel and enterprising; those of the south as revengeful, cunning, but gifted with the capacity for separating truth from falsehood. He further advocates that inhabitants of temperate regions are more talented than those of the north, more energetic than those of the south and they alone possess that prudence necessary for command. Montesquieu, a century later, has explained the same spirit, when he writes that people in cold climates are stronger physically, more courageous, franker, less suspicious and less cunning than those of the south who are 'like old men, timorous, weak in body, indolent and passive'.[1] The hot climate is the cause of immutability of religion, manners, customs and laws in the eastern countries.[2] He was of the opinion that island 'peoples are more zealous of their liberties than those of the continents'.[3] Kant has also described that all inhabitants of hot lands are exceptionally lazy; they are also timid and the same two traits also characterize folk living in the Far North. All the above opinions expressed by earliest thinkers have paved the way for geographers to study the relationship between man and variables of nature.

Growth of Environmentalism/Determinism

Geographers have inherited the idea of environmentalism from Greek philosophers that nature is all-powerful and not only directs but determines all the human activities. With this background, the concept of 'determinism' or 'environmentalism' has been developed in geography. According to Haggett, 'environmentalism is the view that natural environment plays the major role in determining the behaviour patterns of man on the earth's surface'.[4] The nature has been described as a determining factor not only of human activities and behaviour but also its physical structure, mental ability, etc. Humboldt, Ritter, Haeckel, Buckle, Demolins and Semple were the chief advocates of determinism. A brief review of their ideas is necessary in order to understand the concept of environmentalism.

Alexander von Humboldt (1769–1859), a German geographer, was considered as the pioneer of determinism in geography. In his book *Kosmos* (1849), he described the effect of the configuration of the Mediterranean on the evolution of early civilization. He writes: 'The active life of the Greeks, especially of the Ionians, and their early predilection for maritime expeditions found a rich field for its development in the remarkable configuration of the Mediterranean, and in its relative position, to the oceans situated to the South and West.'[5]

He believed that the phenomena on the earth's surface were governed by laws, but that all these would only become apparent when all facts and relationships had been observed in all parts of the earth.

Carl Ritter (1779–1859), also a German geographer, has propounded the influence of environmental factors not only on human activities but also on human character. The relationship between man and nature has been described by him in his book, *Europe in Gernalde*, which he divided into two parts, the first part showing what the land owed to man and the second one depicting what the man owes to nature. Ritter, in his famous volume, *Erdkunde*, has described in detail the man-environment relationship. Both Ritter and Humboldt died in

1859. At the same time, Charles Darwin put forward the theory of origin of species, which seeks to link the origin of species with environment.

Haeckel (1834–1919) has outlined a new science, 'Ecology', the study of the mutual relations of all organisms living in one and the same place and of their adaptation to their environment. Environmentalism has been well illustrated by Buckle in his book, *History of Civilization in England* (Vol. I, 1857; Vol. II, 1861). He was of the opinion that everywhere the hand of nature is upon us, and the history of the human mind can only be understood by connecting it with the history and the aspects of the material universe. According to Buckle:

> Man modifying nature and nature modifying man, yet it is nature itself which determines when and how man shall be so active, in other words, natural powers are in every instance dominant, the determinism is complete.

Demolins' (1852–1907) two volumes, *Essai de geographie sociale, comment la route cree le type sociale* illustrate the strict environmentalism in geography which sprang out of the teachings of Frederic Leplay. According to Demolins, the environment rigidly controls the structure and development of people. He concentrated on the study of environmental forces and pointed out that these forces/factors are responsible for the growth of social organization as well as of civilization of a region.[6]

Frederich Ratzel (1844–1904), another German scholar, may be considered as a leader of environmentalism. His classical work, *Anthropo geographie* is full of illustrations and thoughts concerning environmental impact over man. He proposed the concept of terrestrial unity and described the complex interrelationship between physical variables like altitude, topography, climate, vegetation and also their impact on man.

Ellen Churchill Semple, a leader of deterministic thought in geography, has not only elaborated the concept of Ratzel but also supported her views with numerous examples. In her classical work, *Influence of Geographic Environment* (1911), she noted:

> Man is a product of the earth's surface. This means not merely that he is a child of the earth, dust of her dust, but that the earth has mothered him, fed him, set him tasks, directed his thoughts, confronted him with difficulties that have strengthened his body and sharpened his wits, given him his problems of navigation or irrigation and at the same time whispered hints for their solution.[7]

According to Semple, religious ideas are not exempt from nature's control. She writes that 'Buddha born in the stern Himalayan piedmont, fighting the lassitude induced by heat and humidity, pictured his heaven as *Nirvana*, the cessation of all activity and individual life'.[8] Semple's book is packed with details of examples selected from all parts of the world.

Apart from the above scholars, several other geographers have supported fully or partially the concept of environmentalism. But, with the growth of knowledge, scientific and technical developments, it was realized that man can use nature for his comforts. Several doubts have been raised by scientists/scholars as well as by geographers about total environmental control. Even similar environments do not always evoke the same response. Man's preference for certain items and his skill is also important. The distribution of population, expansion of agriculture, location of towns, industries, mental ability of certain races, etc. have some relation with environment but are certainly not controlled totally by physical factors. Tatham has rightly observed that 'although environment undoubtedly influences man, man in turn changes his environment, and the interaction is so intricate that it is difficult to know when one influence ceases and the other begins'.[9] This discussion has given rise to a new concept in geography known as 'possibilism'.

Possibilism

The concept of possibilism in geography was developed in the beginning of the h century. With the growth of knowledge, industrial revolution, and development in the field of science and technology it has been established that man can use nature according to his wishes. Thus, contrary to deterministic views of all-powerful nature, in possibilism, man has been considered supreme. Haggett has defined it as follows: 'Possibilism, in contrast to environmentalism, stresses the freedom of man to choose alternative patterns of behaviour despite geographic location.'[10]

Febvre was the first French scholar who introduced the concept of possibilism followed by other geographers like Vidal de la Blache, Jean Brunhes, Isaiah Bowman and Carl Sauer in the United States. Febvre was basically a historian and in his book, *Geographical Introduction to History*, has explained the man-environment relationship in a new form, when he writes: 'Man is a geographical agent and not the least. He everywhere contributes his share towards investing the physiognomy of the earth with those "changing expressions" which is the special charge of geography to study.'[11] He further elaborates: 'There are no necessities, but everywhere possibilities, and man as a master of these possibilities is the judge of their use.'[12] In geographical studies, human elements are equally, rather more important. He explained this fact in his statement: 'The geographical elements of the environment are fixed only in the narrow and special sense of the word. The moment we give them human associations, they are as changeful as humanity itself. That is why modern geography has so definitely steered away from determinism and towards a study of types of actually working regional combinations of human and environmental condition.'[13]

Blache is considered as the founder of the concepts of possibilism. In his book *Principles of Human Geography* (1950), he has not only explained the man-environment relationship, but has described the role of man in modification of his environment. According to him, 'nature is never more than an adviser'.[14]

The choice of man is guided and restrained by the mental and social patterns of the group and its level of technology, in short, its culture. Human activity modifies both organic and inorganic features of the earth. Blache was of the opinion that man is not a passive being but an active force and changing his environment. In fact, he considered man as a geographical factor. He has also expressed his views in his many papers contributed to the *Annals de Geographie*, the journal which he founded and has gained an international recognition. Brunhes, a student of Blache, has not only extended the philosophy of Blache but also explained in detail the man-environment relationships in his book, *Human Geography* (1920). Brunhes has pointed out that man is always active in changing the environment. He emphasized the 'terrestrial harmonies' as a basis of relationship between man and his environment. According to him, 'we must add to the group of material forces whose incessant interplay we have seen this new force, human activity; which is not only a material thing, but which also expresses itself through material effects'.[15] The limits set by nature to man's action vary from place to place on the earth's surface and from one historical period to another. 'Thus, everything on the surface of globe is for men, a matter of habit, of sound understanding, of physical facts, and of skilful adaptation of these facts.'[16] With several examples, Brunhes supported the dominant role of a man.

Bowman, in his work *Geography and the Social Sciences* (1934), has illustrated the influence of human activities on the environment. He writes: 'As knowledge of the world spread, the associations of events or conditions with place widened, they became more complex, they had less or more significance with respect to mankind.'[17] He further elaborated that 'the geographical elements of the environment are fixed only in the narrow and special sense of the word. The moment we give them human associations, they are as changeful as humanity itself.'[18]

Several other scholars too have given much importance to man's work and his control over nature. But like determinism, possibilism is also an extreme concept and soon people realized that the impact of nature cannot be ignored. Although man has utilized nature for his benefit, there is a limit to such utilization. There are several regions of the world that are not yet inhabited by man either due to extreme climatic conditions or adverse physiography, etc. Man is helpless before natural hazards like earthquakes, volcanism, floods, landslides, sea storms, etc. which cause heavy damage to man and material wherever they occur. In fact, the whole world and its society is working in complete harmony with nature – this balance is a creative force. This fact has been realized and a new concept, in the form of neo-determinism, has emerged in geography.

Present-Day Determinism or Neo-Determinism

Between the extremes of determinism and possibilism, a new concept has been developed in geography which is known as neo-determinism. In fact, seeds of this concept were present in possibilistic thought when they used terms like 'response', 'influence', 'adjustment', etc. Thus, a mid-way theory has been devised first by Griffith Taylor, who described his point of view as *stop-and-go determinism* in which he admitted the impact of nature on man but also recognized the human skill, mental ability, scientific development and technical know-how, through which man can use nature to a certain extent according to his needs. This, he defines as follows:

> Man is able to accelerate, slow or stop the progress of a country's development. But he should not, if he is wise, depart from the directions as indicated by the natural environment. He is like the traffic controller in a large city, who alters the rate but not the direction of progress, and perhaps the phrase 'stop-and-go determinism' expresses succinctly the writer's geographical philosophy.[19]

According to this concept, nature used to provide opportunity for man to use these possibilities according to their wisdom. But man is not free from the control of nature. Taylor suggests that the goal must be adjustment to nature's plan, the carrying out of nature's programme. The opportunities offered by any environment are not equal, some demand little efforts from man, others, continual struggle, some yield large, others meagre returns. In no environment are the possibilities limitless, and for every choice, price must be paid, but within these the limits of freedom to choose, exist.

Other Thoughts

This neo-determinism is nothing but a compromise between extreme philosophies as Spate has observed: 'In not a few ways and place, and to a significant degrees, human activity is affected by external environment – that is all that neo-environmentalism claims, and it does rule out causative effects of human efforts. But the attempt to the rule out environmentalism, leads to difficulties.'[20] The role of a geographer is basically advisory, he can help to plan efforts in such a way so that harmony with nature and development can be established, which is now called 'sustainable development'. The leader in the Organization of Academic Geography in the United States in the early 20th century, William Morris Davis, also believed very clearly in the importance of environmental influences. In this reference, Francis Bacon's saying is significant that 'nature can be conquered only by obeying her'. The impact of nature should be assessed in its totality rather than 'single factor determinist' point of view. Ecological studies have taught us that all nature is interconnected. While discussing 'environmental causation', John E. Chappell, Jr., has given ten general principles for a more realistic evaluation of environmental impact and limitations:[21]

1. The single factor determinist is merely a straw-man.
2. The fact that all nature is one does not imply that unity cannot be analysed.
3. Causation can indeed be discovered without specifying every link in a mechanistic causal chain.
4. The most profound truths are usually the hardest to discover, to confirm and to quantify.
5. Subjective truths are valued more highly, but objective truths are usually more powerful in the long run.

6. The environment is not stable, and therefore changing culture does not, in itself, prove lack of environmental influence.
7. Cultural diffusion is not the logical antithesis of environmental causation.
8. Environmentally-shaped traits generated in one place may be carried to another place, where they will not be re-shaped immediately.
9. Environmental influences are not limited to material phases of life, they also affect our minds and our habits.
10. Environmental determinism is not a politically reactionary philosophy, and it can be applied towards the solution of large practical problems.

Probablism is another term used by a few geographers. According to Haggett, 'probablism is a compromise position between environmentalism and possibilism that assigns different probabilities to alternative patterns of geographic behaviour in a particular location or environment'.[22] Fleure, Spate, Martin, Woolridge and East, Roxby, Herbertson, etc. have expressed their ideas on man-environment relationship in terms of man's adjustment to nature and have also given due weightage to the modern scientific developments.

The concept of voluntarism is nothing but an expression of above mentioned ideas in a different way. In this concept, due importance has been given to man but he can change the nature up to a certain limit. Similarly, in 'normative approach', relationship between man and nature has been explained in terms of 'optimum location' which provides an opportunity for adjustment and development. In fact, 'ecological approach' is the best way to explain the man-environment relationship. Both environment and man are integral parts of the ecological framework, combined with technological development which is responsible both for development and degradation of the environment. The man-environment relations are based on mutual interaction and mutual transformation. There is a need to humanize nature, as Raza has rightly observed: 'We cannot think of a "non-humanized nature" and "non-natural man".' He (the man) alone among the known species of this planet, just does not adapt himself to nature but actually interacts with it; who is not only a passive element but an active factor in the ecosystem.[23]

Notes

1. Montesquieu, *Spirit of Laws*, Book XIV, Chapter 2.
2. Ibid., Book XIV, Chapter 4.
3. Ibid., Book XVIII, Chapter 5.
4. P. Haggett (1972/1983), *Geography: A Modern Synthesis*, Harper & Row, New York, p. 591.
5. Alexander von Humboldt (1849), *Kosmos*, Vol. 2, Hippolyte Bailliere, London, p. 484.

6. G. Tatham (1957), 'Environmentalism and Possibilism' in G. Taylor (ed.), *Geography in 20th Century*, Methuen, London, p. 138.
7. E.C. Semple (1911), *Influences of Geographic Environment*, Henry Holt & Co., New York, pp. 1–2.
8. Ibid., p. 40.
9. Tatham, op. cit., p. 148.
10. Haggett, op. cit., p. 591.
11. Febvre quoted by Tatham op. cit., p. 153.
12. Ibid., p. 154.
13. Ibid., p. 155.
14. V. Blache (1950), *Principles of Human Geography*, Constable & Co., London, p. 321.
15. J. Brunhes (1920), *Human Geography*, Harrap & Co., London p. 27.
16. Ibid., p. 611.
17. I. Bowman (1934), *Geography in Relation to the Social Sciences*, Charles Scribner's Sons, New York, p. 36.
18. Ibid., p. 37.
19. Taylor, G., *Australia*, p. 445.
20. O.H.K. Spate (1960), 'Quantity and Quality in Geography', *Annals of the Association of American Geographers*, Vol. 60, 1960, p. 389.
21. J.E. Chappell (Jr.) (1981), 'Environmental Causation', in M.E. Harvey and B.P. Holly (eds), *Themes in Geographic Thought*, Croom Helm, London, pp. 163–186.
22. Haggett, op cit., p. 591.
23. Moonis Raza (1983), 'Man in the Ecosystem', *The Geographical View*, Occasional Paper, SSRD, JNU.

3

The Nature of Ecosystems

The whole earth and its parts, viz. land surface, water bodies, atmosphere as well as living organisms are working in an integrated system, which not only directs but also determines the pattern of life and is also an indicator of environmental relationship with living organisms. Each and every part of the earth's surface is having its own ecosystem but at the same time, it is also a part of a larger system, which may cover the whole of the continent, parts of various continents or even the whole of the earth. The single giant ecosystem is the 'biosphere'. The term 'ecosystem' was coined by A.G. Tansley in 1935, who defined it as 'the system resulting from the integration of all the living and non-living factors of the environment'. He further stated that 'the whole system, includes not only the organism complex but also the whole complex of physical factors forming what we call the environment of the biome – the habitat factors in the widest sense. It is the systems so formed which are the basic units of nature on the face of the earth'.

Although the concept of ecosystem is much older than the term itself, many terms parallel to it have been used by bio-scientists, such as *biocoenosis* (Mobius, 1877), *microcosm* (Forbes, 1887), *holocoen* (Friederichs, 1930), *biosystem* (Thienemann, 1939), *geobiocoenosis* (Sukachev, 1944), *ecosphere* (Cole, 1958), etc. However, the term 'ecosystem' is now accepted and most preferred, where 'eco' implies the environment, and 'system' means interdependent, interacting complex.

According to Monkhouse and Small, ecosystem is 'an organic community of plants and animals viewed within its physical environment or habitat'.[1] The habitat or physical environment controls the whole world of organic community, viz. plants and animals. With the change in physical environment, the types of plants and animals too have changed. Similar views have been expressed by Haggett: 'Ecosystems are ecological systems in which plants and animals are linked to their environment through a series of feedback loops.'[2]

Strahler has defined ecosystem as 'the total assemblage of components entering into the interactions of a group of organism'. He further elaborated that 'to the geographer, ecosystems are the part of the physical composition of the life layer'.[3]

Ecosystem has also been defined as:

> a unit that includes all the organisms (biological factor) in a given area interacting with the environment (physical factors) so that a flow of energy leads to a clearly defined trophic (nutrient requiring) structure, biotic diversity, and material cycles (i.e. exchange of materials between living and non-living sectors).[4]

In fact, our earth has a giant ecosystem in the form of the biosphere, where abiotic and biotic components are constantly acting and reacting upon each other, bringing forth structural and functional changes in it. The biosphere is a vast ecosystem and difficult to study as a whole, therefore, for the convenience of study, it can be divided into the following major sub-divisions:

1. *Natural Ecosystems:* These are of two types:
   (i) *Terrestrial* as forest, grassland, desert, mountains, etc.
   (ii) *Aquatic* (water-open): (a) *fresh water*, which may be *lotic* (running-water like spring, stream, rivers) or *lentic* (standing water like lake, pond, swamp, pools, etc.); and (b) *marine*, such as deep sea or shallow sea, delta, etc.
2. *Artificial (man-engineered) Ecosystems*: These are maintained by man such as cropland ecosystems.

Components of the Ecosystem

The two major components of ecosystem are:

- the biotic components comprising all living organisms, and
- the abiotic components including the physical (non-living) environment.

Biotic Components

In an ecosystem, organisms are usually divided into two categories on the basis of their nutritional (trophic) relationship. These are: (a) autotrophic components, and (b) heterotrophic components.

1. *Autotrophs* are those organisms which can produce their own food. These are green plants (with chlorophyll) and certain bacteria, which obtain their energy from the sun and through photosynthesis produce food. Since these organisms produce food for all the other organisms, they are known as producers.
2. *Heterotrophs* are those organisms which depend directly or indirectly upon the autotrophs for their food, thus known as consumers. The consumers are of two types:
   (i) *Macroconsumers* also known as herbivores, carnivores or omnivores. Herbivores are the plant eaters, while carnivores are animal eaters and omnivores eat all kinds of food. They all are phagotrophs which include chiefly animals that ingest other organic and particulate organic matter. These are also known as primary consumers.

(ii) *Microconsumers* are also known as decomposers or osmotrophs. They are decomposers because of their role in decomposition of dead organic matter. They secrete digestive enzymes to break down food into simpler substances and then absorb the digested food.

Functionally, the biotic components can be divided into three types, viz. producers, consumers and decomposers. Producers produce their own food like plants, which are primary producers, while man and animals obtain food from plants. Similarly, consumers are also primary, secondary or tertiary. Man, cow, goat, etc. are primary consumers, while meat eaters are secondary consumers. Tertiary consumers are those animals which eat meat such as snakes eat frogs, peacocks eat snakes, etc. Among decomposers those bacteria, fungi, etc. have been included which are responsible for decomposition of a product and release inorganic nutrients in environment.

Abiotic Components

The physical environment with its several interacting variables constitutes the abiotic component of ecosystem. It consists of (i) the lithosphere, i.e. the solid mineral matter of the earth, (ii) the water in the oceans, rivers, lakes, etc. or in other words, the hydrosphere, (iii) the gaseous mixture in the air (the atmosphere), and (iv) the radiant solar energy.

Figure 3.1

Ecosystem Model Showing its Structure and Function

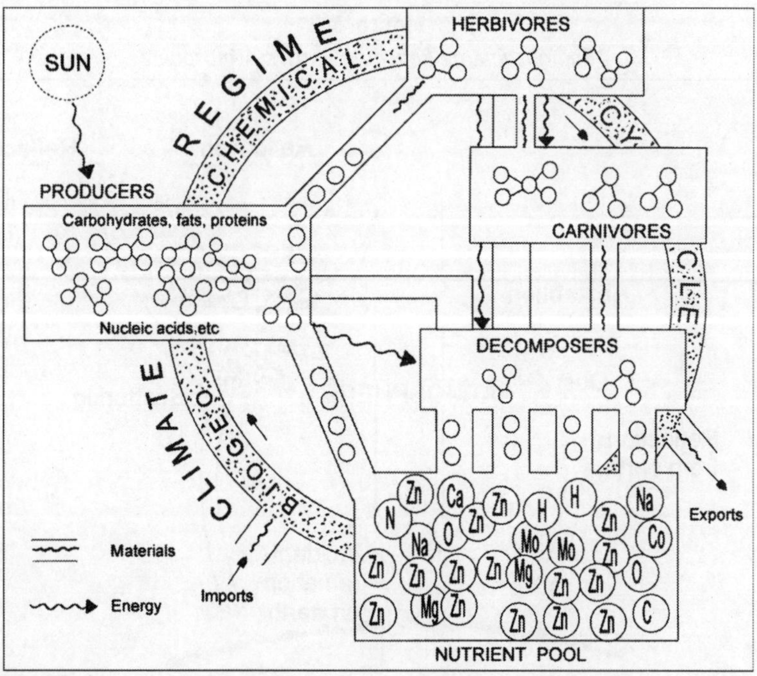

Figure 3.1 is a generalized depiction of the structural elements of an ecosystem and their functional linkages.

Ecosystem Functioning

The functional pattern of an ecosystem is a very important and basic aspect of study, because all its components are dynamic in one way or the other and are responsible for the creation of a unique state of man-environment relationship as well as a habitat, different from other regions. Ecosystem functioning can be studied under three heads, viz. energy flow, productivity and biogeochemical cycling.

Energy Flow in an Ecosystem

Energy is defined as the ability to do work and is the essence of life. Without energy transfers, there could be no life and no ecological systems. The energy used for all the life processes is derived from solar radiant energy. The behaviour of energy is described by the laws of thermodynamics: (i) energy may be transformed from one type into another but is never created or destroyed, and (ii) no process involving an energy transformation will spontaneously occur unless there is a degradation of the energy from a concentrated form into a dispersed form, etc.

The radiant energy produced in the sun travels through space in the form of waves. But only a small fraction of solar radiation reaches the earth to provide energy for the biotic components of the ecosystem. Figure 3.2 indicates the fate of radiant energy reaching the earth's atmosphere. It becomes clear from the figure that most of the radiation is lost in space by processes of reflection, absorption as well as through scattering. Its energy is greatly altered as it passes through cloud cover, water and vegetation. The daily input of

Figure 3.2

Radiant Energy Reaching Earth's Atmosphere

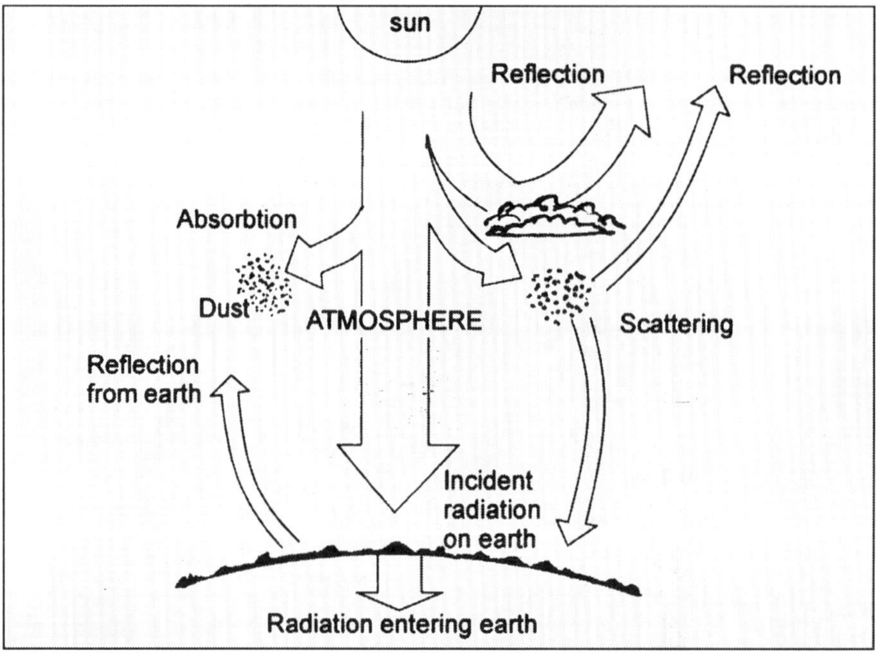

sunlight to autotrophic layer of an ecosystem varies mostly between 100 and 800 to 300–400 gcal per cm^2 (= 3000 to 4000 kcal per m^2) in the temperate zone. The total radiation flux within different strata of the ecosystem varies from season to season as well as with nature of the earth's surface, thus controlling the distribution and response of organisms accordingly.

The energy reaching the earth's surface is transformed and/or absorbed by plants and other organisms. It is used by the green plants during photosynthesis by converting the light energy to chemical energy and making it available to other organisms as food. Such energy transfers along the food chain were studied for the first time by Lindeman in 1942. He stated that all the functions of an ecosystem could be explained in terms of energy by the knowledge of two attributes of each trophic level – the level of energy storage and the efficiency of energy transfer. Odum (1957), Golley (1960), Slobodkim (1959, 1960, 1962), Teal (1962) and Kozlovsky (1968) have made further detailed investigations. The process of energy flow in an ecosystem has been depicted by Odum in a simplified diagram as shown in Figure 3.3, where the 'boxes' represent the trophic levels and the 'pipes' depict the energy flow in and out of each level. It becomes clear from this simplified model that the energy flow is greatly reduced at each successive trophic level from producers to herbivores and then to carnivores.

Figure 3.3

Energy Flow in a Food Chain

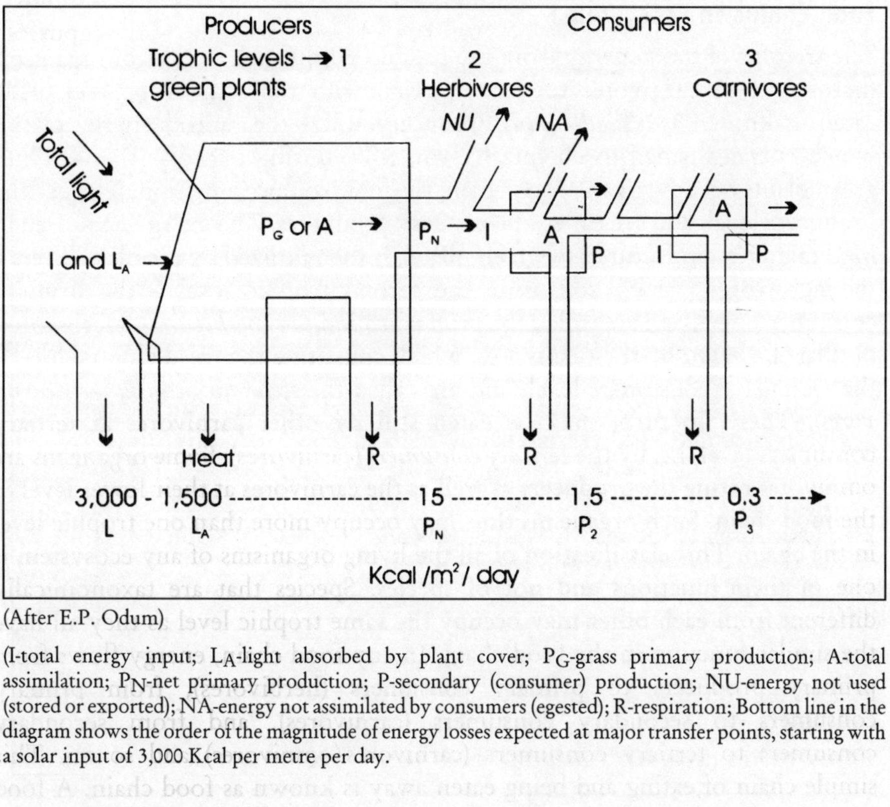

(After E.P. Odum)

(I-total energy input; L_A-light absorbed by plant cover; P_G-grass primary production; A-total assimilation; P_N-net primary production; P-secondary (consumer) production; NU-energy not used (stored or exported); NA-energy not assimilated by consumers (egested); R-respiration; Bottom line in the diagram shows the order of the magnitude of energy losses expected at major transfer points, starting with a solar input of 3,000 Kcal per metre per day.

Productivity of Ecosystem

The productivity of an ecosystem refers to the rate of production, i.e. the amount of organic matter accumulated in any unit of time. It is of primary, secondary and net productivity.

1. *Primary Productivity* is defined as 'the rate at which radiant energy is stored by photosynthetic and chemosynthetic activity of producer organisms (chiefly green plants) in the form of organic substances which can be used as food materials'.[5] Primary productivity is of two types:
   (i) *Gross Primary Productivity* is the total rate of photosynthesis including organic matter used up in respiration during a particular period.
   (ii) *Net Primary Productivity* is the rate of storage of organic matter in plant tissues in excess of respiratory utilization by plants during the measurement period.
2. *Secondary Productivity* is the rate of energy storage at consumer level. It actually remains mobile from one organism to another and does not live *in situ* like primary productivity.
3. *Net Productivity* refers to the rate of storage of organic matter not used by the consumers. Thus, it is the rate of increase of biomass of the primary producers which has been left out by the consumers.

Food Chains in Ecosystems

The transfer of food energy from the producers, through a series of organisms (herbivores to carnivores to decomposers) with repeated eating and being eaten, is known as a *food chain*. Producers utilize the radiant energy of sun which is transformed to chemical form, ATP during photosynthesis. Thus, green plants occupy, in any food chain, the first trophic (nutritional) level – the producers level, and are called the primary producers. The energy, as stored in food matter manufactured by green plants, is then utilized by the plant eaters – the herbivores, which constitute the second trophic level – the primary consumers level, and are called the *primary consumers* (*herbivores*). Herbivores, in turn, are eaten by the carnivores, which constitute the third trophic level – the secondary consumers level, and are called the *secondary consumers* (*carnivores*). These, in turn, may be eaten still by other carnivores at tertiary consumers level, i.e. by the *tertiary consumers* (carnivores). Some organisms are omnivores eating the producers as well as the carnivores at their lower level in the food chain. Such organisms thus may occupy more than one trophic level in the chain. This classification of all the living organisms of any ecosystem is one of their functions and not of species. Species that are taxonomically different from each other may occupy the same trophic level as they all have the similar function in the food chain. In any food chain, energy flows from primary producers to primary consumers (herbivores), from primary consumers to secondary consumers (carnivores), and from secondary consumers to tertiary consumers (carnivores/omnivores) and so on. This simple chain of eating and being eaten away is known as food chain. A food chain in grassland ecosystem starts with grasses and herbs and goes through grasshoppers, the frogs, the snake, the hawk in an orderly sequential arrangement based on the food habits, whereas, in a pond the order would start with phytoplanktons, going through water fleas, smaller fish, bigger fish, birds, larger animal and so on.

In nature, we generally distinguish two general types of food chains:

Grazing Food Chain

This type of food chain starts from the living green plants, goes to grazing herbivores (that feed on living plant materials with their predators), and on to carnivores (animal eaters). Ecosystems with such type of food chain are directly dependent on an influx of solar radiation. This type of chain thus depends on autotrophic energy capture and the movement of this captured energy to herbivores. Most of the ecosystems in nature follow this type of food chain. From energy standpoint, these chains are very important. The phytoplanktons-zooplanktons-fish sequence or the grasses-rabbit-fox sequence are the examples of grazing food chain.

Detritus Food Chain

This type of food chain goes from dead organic matter into microorganisms and then to organisms feeding on detritus (detritivores) and their predators. Such ecosystems are thus less dependent on direct solar energy. These depend chiefly on the influx of organic matter produced in another system. For example, such type of food chain operates in the decomposing accumulated litter in a temperate forest. A good example of a detritus food chain is based on mangrove leaves described by Heald (1969) and Odum (1970). In the brackish zone of southern Florida, leaves of the red mangrove – *Rhizophore mangle* fall into the warm, shallow waters. The fallen leaf fragments (acted on by such saprotrophs as fungi, bacteria, protozoa, etc. and colonized mainly by phytoplanktonic and benthicalgae) are eaten and re-eaten (coprophagy) by a key group of small animals. These animals include crabs, copepods, insect larvae, grass shrimps, mysids, nematodes, amphipods, bivalve molluscs, etc. All these are detritus consumers. These animals are in turn eaten by some minnows and small game fish etc. i.e. the small carnivores, which in turn serve as the main food for larger game fish in particular. The mangroves considered generally as of less economic value make a substantial contribution to the food chain that support the fisheries, an important economy in that region. Similarly, detritus from seagrasses, saltmarsh grasses and seaweed support fisheries in many estuarine areas.

But, under natural situations, a system must always be self-sufficient. In fact, this type of food chain (detritus type) is simply a sub-component of another ecosystem. And, the above said two types of food chains in nature are indeed linked together belonging to the same ecosystem.

Food Webs: Interlocking Pattern of Organisms

However, food chains in natural conditions never operate as isolated sequences, but are interconnected with each other forming some sort of interlocking pattern, which is referred to as food web. Under natural conditions, the arrangement of food chain hardly occurs and these remain indeed interconnected with each other through different types of organisms at different trophic levels. For example, in grazing food chain of a grassland, in the absence of rabbit, grass may also be eaten by mouse. The mouse in turn may be eaten directly by hawk or by snake first which is then eaten by hawk. Thus, in nature, there are found alternatives, which all together constitute some sort of interlocking pattern, known as the *food web*.

In such a food web in grassland, as shown in Figure 3.4, there may be seen as many as five linear food chains, which in sequences are:

1. Grass→Grasshopper→Hawk
2. Grass→Grasshopper→Lizard→Hawk
3. Grass→Rabbit→Hawk (or vulture or fox or even man, if present)
4. Grass→Mouse→Hawk
5. Grass→Mouse→Snake→Hawk

Figure 3.4
Food Web in a Grassland Ecosystem

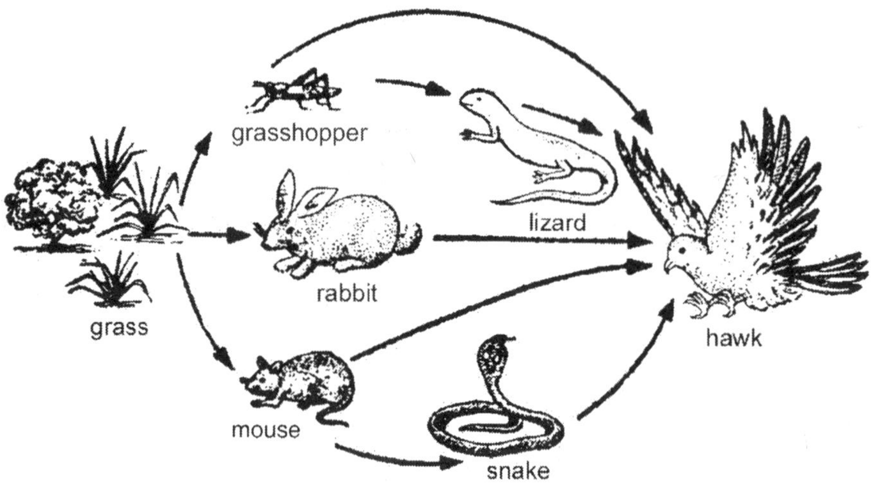

Besides those shown in Figure 3.4, there may also be present some other consumers as vultures, fox and man in grasslands, and if so, the food web may be even more complex than shown here. However, these all five chains are interlinked with each other at different points, forming food web. A similar food web in a pond, with different interlinked chain is shown in Figure 3.5. The food webs are very important in maintaining the stability of an ecosystem in nature. For example, decrease in the population of rabbit would naturally cause an increase in the population of alternative herbivore, the mouse. This may decrease the population of the consumer (carnivore) that prefers to eat rabbit. Thus, alternative (substitutes) serve for maintenance of stability of the

Figure 3.5
Food Web in a Pond

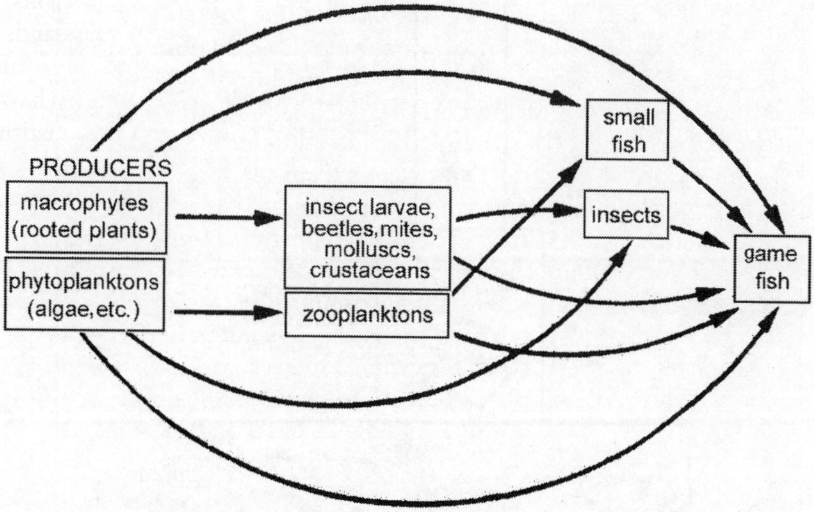

ecosystem. Moreover, a balanced ecosystem is essential for the survival of all living organisms of the system. For instance, had primary consumers (herbivores) not been in nature, the producers would have perished due to overcrowding and competition. Similarly, the survival of primary consumers is linked with the secondary consumers (carnivores) and so on. Thus, each species of any ecosystems is indeed kept under some sort of a natural check so that system may remain balanced.

The complexity of any food web depends upon the diversity of organisms in the system. It would accordingly depend upon two main points:

1. Length of the food chain. Diversity in the organisms based upon their food habits would determine the length of food chain. More diverse the organisms in food habits, more longer would be the food chain.
2. Alternatives at different points of consumers in the chain. More the alternatives, more would be the interlocking pattern.

In deep oceans, seas, etc. where we find a variety of organisms, the food webs are much complex.

Ecological Pyramids: Graphic Representation of Trophic Structure

Trophic structure, i.e. the interaction of food chain and the size metabolism relationship between the linearly arranged various biotic components of an ecosystem is characteristic of each type of ecosystem. The trophic structure and function at successive trophic levels, i.e. producers – herbivores – carnivores can be shown by *ecological pyramids* where the first or producer level constitutes the base of the pyramid and successive levels, the tiers making the apex. The three types of ecological pyramids are: (i) Pyramid of numbers, (ii) Pyramid of biomass, and (iii) Pyramid of energy.

Figure 3.6
Pyramid of Numbers in a Grassland Ecosystem

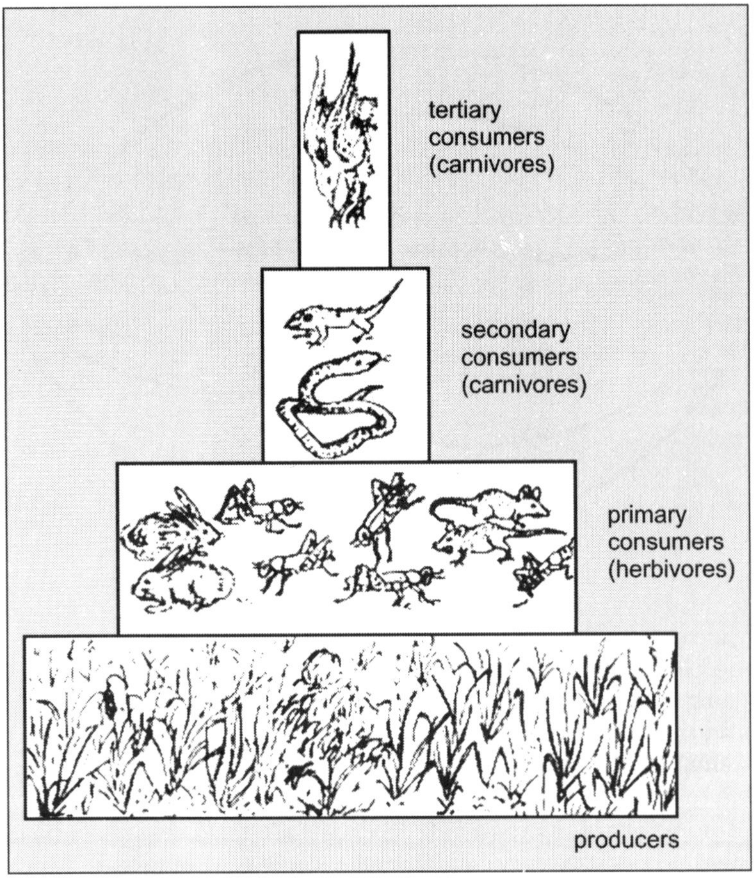

1. *Pyramid of Numbers:* They show the relationship between producers, herbivores and carnivores at successive trophic levels in terms of their numbers. The pyramid of numbers in a grassland ecosystem is shown in Figure 3.6. In a grassland ecosystem, the producers mainly grasses, are maximum in numbers. This number then shows a decrease towards apex, as the primary consumers like rabbits, mice, etc. are lesser in number than grasses; the secondary consumers, snakes and lizards are lesser in number than rabbits mice etc. Finally top (tertiary) consumers hawks or other birds are least in number. Thus, this pyramid becomes upright. The shape of the pyramid of numbers varied in forest ecosystem and parasitic food chain.
2. *Pyramid of Biomass:* The pyramid of biomass in a grassland ecosystem is upright (Figure 3.7), while it is inverted in a pond ecosystem; where producers are small organisms, their biomass is least, and this value gradually shows an increase towards apex of the pyramid, thus making a pyramid *inverted* in shape (Figure 3.8).

Figure 3.7

Pyramid of Biomass in a Grassland Ecosystem

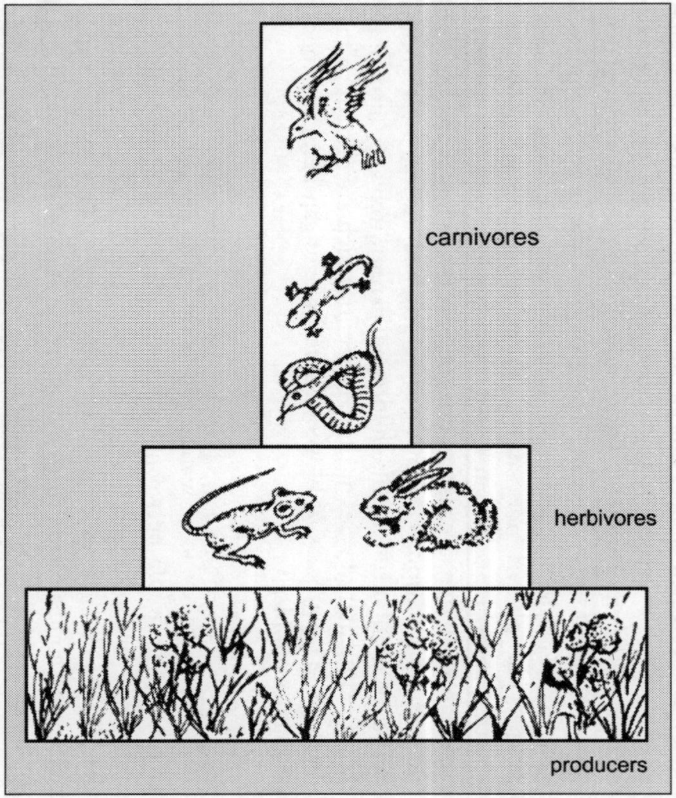

Figure 3.8

Pyramid of Biomass in a Pond Ecosystem

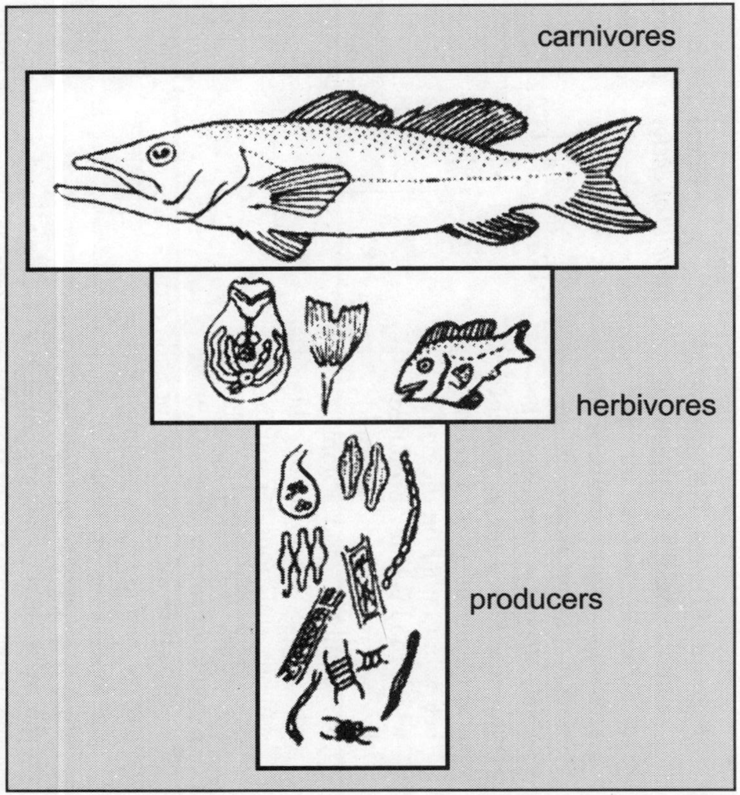

3 *Pyramid of Energy:* The energy pyramids give the best picture of overall nature of the ecosystem. Here, number and weight of organisms at any level depends not on the amount of fixed energy present at any one time in the level just below but rather on the rate at which food is being produced. In contrast with the pyramids of numbers and biomass, which are pictures of the standing situations (organisms present at any moment), the pyramid of energy (Figure 3.9) is a picture of the rates of passage of food mass through the food chain. In shape, it is always *upright*, as in most of the cases there is always a gradual decrease in the energy content at successive trophic levels from the producers to various consumers.

Figure 3.9
Pyramid of Energy in a Ecosystem

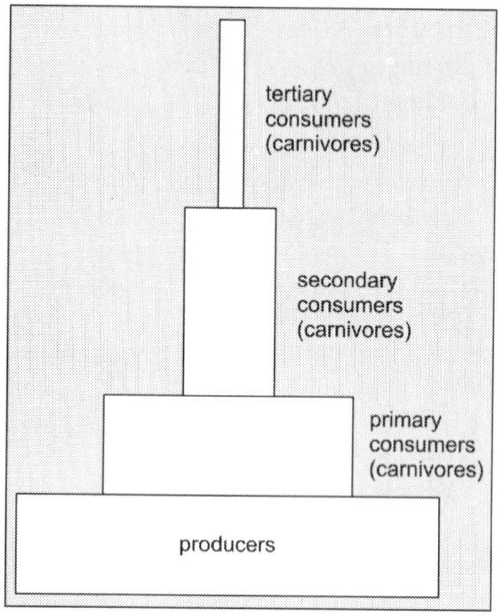

Biogeochemical Cycles

The chemical elements, including all the essential elements of protoplasm, tend to circulate in the biosphere in circular paths from environment to organism and back to the environment. These move in a more or less circular path, thus known as *biogeochemical cycles.*[6] The growth of life process requires about 40 elements. Hydrogen, carbon and oxygen are the basic elements, while nitrogen, phosphorus, potassium, calcium, sulphur, magnesium and iron are the other important elements. All the elements continue to cycle at the ecosystem level, they are also an integral part of the larger and/or global cycle. The cycling of various elements can be grouped into two categories, viz. (i) gaseous cycles of elements like hydrogen, carbon, oxygen, nitrogen, sulphur; and (ii) sedimentary cycles of those elements which are present in the soil and organisms. The process of biogeochemical cycles can be understood through following examples:

Hydrological Cycle

The hydrological cycle or water cycle is the most important one. It involves interchange of water between earth's surface and the atmosphere through rainfall and evapotranspiration. The water from water bodies like oceans, lakes, rivers, etc. gets evaporated by solar energy. These water vapours after cooling and condensation form clouds, result in rainfall, snowfall, etc. A large part of the rainfall occurs over oceans, and a sizeable part of the water vapours goes back to the oceans via rivers and streams. Some part of it infiltrates into the soil to become soil or underground water. A small quantity of water is also absorbed by the plants and consumed by other animals. The same is released during respiration and transpiration of plants. Thus, there is a continuous cycling of water as shown in Figure 3.10.

Figure 3.10
Hydrological Cycle

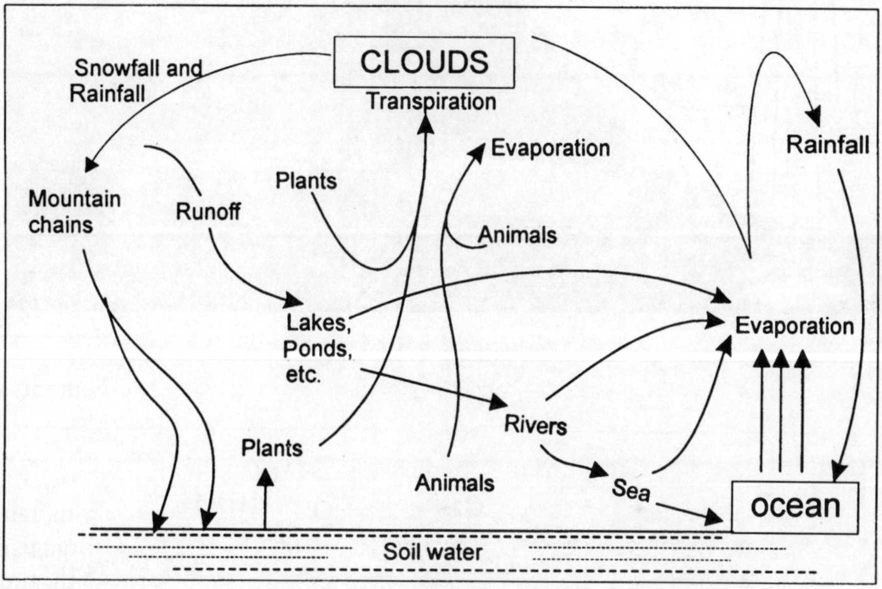

Carbon Cycle

Like the hydrological cycle, carbon cycle is also very important for man. The small amount of carbon dioxide (CO_2 – 0.03%) in the atmosphere is the only source of all carbon that passes through the organisms along the food chains. As shown in Figure 3.11, carbon moves from the atmospheric pool to green plants (producers), then to animals (consumers) and finally to bacteria and other microorganisms (decomposers) that return into the atmosphere, through decomposition of dead organic matter. The carbon dioxide has the unique property of absorbing infra-red radiation and its small quantity helps in keeping the earth warm. A part of this cycle also operates in the ocean.

Oxygen Cycle

The oxygen constitutes 21 per cent of the atmosphere. It also occurs in a bound state as oxides and carbonates in rocks and in water. The plants release oxygen in photolysis of water during photosynthesis. Again, gaseous oxygen is used in respiration of all the organisms and in the oxidation of the organic matter. Another phase of the oxygen is the ozone layer of the outer stratosphere which serves an important function of protecting the life from the ionizing ultra violet waves. The pattern of oxygen cycle has been depicted in Figure 3.12.

Figure 3.11
A Simplified Carbon Cycle

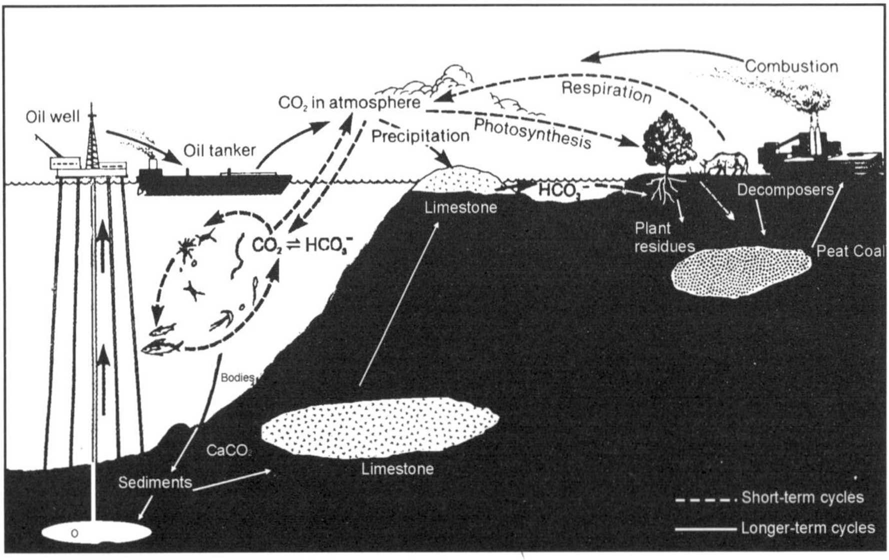

Figure 3.12
Oxygen Cycle

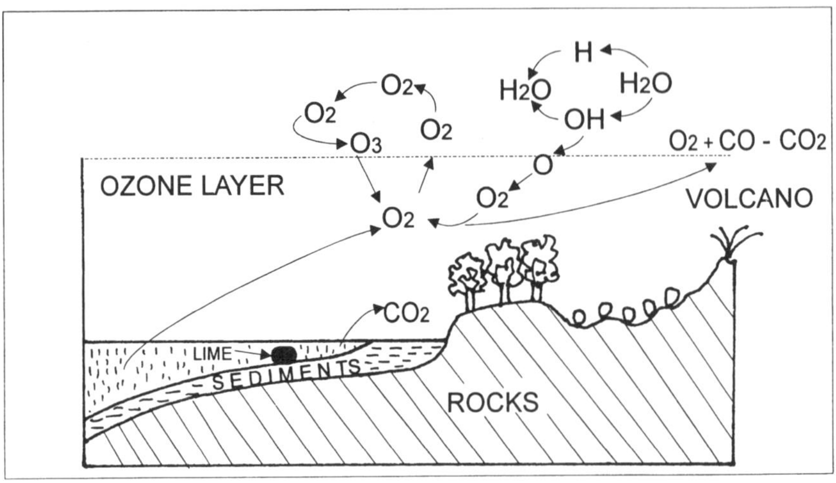

Nitrogen Cycle

Nitrogen is an essential element of all forms of life and in atmosphere, its percentage is 79. But nitrogen is never taken directly from the atmosphere. The chief sources of nitrogen for plants are nitrates in the soil. During the cycle, the reduced and oxidized forms are involved at one or other stage. The chief symbiotic nitrogen fixers are bacteria belonging to the genus *Rhizobium* found in the root nodules of legumes, while asymbiotic nitrogen fixers are some blue-green algae. Plants in turn are eaten by animals. The dead organic matter formed due to death of plants and animals is decomposed by bacteria, fungi, etc. which releases nitrogen either in free state to atmosphere or as ammonia gas. Ammonia gas may also reach the soil as nitrates through the activity of nitrifying microbes. The whole pattern of the nitrogen cycle becomes clear from Figure 3.13.

Figure 3.13

Nitrogen Cycle

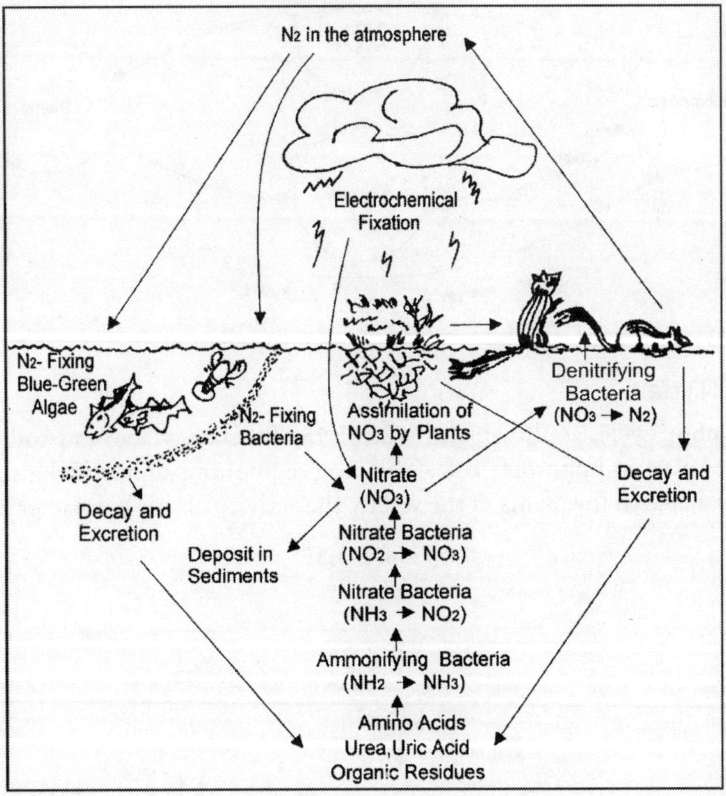

Phosphorus Cycle

This is a sedimentary cycle and a simpler one, in comparison to nitrogen cycle. As shown in Figure 3.14, phosphorus, an important and necessary constituent of protoplasm, tends to 'circulate', the organic compounds being broken down eventually to phosphates which are again available to plants. The great reservoir of phosphorus are rocks or other deposits which have been formed in past geological ages. These erode and release phosphates to ecosystems. A major portion of phosphates goes into the sea in the form of sedimentation. Sea birds also have a role in returning phosphorus to the cycle from sea to land. The general pattern of phosphorus cycle is more or less similar to that of other minerals.

Figure 3.14
Phosphorus Cycle

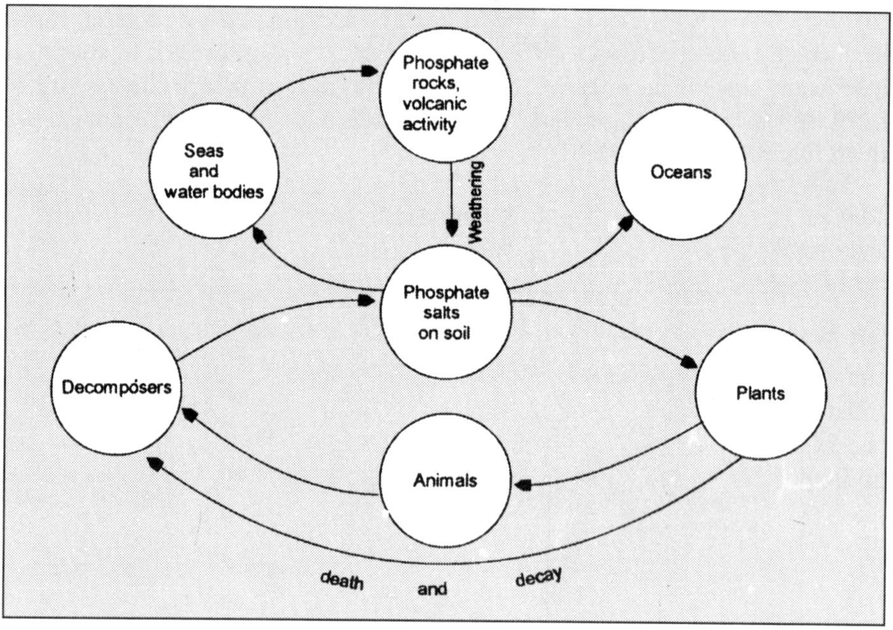

Sulphur Cycle

The sulphur cycle, as shown in Figure 3.15, links air, water and soil, where microbes play an important role. The reserve pool of sulphur is the soil. It is made available to the plants in the soil by the activity of sulphur bacteria. Some

Figure 3.15
Sulphur Cycle

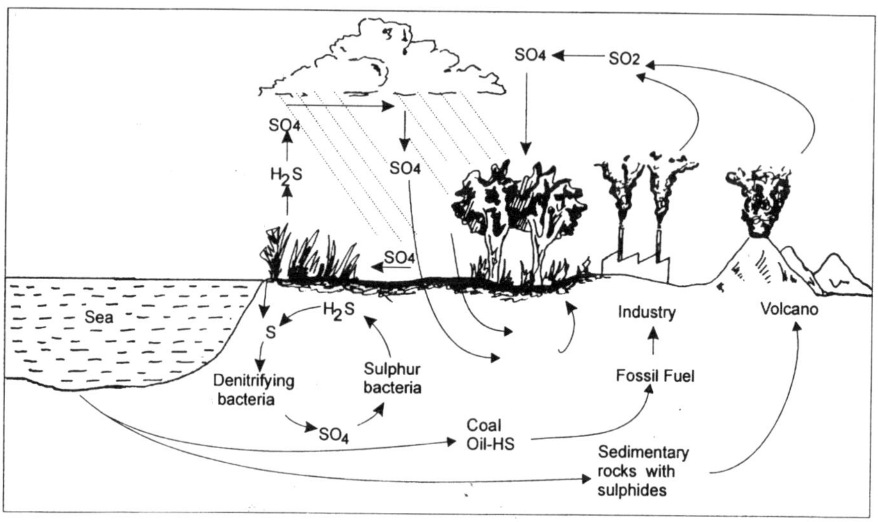

quantities are added to the atmosphere by the burning of the fossil fuels. Later, sulphur dioxide and hydrogen sulphide return to the soil as sulfates or sulphuric acid, with the rain. In tissues of organisms, sulphur is in the form of proteins. The decomposition of proteins releases sulphur. The sulphur cycle is a good example of the interaction and complex biological and chemical regulation between different mineral cycles.

Apart from above mentioned cycling of elements, other elements too follow an almost similar pattern of cycling. The cycles of these elements are usually restricted to the same ecosystem but some of them are widespread and a part of the global system.

Major Ecosystems

In the world, there are several ecosystems working at macro or micro level. As pointed out earlier, the biosphere is the biggest ecosystem which combines all the ecosystems of the world. But with the difference in physiography, climate, natural vegetation, soil and water bodies, separate ecosystems have been developed. The biosphere can be divided into following major ecosystems (Figure 3.16):

Figure 3.16

Basic Types of Ecosystems

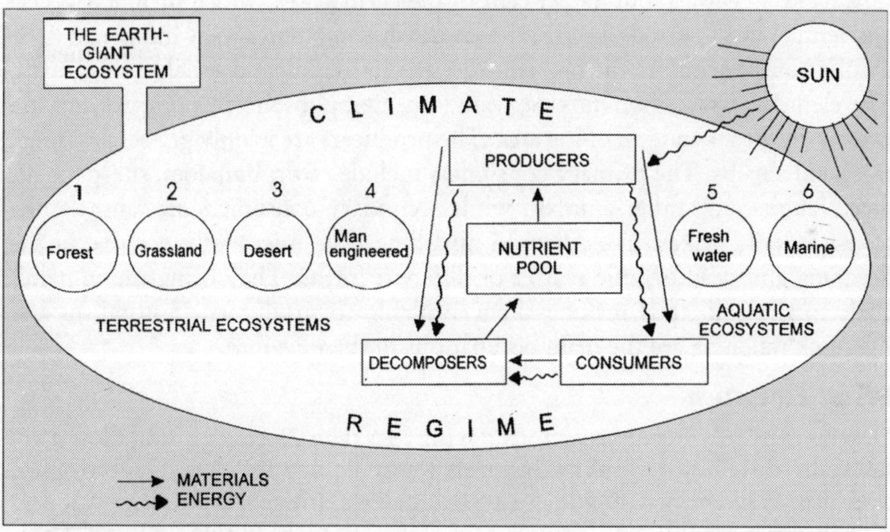

(A) *Natural Ecosystems*

1. Terrestrial Ecosystems
   (i) Forest (ii) Grassland
   (iii) Desert
2. Aquatic Ecosystems
   (i) Fresh water (ii) Marine

(B). *Man-Engineered or Artificial Ecosystems*

(i) Cropland

The major ecosystems are described as under:

Forest Ecosystem

About 30 per cent of the land area of the earth is under forest cover, but due to man's intervention this area is gradually becoming smaller. But still, forest ecosystem is very important. The coniferous forests stretch as broad belts across North America and Eurasia. Temperate deciduous forests occupy areas in eastern North America, parts of Europe, Japan and also in Australia. Among forests the tropical evergreen forests are found in the tropical regions.

In the forest ecosystem, the abiotic components are the organic substances present in the soil and atmosphere and also minerals present in dead organic debris. The producers are trees of different species. The primary consumers are animals of various types, while secondary consumers include carnivores like lions, tigers, snakes, birds, lizards, foxes, etc. The forest ecosystems are of great concern from the environmental point of view. The rate of exploitation of forest by man is growing day by day, thus causing a great concern to all nations of the world because of its impact over global climate and on several animal species.

Grassland Ecosystem

Grasslands occupy about 19 per cent of the earth's area, which include tropical and temperate grasslands. In this, the savannah ecosystem is very important. The abiotic components are the nutrients present in the soil and aerial environment. The elements like carbon dioxide, water, nitrates, phosphates, sulphates, etc. are present in the air and soil of the area. The producers are mainly grasses and small trees and shrubs. The primary consumers include cows, buffaloes, sheep, goats, deer, rabbits, and other animals, while secondary consumers are animals like foxes, jackals, snakes, frogs, lizards, birds, etc. The microbes are active in the decaying and dead organic matter of different forms. They bring the minerals back to the soil, thus making them available to producers. Pastoralism and livestock ranching are the main occupations in these regions.

Desert Ecosystem

Deserts generally occur in regions having less than 25 cms of rainfall and are unevenly distributed. Scarcity of rainfall may be due to: (i) high sub-tropical pressure, as in the Sahara and Australian deserts, (ii) geographical position in rain shadows, or (iii) high altitude as in Tibetan, Gobi, Bolivia deserts. There are three life forms of plants that are adapted to deserts: (i) the annuals, which avoid drought by growing only when there is adequate moisture, (ii) the succulents, such as the cacti, which store water, and (iii) the desert shrubs, which have numerous branches and a special root system which help them to adapt to desertic conditions. The most common animals apart from camels and goats, are reptiles and insects, able to live under xeric conditions. Due to poor vegetation, dead organic matter is less and some fungi and bacteria act as decomposers. Apart from hot desert, there are also widespread cold deserts encircling the north and south poles, where permanent ice caps have developed an entirely different type of ecosystem, sustainable to cold climatic conditions.

Freshwater Ecosystem

Freshwater habitats can be divided into two categories:

1. *Standing water or lentic (calm):* lake, pond, swamp or bog.
2. *Running water or lotic (washed):* river, spring, stream.

Although freshwater habitats occupy a small portion of the earth's surface, they are of great importance to man because they provide drinking water as well as water for domestic and industrial needs. Odum has defined freshwater ecology as:

> Freshwater ecology emphasizes the organisms in environment relationship in the freshwater habitat in the context of the ecosystem principle.[7]

A pond is a good example of freshwater ecosystem, which exhibits a self-sufficient, self-regulating system. A pond is a place where living organisms not only live but interact with abiotic and biotic components, thus forming an ecosystem which is different from other systems. Similarly, lakes, swampy regions and delta regions of rivers have their own ecosystems in which producers, consumers and decomposers interact and are responsible for the unique ecosystem of each.

Marine (Ocean) Ecosystem

The marine ecosystem is different from freshwater ecosystem mainly because of its salty water and also because (i) the sea covers 70 per cent of earth's area, (ii) the sea is deep, (iii) the sea is continuous, and (iv) the sea water is in continuous circulation. According to Odum, 'marine ecology emphasizes the totality or pattern of relationship between organisms and the sea environment.'[8] In a marine ecosystem, the ecology of shallow and deep waters as well as estuarine part are different from each other. Each ocean also represents a very large and stable ecosystem.

Cropland Ecosystem

Apart from above mentioned natural ecosystems, there are also man-engineered ecosystems in which man is very much involved in their operation. One such is the cropland ecosystem, in which man has developed croplands after considering the soil, climatic and other environmental conditions. These are ecosystems of dominant crop species like wheat, maize, jowar, paddy, sugarcane, cotton, tea, coffee, etc.

The abiotic components of this ecosystem include climatic conditions and mineral contents of the soil. In case of any deficiency, man used chemical fertilizers and/or water for irrigation, etc. The various types of foodgrains, pulses and commercial crops are grown in these croplands, which provide food and fodder not only to man and animals, respectively, but in the fields several types of animals like birds, rats, rabbits, and other smaller reptiles insects, etc. also survive. The decomposition of dead organic matter of plants and animals make the minerals available again.

There is a need that every ecosystem should be studied in detail. Apart from above mentioned macro or global ecosystems, there are several meso and micro regions existing in each region. For example, India has the following major ecosystems:

1. Himalayan mountain ecosystem
2. Plain ecosystem
3. Desert ecosystem
4. Central India plateau ecosystem
5. Peninsular (Deccan plateau) ecosystem
6. Coastal plain ecosystem
7. Island ecosystem

These ecosystems have been identified on the basis of physiography, which also controls climate, natural vegetation, soil and also man's adoption to nature. Each major region can be divided into sub-ecosystem regions, for example, Rajasthan has the following ecosystems:

1. Arid ecosystem
2. Irrigated arid ecosystem
3. Sub-arid ecosystem
4. Aravalli hills ecosystem
5. Eastern plains ecosystem
6. Hadoti plateau ecosystem

Further identification of micro-ecosystems will help not only in understanding the ecosystem but also in regional development and planning. The goal of sustainable development can only be achieved, when our developmental policies are based on the micro-ecosystem of each region.

Notes

1. F.J. Monkhouse and J. Small (1978), *A Dictionary of Natural Environment*, John Wiley, New York, p. 38.
2. P. Haggett (1972), *Geography : A Modern Synthesis*, Harper & Row, New York, p. 111.
3. A.N. Strahler and A.H. Strahler (1977), *Geography and Man's Environment*, John Wiley, New York, p. 271.
4. *Encyclopaedia Britannica*, Vol. III, 1980, p. 780.
5. E.P. Odum (1971), *Fundamentals of Ecology*, W.B. Sounders, Philadelphia, p. 43.
6. Ibid., p. 86.
7. Ibid., p. 295.
8. Ibid., p. 324.

4

Biomes: World Ecosystem Types

Biomes are the regional-scale assemblages of ecosystem on the surface of the earth and occupied by distinctive plant and animal communities. The pattern of biomes is essentially controlled by primary production, which provides the energy base for the biological communities that make up each biome. There are wide variations in productivity, and thus, variations in the types of plants. These are controlled by a series of abiotic environmental factors. The world pattern of terrestrial biomes is closely related to global climates. Human influences are becoming increasingly important elements affecting the nature of biomes.

Some of the definitions of biomes are as follows:

> A mixed community of plants and animals (a biotic community) occupying a major geographical area on a continental scale. (Goudie et al., 1995: 60)

> A major regional ecological community of plants and animals extending over a large area, with a uniform climate and vegetation. (Simmons, 1979: 347)

> A biome is a major ecological or biotic community of living plants and animals that occupies an extensive are. (Mathur, 1988: 177)

> Biomes are groups of ecosystems that are similar in structure and function. Wherever a particular biome is found, climate and soils tend to be similar and organisms tend to have a similar range of life. (Kerski and Ross, 2005: 24)

Each biome is characterized by similarity of vegetation structure and to a great extent, in species composition and is usually related to climate. Biomes are usually recognized by the structure of the vegetation. Because it is relatively stationary, vegetation is more easily observed than the animals of a biome. To understand biomes, it is also important to consider the characteristics of the regional climate and the limitations or requirements imposed upon life by specific temperature and precipitation patterns.

Table 4.1

Major Environmental Zones/Biomes of the World

Type	*Zone Name*	*Percentage of Land Surfaces*	*Main Areas*	*Dominant Vegetation Cover*	*Main Man-made Changes in Vegetation*	*Main Precipitation Characteristics*	*Main Thermal Characteristics*
Forested	Equatorial	8	Amazon Basin in S. America, Indonesia, S.Asian Peninsulas, Congo Basin in Africa	Natural broad-leaved evergreen forest; wide variety of species; swamp forests on floodplains and coast	Very variable: ranges from extensive clearing and cultivation (e.g. in Java) to little impact (e.g. in Amazonia); low to very high population densities	High rainfall (over 100 cm or 39 in) throughout the year; heaviest at Equinoxes	Uniformly high temperatures; little seasonal variations
	Midlatitude (Peripheral)	7	Europe, E.China, E. United States	Broad-leaved, deciduous, and mixed forests, merging with warm temperate, evergreen forests on eastern periphery	Very extensive clearing and cultivation; medium-to-high population densities throughout	Moderate precipitation (75–100 cm or 30–39 in) all year; heaviest in winter or autumn on western periphery, in summer maximum in eastern warm temperate area	Cool to warm temperate; seasonal range increases with continentality
	Boreal	14	Russia and Scandinavia, Canada, Alaska, NW United States	Needle-leaf forests; relatively uniform stands with small number of species (e.g. spruce, fir, pine, larch)	Limited clearing on equatorward fringes; very low population densities	Light precipitation (25–50 cm or 10–20 in), mainly in summer	Short, cool summers; very large annual temperature range
Inter-mediate	Savannah	24	African tropics, S.America, S.E.Asia	Ranges from open, tall-grass savannah to deciduous monsoon forest; gallery forest along stream systems	Burning, grazing, variable clearing, and cultivation; high population densities on floodplains in monsoon Asia, otherwise low	Variable rainfall (25–200 cm or 10–79 in) with pronounced spring or summer maximum	Warm; small seasonal variations

Contd...

...Contd

	Mediterranean	1	Lands around Mediterranean Sea, S.Australia	Evergreen drought-resistant hardwoods and shrubs	Extensive clearing and cultivation, especially in lands around Mediterranean Sea; variable population density	Low to moderate rainfall (50–75 cm or 20–30 in) with pronounced summer drought	Warm temperate; moderate annual range
	Midlatitude grasslands	9	C.Asia and E.Europe, C.North America, E.Australia	Grasslands varying from tall-grass prairies to short-grass steppes with decreasing humidity	Hunting and grazing; main settlement and cultivation in last 150 years; low population density	Low to moderate rainfall (30–60 cm or 12–24 in), mainly in spring and summer; considerable year to year variation	Very strong seasonal variations; cold winters dominated by invasions of polar air
Barren	Arid and semi-arid	21	C.Asia, Sahara and S.W.Asia, C. and W.Australia	Widely dispersed, drought-resistant shrubs; salt flats, plantless sand, and rock deserts	Little impact outside small irrigated areas	Very low rainfall (0–25 cm or 10 in) with considerable year to year variation	Very high summer temperatures; seasonal variations range from moderate in tropics to very high in mid latitudes
	Tundra	5	N.Canada and Alaska, Russia and N. Scandinavia	Low herbaceous plants, mosses, and lichens	Little impact	Low annual totals (10–40 cm or 4–16 in) with late summer or autumn maximum; light winter snowfall	Severe cold; short, cool summers
	Polar	11	Antarctica, Arctic	Ice caps; no plant life	No impact	Low annual totals; little detailed precipitation data	Extreme cold; no months with above-freezing average temperatures

The three groups of major biomes, viz. Forest Biomes, Grassland and Mediterranean Biomes and Barren (Arid & Tundra) are depicted in Figures 4.1, 4.2 and 4.3.

Figure 4.1
Forest Biomes of the World

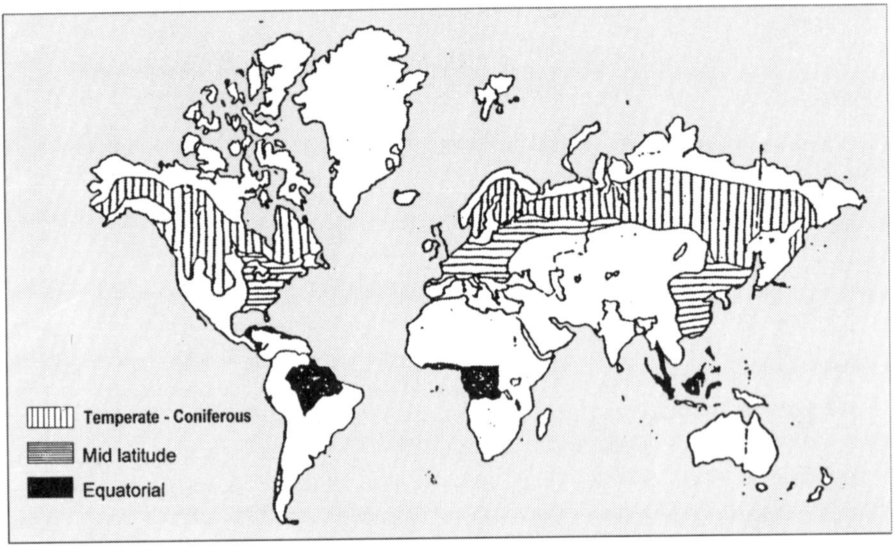

Figure 4.2
Grassland and Mediterranean Biomes

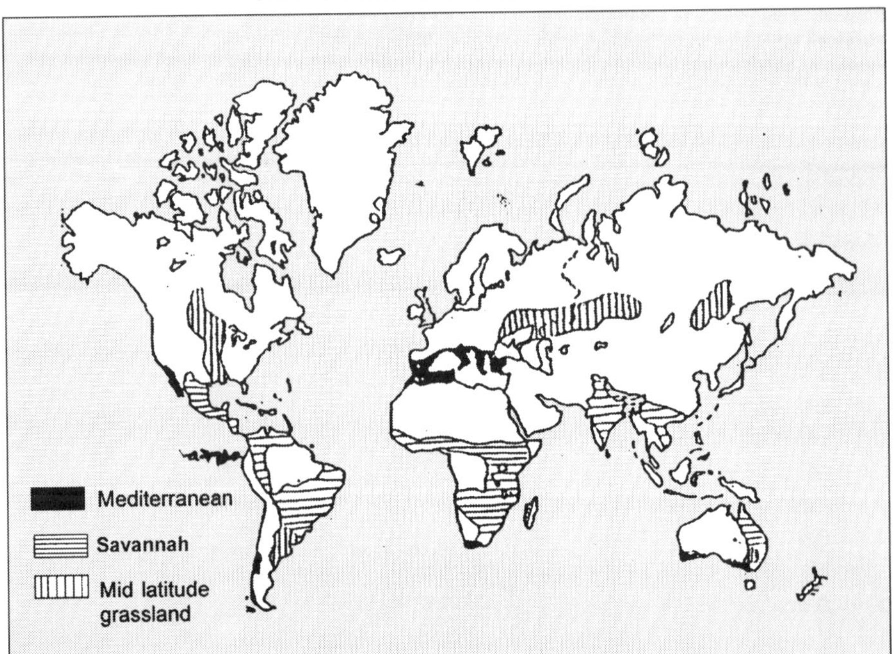

Figure 4.3
Barren Areas Biomes

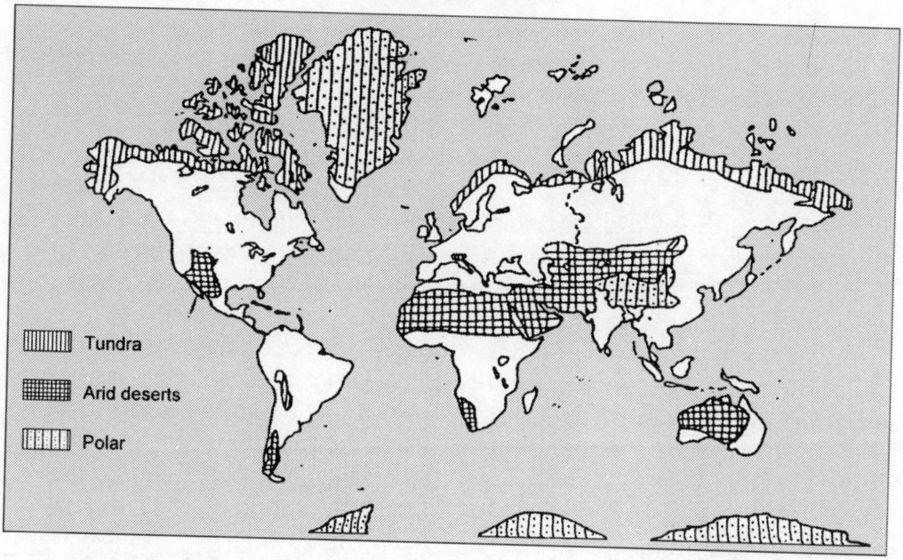

Major Biomes of the World

The global pattern of biomes is varied due to difference in physiography, climate, natural vegetation, soil and water bodies. Peter Hagget, in his book, *Geography: A Modern Synthesis* (1975) has described the major ecosystems/biomes as given in Table 4.1.

Joseph Kerski and Simon Ross in their publication, *The Essentials of the Environment* (2005) has identified two broad categories of the biomes of the world, viz. terrestrial biomes and aquatic/marine biome and sub-divided them according to their characteristics as given in Table 4.2.

In fact, major biomes of the world are the terrestrial biomes, which are based on climate and natural vegetation and spread in various regions of the world as shown in Figure 4.4.

A brief description of major biomes of the world is presented here under the following heads:

1. Humid and sub-humid tropics – Tropical rainforest, Savannahs and some other grasslands
2. Hot deserts – Desert and semi-desert
3. Middle latitudes – Deciduous forests, Temperate grasslands and Mediterranean biomes
4. Continental interiors of middle latitudes and coasts of the subarctic – Northern coniferous forest biome
5. Arctic and mountain biomes.

Table 4.2

Major Biomes of the Earth

Terrestrial biome	*Mean annual Temperature (°C)*	*Mean annual Precipitation (cm)*	*Producers*	*Typical Primary Consumers*	*Net Production $g/m^2/year$*	*Other Characteristics*
Temperate forest	–3–20	6–250	Deciduous trees, shrubs, annual and perennial herbs	Browsers, frugivores	1,200	Highly seasonal, cold winters, hot summers
Temperate grassland	–5–20	50–130	Grasses	Grazers, seed-eaters	600	Fire-, drought-, wind influenced, extremes, burrowers
Tropical savannah and scrub	17–30	50–150	Scattered trees over grassland	Grazers, browsers	900	Fire-prone
Taiga (boreal coniferous forest)	–5–5	5–200	Needle-leaved trees	Browsers, seed-eaters	800	Minimal under-story, high litter mass
Tundra	–15–5	10–100	Sedges, grasses	Grazers (lemmings, musk ox)	140	Permafrost, many breeding shorebirds
Chaparral	–5–20	50–75	Hard-leafed scrub	Browsers, seed and fruit predators	700	Summer drought, winter rain, fire-prone
Desert	–5–30	0–50	Stem- or leaf-succulents	Browsers, seed predators	90	Water-rich tissues protected by spines, etc.
Tropical rainforest	17–30	25–450	Tall buttressed trees, vines, epiphytes	Specialized herbivores, tree animals	2,200	Little seasonality
Lotic (flowing) waters			Rooted aquatic plants	Invertebrates, fish	250	Well-aerated
Lentic (lakes, ponds)			Floating plants	Invertebrates, fish phytoplankton	250	Poorly aerated

Contd...

...Contd

Swamps – wetland with trees	Rooted plants, trees and graminoids	Invertebrates, fish, some mammals	< 2,000	High productivity
Marshes	Herbaceous (graminis) wetlands	Invertebrates, fish, specialized rodents	< 2,000	
Estuaries	Rooted plants, phytoplankton	Invertebrates, fish	< 1,500	High productivity
Mangrove swamp	Mangroves (various species)	Invertebrates, fish	1,100	Littoral or estuarine
Littoral (shoreline)	Phytoplankton	Invertebrates		Moderate productivity
Neritic (continental shelf)	Phytoplankton	Invertebrates, fish		Moderate productivity
Reefs	Algae symbiotic with corals	Fish, corals and other invertebrates	< 2,500	Tropical, high species richness, high productivity
Pelagic (open ocean)	Phytoplankton	Invertebrates, fish	125	Low productivity

Figure 4.4
Major Biomes of the World

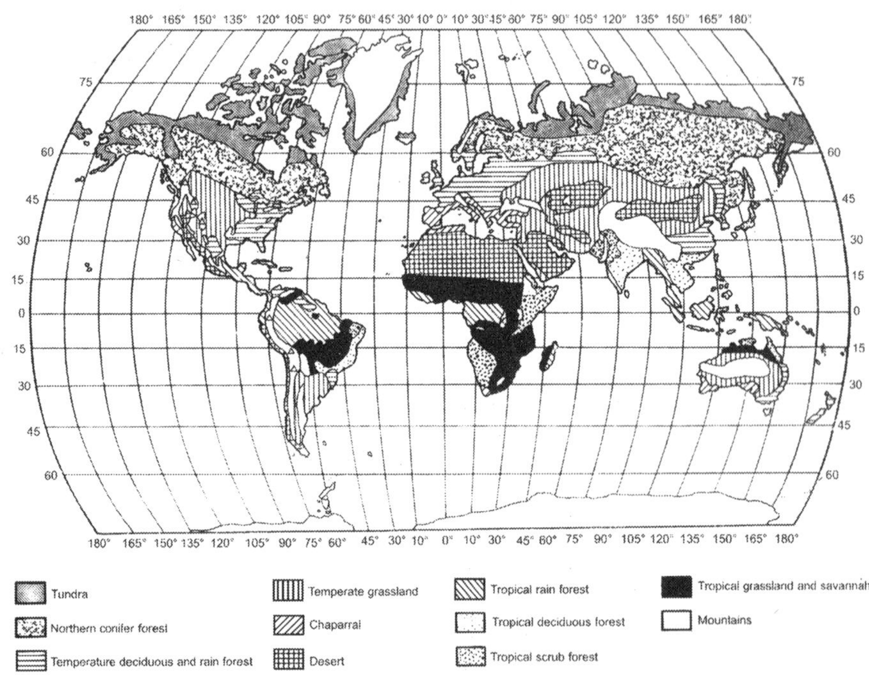

Major biomes of the world

Biomes of Humid and Sub-humid Tropics

Two major biomes, viz. tropical rainforests and Savannahs are included in this zone. The distribution of these types is shown in Figure 4.5.

The *tropical rainforest biome* is found in the tropical regions of the world. The Amazon region is the largest tract of equatorial and tropical rainforest. The other regions included are Congo Basin, Coastal Gulf of Guinea, margins of Madagascar, South-East Asia, the Pacific Coast of Equador and Colombia, and the East Coast of central America.

In the equatorial region, owing to intense heat and high moisture, the growth of trees is dense and luxuriant. Many of the trees grow to tremendous height and in the struggle to reach the sunlight. Epiphytes are found abundantly in these forests. The wild rubber tree is also a native of the tropical rainforest. These forests are having mixed types of trees. In many places, there are more than 50 species of trees per hectare. Therefore, logging is difficult and selective cutting is required.

These forests have the highest primary productivity, the greatest biodiversity and the most structural complexity of any terrestrial biome. The richness in biodiversity extends to all trophic levels and all types of organisms. According to an estimation, about 9,50,000 species of insects have been

Figure 4.5
Distribution of Tropical Forests and Savannah Grassland

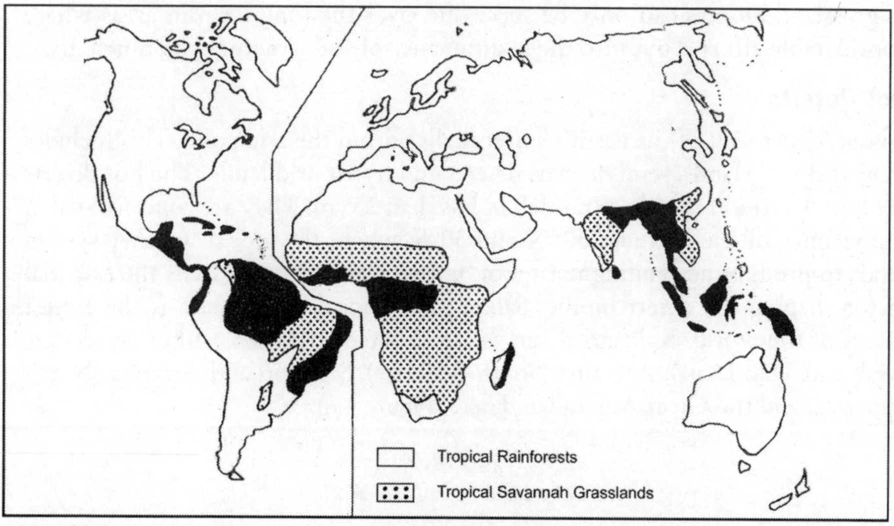

identified and probably some 8,000,000 species are in existence. Majority of these unidentified species are in humid tropics. The three-dimensional mosaic of plant species, large and small, is the driver of both the very high rates of primary productivity and the high indices of biodiversity.

Animal species in tropical rainforests are also very diverse, with insects, amphibia, reptiles and birds. Co-evolutionary relationships appear to be more highly developed in this biome than in any other. Tropical forest soils (oxisols) are in general exceedingly thin and nutriently poor. The leaching process leaves behind large residue of insoluble iron and aluminum oxides in the upper levels of tropical forest soils. Tropical rainforests are being destroyed more rapidly than in any other biome.

Tropical Savannah Grassland Biomes are found between the wet forests and the hot deserts. Africa has the largest region of this biome, including the famous Serengeti plains and the Sahel region. Sections of Australia, Peninsular India, parts of Brazil, Argentina and Venezuela are all parts of Savannah biome. Some local names for these lands include the Leanos in Venezuela, the Campo Cerrado of Brazil and Guinea and Pantanal of South-Western Brazil.

Savannahs are although less productive but sufficient to support the greatest concentration of large grazing animals. These grasslands contrast strongly with rainforests in two ways other than obvious factors of lower primary productivity and dominance of most of the vegetation cover by grasses. Soil has more available nutrients. Grazing in Savannahs may be dominated by large vertebrates. Large animals typical to grasslands include bison (buffalo), pronghorn antelope in North America, wild horses in Eurasia, large kangaroos in Australia, and zebras, giraffes, white rhinoceros and a vast diversity of antelopes in African Savannahs. The Savannahs support the richest fauna of large grazers.

The grassland biome has a higher concentration of organic matter in its soil than other biomes. The richness of grassland soil has led to the development of agriculture in these regions as in the American Prairies. Agriculture in grassland biomes can only be successful over the long term in areas where considerable effort is put into the maintenance of soil structure and nutrients.

Hot Deserts

About 30 per cent of the earth's surface falls within the arid zone. This includes semi-arid rangelands, semi-deserts, deserts and hyper-arid lands. The hot deserts are found in the areas where rainfall in less than 25 cm. They are concentrated in the vicinity of the latitudes 30°N and 30°S, where the global weather system tends to produce descending masses of dry air. Lack of moisture is the essential factor shaping the desert biome. Sahara desert located in Africa is the largest desert of the world. Kalahari desert is also located in Africa. Other deserts are Arab and Thar (Asia), Atacama (South America), Sonoran and Arizona (North America) and the Great Australian desert (Figure 4.6).

Figure 4.6
Hot Desert Biomes of the World

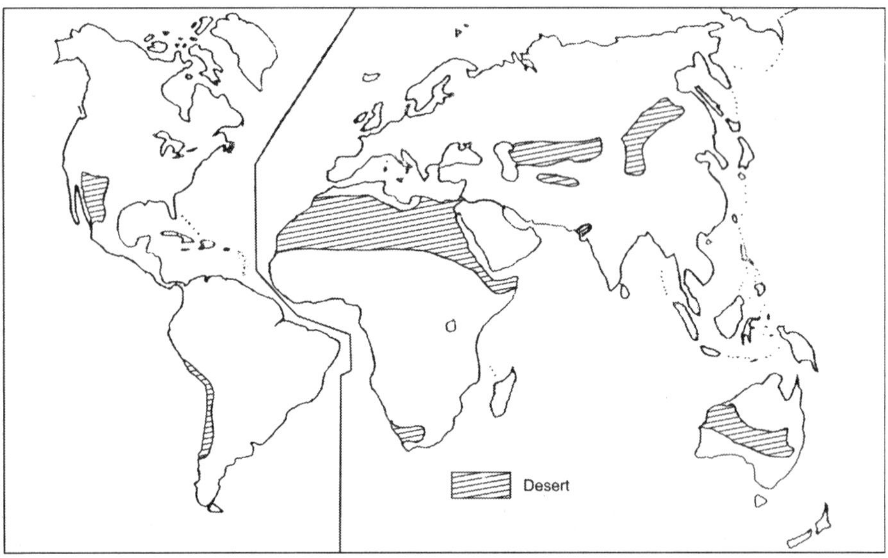

Biomes of the arid zone are characterized by low levels of primary biological productivity. This is the inevitable result of the scarcity of water. Inspite of low total primary productivity, the biodiversity of these biomes is not as low as might be expected. The higher trophic levels are inevitably influenced by the harsh thermal environment and small amount of energy available for consumption. There is a tendency for biodiversity to be at its highest in locations in which there are some limited water supplies and for biodiversity to decrease as water availability increases, and more competitive plants can be sustained.

Scarcity of water is the limiting factor. Humans have tried to offset this since the dawn of civilization. Indeed, it is notable that among the earliest so-called 'hearths' of civilization are the desert areas traversed by great rivers (e.g. the lower Nile and Mesopotamia, the zone between the Tigris and Euphrates rivers in present-day Iraq). Here some of the first plant domestication took place. The light, easily cultivated soils provided an ideal habitat, given water, for disturbance-tolerant grasses, which are the ancestors of our most important cereals. Water was available from the great rivers. Natural flood regimes were soon supplemented by basic irrigation mechanisms. In the twentieth century, huge areas of the margins of the Sahara, the deserts of Central and South Asia and in the southwest of the United States, as well as many other parts of the world, have been made into highly productive, intensive agricultural areas by large-scale irrigation schemes.

The Middle Latitudes

In this climatic region, most of the developed world is located. It is perhaps that group of biomes, most modified by humans. Figure 4.7 depicts the distribution of the three sub-categories of middle latitudes, i.e. deciduous forests, temperate grasslands and the Mediterranean biomes.

Deciduous Forests Biome is also known as temperate coniferous forest or boreal or taiga forest biome. These forests once covered most of the middle latitudes and were distributed in North America and Eurasia. Besides, there are small patches of natural coniferous forests at higher altitudes in Germany, Poland, Switzerland, Austria and other parts of Europe and on the high rocky mountains of North America.

Figure 4.7

Distribution of Deciduous (Temperate) Forests, Temperate Grasslands and Mediterranean Biomes

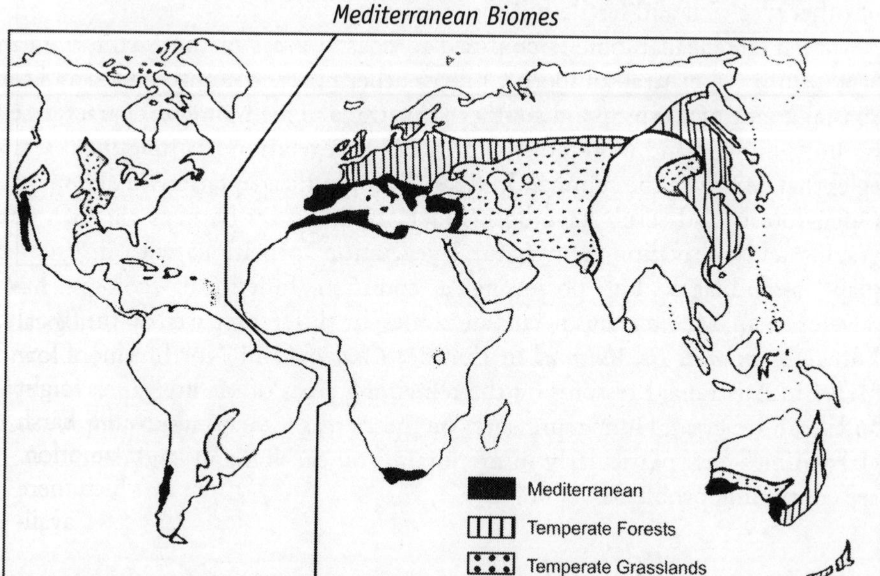

The coniferous trees are the most dominant members of this biome. These trees form dense cover of forests which are the richest sources of softwood in the world. The pine, fir, spruce and larch are the main types of trees found in this biome. The characteristics of these forests contrast with tropical rainforests. Temperate deciduous forests have significantly lower primary productivity. This lower level of productivity is largely caused by the dormancy of trees during the cold winter. Biodiversity is lower and vegetation structure, less complex. Soil conditions too, are different. Lower rates of decomposition and uptake mean that proportion of plant nutrients which are in available form in the soil is smaller. But soils are fertile; because of this reason much of this biome has been converted into intensive form of agriculture.

Temperate Grassland Biome is located in interiors of the continents which come in the westerly wind belt but because of their more interior locations they do not get sufficient rainfall and hence the grasslands are particularly treeless. The temperate grassland biomes are widely distributed in various continents and are known by different names such as Steppes (Russia), Prairies (North America), Downs (Australia), Velds (South Africa) and Pampas (Argentina). These grasslands are now extensively cultivated areas and are famous for growing cereals. In these regions, the only naturally occurring trees are deciduous broadleaves along streams and other limited sites. These regions are called grasslands because of the original predominance of grass-like plants. No other natural ecosystem or biome tells the story of the impact of human activities better than the temperate grassland biomes of the world. The large-scale removal of vegetation for agricultural purposes has resulted in the loosening of the soil cover because of the destruction of dense network of root systems of grasses. The change in the soil cover has resulted into accelerated rate of soil erosion and therefore loss of otherwise rich and fertile soil.

Mediterranean Biome is confined to coastal areas of the Mediterranean Sea, California, central-southern Chile, southern parts of South Africa and the coastal zones of western and southern Australia. This biome is characterized by hot summers and mild-moist winter. Most vegetation has functional strategies that deal with the summer drought. Xerophytic adaptations are common. Open forests and scrub have been much modified by humans through fire grazing and cultivation. Semi-natural vegetation formations with distinctive plant assemblages, but possessing a common functional ecology, have developed in Mediterranean climate zones in different parts of the world, known as *Maquis* or *Mattoral* in Europe, *Chaparral* in North America and *Mallee* in Australia. Pressures on the remaining areas of Mediterranean forests and scrub are great. Human pressures on these areas have been substantial since classic times and, particularly in mountain, soil erosion and land degradation are continuing problems.

Continental Interiors of Middle Latitudes and Coasts of Sub-Arctic

In parts of the middle latitudes, extending in the northern hemisphere to 60° N, winter climate is severe, with temperatures below 0°C for extended and continuous periods, and with short though warm summers, in some cases with little more than 120 days when temperatures are above the threshold for growth. Typically these areas are found in the interiors of continents, far from oceanic moderating influences. Mostly such areas are quite dry and some merge into semi-desert, though generally there is enough rainfall to sustain tree growth. However, primary productivity is low, typically less than one-sixth of that in tropical forests. Correspondingly, biological water demand is much lower and evapotranspiration rates over much of the year are low. The limiting factor controlling vegetation here is the short thermal growing season. Similar thermal though moisture conditions may be found in the coastal areas of the subarctic, such as southern Alaska, southern Chile or western Scandinavia. In all of these locations, conditions are ideal for trees, particularly of a handful of species of large, slow-growing conifers. This is the boreal forest biome, the extent of which is shown in Figure 4.8.

These forests are dominated by a handful of species of conifers, pine, spruce and larch species. Structure is simple, with often only two or at most three layers of vegetation, and biodiversity is low. Both plant and animal life must be adopted to long and cold winter. There are few tall trees, which utterly dominate the biome and grow slowly. In the winters, the birds and mammals migrate or become dormant. The soil and decomposing systems in these forests are quite different from those in tropical rainforest and temperate deciduous

Figure 4.8

Northern Coniferous Forest Biome

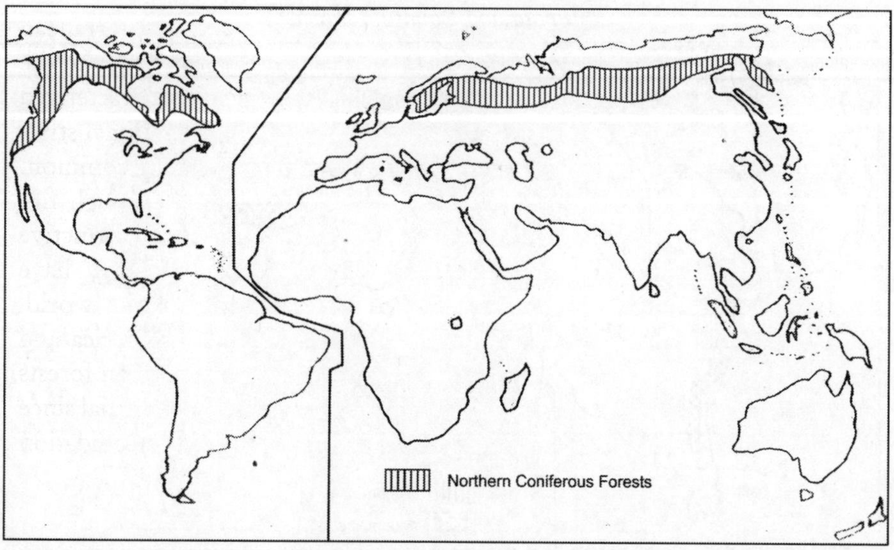

forests. There are large amounts of unavailable nutrients in soil litter. Boreal forests are not immune to human impacts, however the trees themselves are valuable raw materials which have been ruthlessly exploited.

Arctic and Mountain Biomes

Arctic or Tundra biome includes parts of Alaska, extreme northern parts of Canada, the coastal strip of Greenland, and the arctic seaboard regions of European Russia and northern Siberia. Besides, Tundra biome has also developed over arctic islands. Vegetation rapidly changes to the north of tree line because of increasing severity of climate. Thus, based on variations in the general characteristics of vegetation in the arctic tundra, tundra biome is divided into sub-divisions, e.g. Arctic Tundra Biome and Alpine Tundra Biome, where the latter is found over high mountains of tropical to temperate areas. Alpine tundra occurs in the Andes, the white mountains of California, the Rockies, the Alps, the Himalayas and the Mount Kilimanjaro of equatorial Africa (Figure 4.9).

The climatic conditions are the controlling factor of this biome. There is general absence of insolation and sunlight and temperature is very low throughout the year. This severe climatic condition does not favour much vegetative growth and hence most of the Tundra remains a barren land. Primary biological productivity in these biomes is at about the same level as terrestrial deserts or oceanic waters. Only a few species of small plants can survive the cold climate in which temperatures in the short growing season rise little above the threshold for growth. Vegetation is dominated by hardly grasses, mosses and a few shrubs. The animals of arctic tundra biome are grouped into two categories, viz. resident animals and migrant animals. Most

Figure 4.9
Tundra and Mountain Biome

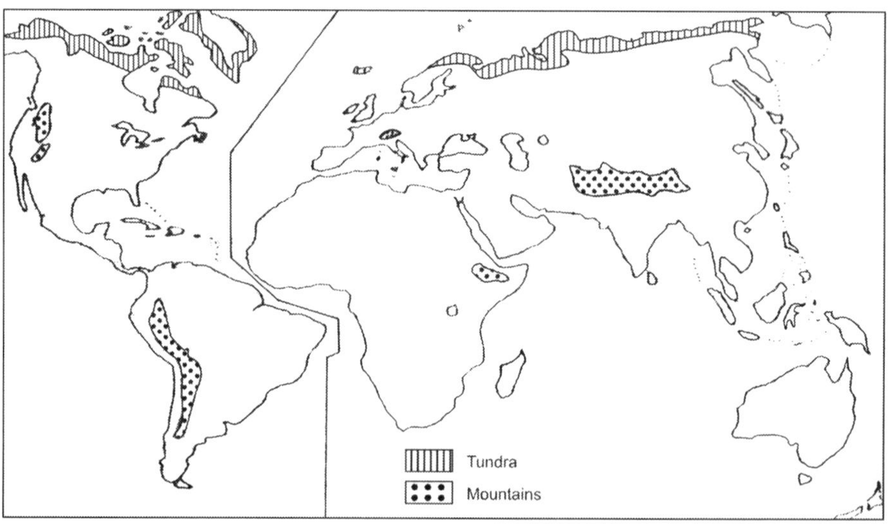

of the animals leave Arctic Tundra and migrate southward during winter season. On the other hand, Alpine biome features grasses and stunted shrubs, such as willows and heaths. Man is closely associated with the biota of the tundra biome because even his very existence depends upon animals of both terrestrial and aquatic habitats. This is a fragile environment and its ecosystems are easily damaged by even low levels of human impact.

References

Goudie, Andrew et al. (eds) (1995), *The Encyclopedic Dictionary of Physical Geography*, Blackwell, Oxford, p. 60.

Simmons, I.G. (1979), *Biogeography*, Edward Arnold, London, p. 347.

Mathur, H.S. (1988), *Essentials of Biogeography*, Pointer Publishers, Jaipur, p. 177.

Kerski, Joseph and Ross, Simon (2005), *The Essentials of the Environment*, Hodder Arnold, London, p. 24.

Haggett, P. (1972), *Geography: A Modern Synthesis*, Harper and Row, New York, p. 111.

5

Human Impact on the Natural Environment

The human impact on natural environment is one of the most pressing issues of contemporary times and a subject of discussion in geographical literature. The ways in which human beings have changed and are changing the face of the earth and the human role in the natural processes and systems, have drawn the attention not only of natural scientists but also of social scientists as well as of planners and policy makers. The geographical views regarding man-environment relationship has been described in Chapter 2. The basic law of ecology is that everything is connected to everything else and that one cannot change just one thing in nature. George Perkins Marsh published *Man and Nature* in 1864, in which he dealt with human influence on the woods, the waters and the sands. This relationship has been described in depth in their own way by geographers like Brunhes, Blache, Semple, Ratzel, Barrows, Sauer, Gerasimov and by many Anglo-Saxon and American geographers. The subject of the human impact on environment, dealing as it does with such matters as environmental degradation, pollution and desertification, etc. has also been discussed in many textbooks and research monographs on geography. Although the entire text of this book is in some way or other related with this aspect, the present chapter aims to discuss the human impact on climate, vegetation, soil, geomorphology, etc.

Human Impact on Climate and Atmosphere

The increasing human population and the rising level of technology, both have become significant factors in the variations in world climate and are responsible for the various changes in atmospheric conditions as well as air pollution. The human influence on global climate is due to the following mechanisms:

— gas emissions
— CO_2 – industrial and agricultural
— ethane
— chlorofluorocarbons (CFCs)

— nitrous oxide
— krypton 85
— water vapour
— miscellaneous trace gases
— aerosol generation
— thermal pollution
— albedo change
— dust addition to ice caps
— deforestation
— overgrazing
— extension of irrigation
— alteration of ocean currents by constricting straits
— diversion of fresh waters into oceans

The problem of carbon dioxide (CO_2) emission has become a major environmental concern. Since the beginning of the Industrial Revolution, humans have been taking stored carbon out of the earth in the form of coal, petroleum and natural gas, and burning it to make carbon dioxide, heat, water vapour and small amounts of sulphur dioxide (SO_2) and other gases, which are responsible for air pollution, green house effect, increase in surface temperature, or in other words, global warming. By 2050, it is possible that the increase in global surface temperatures ranges between 1.4 and 2.2 degree celsius. All other aspects of atmospheric pollution will be discussed under the heading, 'Air Pollution'.

The impact of human influence on the atmosphere is more because the atmosphere acts as a major channel for the transfer of pollutants from one place to another. It is in this way that harmful substances have been transferred long distances from their sources of emission. A second example of the possible widespread and ramifying ecological consequences of atmospheric pollution is provided by acid rain.

In recent years, the greatest attention has been focused on the role of CFCs, the production of which has been rising in last few decades. These gases may diffuse upwards into the stratosphere where solar radiation causes them to become dissociated to yield chlorine atoms which react with and destroy the ozone present there. The Antarctic ozone hole has been identified through satellite monitoring. In fact, air pollution and its variability both in space and time is more due to human action.

The main cause of global warming is now termed as anthropogenic, which means it has been caused primarily by human activities. The scientific consensus is that human activities have been warming the world and it will continue to get hotter irrespective of whatever we choose to do now. Predicions suggest that if we change our behaviour for the better immediately, then the increase in warming will be kept under control to some extent, but if we carry on in the same way as we do now, then the warming will increase and the consequences will be dire, at least for the large proportion of the world's population.

Human Impact on Vegetation

The human impact on vegetation is greater than on any of the other components of the environment. The nature of whole landscape has been transformed by human-induced vegetation change as depicted in Figure 5.1.

Figure 5.1

Human Impact on Vegetation

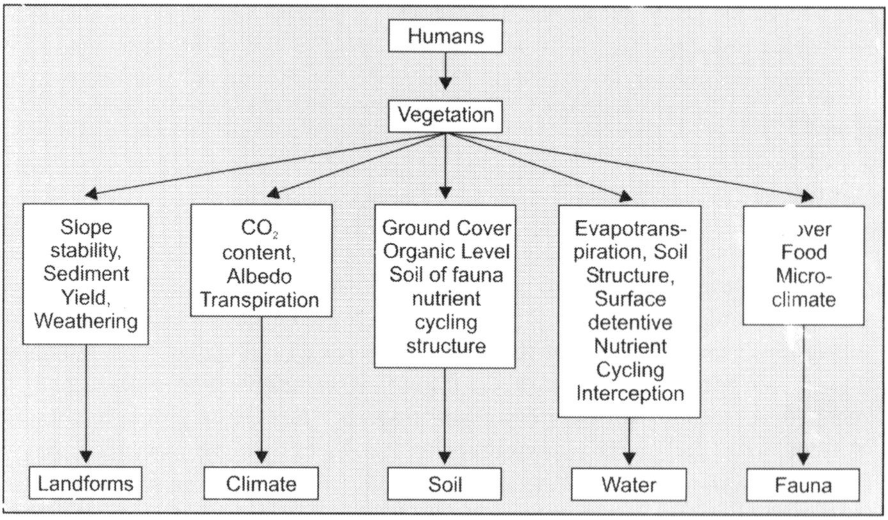

The extent of human influence on vegetation is provided by Westhoff (1983) in the following four-part scheme:

1. *Natural*: a landscape or an ecosystem not influenced by human activity.
2. *Sub-natural*: a landscape or ecosystem partly influenced by humans, but still belonging to the same formation type as the natural system from which it derives.
3. *Semi-natural*: a landscape or ecosystem in which flora and fauna are largely spontaneous, but vegetation structure is altered so that it belongs to other formation type such as pasture, moorland, etc.
4. *Cultural*: a landscape or ecosystem in which flora and fauna have been essentially affected by human agencies in such a way that dominant species may have been replaced by other species as in case of arable land.

Man has used fire for the clearance of forest cover, mainly for the use of land for cultivation or for habitation. This practice was common during the early stage of the development of civilization. But, this practice is still prevalent among many tribals. Although, controlled burning is a part of forest management, uncontrolled and deliberate fire is the cause of destruction not only of vegetation but also of animals and other forest creatures. Similarly, uncontrolled and heavy grazing is not only a cause of the disappearance of vegetation cover but is also responsible for desertification and other environmental problems.

The deliberate removal of forest or deforestation is one of the most long-standing and significant ways in which humans have modified the environment, whether achieved by fire or cutting. Sometimes, forests have been cut down to allow agriculture, at other times, to provide fuel for domestic purposes or to provide charcoal or wood for construction, etc. One of the great phases of forest clearance occurring at the present time is in humid tropics. The annual regression rate of tropical moist forest is about 11 million hectares. The spread of desert-like conditions (desertification) in arid and semi-arid areas is always due to man's influence. Some of the air pollutants, that humans have released into the atmosphere, have had detrimental impact on plants – sulphur dioxide, for example, is toxic to them. Local concentrations of industrial fumes also kill vegetation. Photochemical smog is also known to have adverse effect on plants both within cities and also on the its outskirts. The adverse effects of pollution on plants are not restricted to air pollution, for water and soil pollution can also be serious. Excessive amounts of heavy metals may prove toxic to them. Salt marshes, mangrove swamps and other kinds of wetlands are particularly sensitive to oil spills. In fact, human activities have become a prime cause for the gradual removal of forest cover in many parts of the world and has become a main cause of the degradation of the environment.

Human Impact on Animals

The range of impact that humans have had on animals, though large, can be grouped conveniently into five main categories: domestication, dispersal, extinction, expansion and contraction. The extinction of animals by human predators has been extensive over the past 20,000 years, and in spite of recent interest in conservation, it continues at a high rate.

The extreme effect of human interference with animals is extinction. The problem of extinction of several animal species has become a world-wide ecological problem. This decline is partly due to intentional killings for subsistence and commercial purposes, but much wildlife decline occurs due to pollution and also due to change in ecosystem and natural habitat. For example, the use of DDT is one of the causes for the extinction of several types of insects, birds and other small animals. The polluted water containing heavy metals and methyl mercury is also harmful. Oil pollution is an increasingly serious problem for marine and coastal fauna and flora. Sea birds are especially vulnerable since oil clogs their feathers. Other industrial pollutants have a clear impact on aquatic systems. Pollution is a serious problem for coral reefs.

The list of other indirect causes of wildlife decline is a vast one. Modern society is particular about the conservation of wildlife and efforts have been made by national and international agencies and also by NGOs to protect them, but, due to selfish and commercial motives, the killing of these animals is still going on. Various type of habitats have also started disappearing due to adverse human impact. On the global basis the loss of wetland habitats is a cause of considerable concern. Such areas are important for the breeding of several animals. The other threats include drainage, dredging, mining, filling, pollution and channelization.

Briefly, the human impact on animal extinction can be grouped in three categories:

1. The blitzkrieg effect, which involves rapid development of human populations with big-game hunting technology so that there is very rapid demise of animal population;
2. The innovation effect, whereby long-established human population groups adopt new hunting technologies and erase fauna that have already been stressed by climatic changes; and
3. The attrition effect, whereby extinction takes place relatively slowly after a long history of human activity because of loss of habitat and competition for resources.

Human Impact on Soil

Soil is the most vulnerable of human resources and is one on which humans have had a very major impact, because they live close to and depend on the soil. Impact on soil can occur with great rapidity in response to land use change by new technologies. The major changes brought about by humans are chemical, structural and hydrological changes, while soil erosion is perhaps the most important.

Salinity is a natural characteristic in some semi-arid and arid soils. But humans have increased the extent and degree of salinity in different ways. The extension of irrigation and different techniques used for water abstraction, can lead to a build-up of salt levels in the soil through the mechanism of raising ground water level. The construction of large dams and barrages to control water flow and to give a head of water creates large reservoirs from which further evaporation can take place. The seepage of water is also responsible for upward movement of ground water. In coastal areas, salinity problems are created by sea water incursion brought about by over-pumping. The clearance of native forest vegetation is also a cause of increase in salinity. Nowadays, measures have been taken for reclamation of salt affected areas through eradication, conversion and control techniques; but soil salinity is still a grave problem in many parts of the world.

Laterization is a chemical process in the soil of the tropics where extensive sheets of a material called laterite has been deposited either through natural process or by human actions. Laterite is a *paleological leprosy*. Man's activities aggravate the dangers of laterization when started by negligent removal of the forest.

Human activities are also responsible for the structural changes in soil. There are many ways in which humans can alter this, especially by compacting it with agricultural machinery and by changing its chemical character through irrigation. Grazing is another activity that can damage soil structure through trampling and compaction. Soil drainage is a slow and gradual process and is responsible for certain environmental changes and also a cause for long-term damage to soil quality.

The introduction of chemical fertilizers has also changed the chemistry of soils. Sometimes these create environmental problems such as water pollution, while their substitution for more traditional fertilizers may accelerate soil structure deterioration and soil erosion. Nowadays, the use of three main fertilizers – nitrates, phosphates and potash – has greatly increased agricultural productivity but has also changed the basic structure of the soil.

Soil erosion is the result of natural processes but at the same time it is also a result of human activities. Myers (1988) has summarized the scale of accelerated soil erosion that has been done by human actions. 'Since the development of agriculture some 12,000 years ago, soil erosion is said by some to have ruined 4.3 million km^2 of agricultural lands, or an area equivalent to rather more than one-third of today's croplands – the amount of agricultural land now being lost through soil erosion, in conjunction with other forms of degradation, can already be put at a minimum of 200,000 km^2 per year.' In fact, soil erosion is a major and serious aspect of the human role in environmental change. The processes of construction, urbanization, mining, etc. are also accelerating this problem, although the prime causes of soil erosion are deforestation and agriculture. Because of the adverse effects of accelerated erosion several techniques have also been used to conserve soil resources. Some of them are – revegetation, stoppage of full enlargement, crop management, slope run-off control, suppression of wind erosion, terracing, contour ploughing, etc.

Human Impact on the Waters

Water is the source of life and right from the origin, human beings are using it for various purposes. The ancient civilizations have developed in river valleys. This is also true about medieval townships and other developments, and all modern developments are related directly or indirectly to water. The main concern is that by using water, humans have influenced both its quantity and quality. Earlier, the influence of human activity on water resources was limited but now this has become a major problem of environmental degradation throughout the world. However, there are many ways in which humans influence water, for example, by direct channel manipulation, modification of watershed characteristics, urbanization and pollution.

The construction of dams and reservoirs is widespread throughout the world for irrigation purpose, to generate power or to provide a reliable source of water. More than 700 dams have been built worldwide and their number is increasing year after year. In countries like Brazil, Argentina, Canada, India, Japan, Turkey, Spain and China, run-off water has been regulated by water reservoirs. The impact of human activities is especially marked in Africa and North America where about 20 per cent of total run-off is now controlled. The impact of dam construction is apparent in the form of change in ecosystem and environmental conditions. Bigger the dam more will be its impact. Dams like Hoover (USA), Kiev (Russia), Aswan (Egypt), Akosombo (Ghana), and Bhakra Nangal, Gandhi Sagar, Damodar and Rihand (India) have had their impact on regional ecology. The recent environmental controversy over Tehri and Sardar Sarovar dams show people's awareness against construction of dams and their possible damaging environmental impact. The environmental consequences of dams include subsidence, earthquake triggering, transmission and expansion in the range of organism, the build-up of soil salinity, changes in ground water levels, waterlogging, deforestation, etc.

The process of urbanization has a considerable impact, both in terms of controlling rates of erosion, the delivery of pollutants to rivers and in terms of influencing the nature of run-off and other hydrological characteristics. The impact of urbanization can be summarized as follows:

1. removal of trees or vegetation,
2. construction of houses and sewerage lines,
3. construction of septic tanks,
4. diversion of nearby streams for public supply,
5. accelerated land erosion,
6. larger quantities of untreated waste into local streams,
7. pollution of streams and wells,
8. drilling of deeper, large capacity industrial wells, and
9. increased use of water for domestic and commercial purposes, etc.

Deforestation is a human activity that gives rise to floods, annual run-off levels and also affects on-stream flow. In many parts of the world, humans obtain water supplies by pumping from groundwater. This has two main effects, i.e. the reduction in the level of water table and the replacement of coastal areas, of fresh water by salt water. Environmental consequences of these two phenomena include ground subsidence and soil salinization.

Water pollution has now become a common problem and is a result of human use and misuse of water for various purposes. The causes and forms of water pollution created by humans are many and can be classified into the following groups:

1. Sewage and other oxygen demanding wastes
2. Infectious agents
3. Organic chemicals
4. Other chemical and mineral substances
5. Sediments
6. Radioactive substances
7. Heat (thermal pollution)

Many human activities can contribute to changes in water quality including agriculture, fire, urbanization, industry, mining, irrigation, etc.

Impact of Human Agencies on Geomorphology

Apart from above mentioned impact of human activities on the natural environment, the human role in creating landforms and modifying the operation of geomorphological processes such as weathering, erosion and deposition, also needs consideration. The range of the human impact on both forms and processes is considerable. Haigh (1978) has classified anthropogenic landforming processes as follows:

1. Direct Anthropogenic Processes
   (i) Constructional:
   tipping – loose, compacted, molten
   graded – moulded, ploughed, terraced.
   (ii) Excavational:
   digging, cutting, mining, blasting, etc.
   (iii) Hydrological interference:
   flooding, damming, canal construction, dredging, channel modification, draining, coastal protection.
2. Indirect Anthropogenic Processes
   (i) Acceleration of erosion and sedimentation, agricultural activity, clearances of vegetation, engineering, road construction, urbanization.
   (ii) Subsidence:
   collapse, settling, mining, hydraulic, thermokarst.
   (iii) Slope failure:
   landslide, flow, accelerated creep, loading, undercutting, etc.
   (iv) Earthquake generation:
   loading (reservoirs)
   lubrication (fault plane).

Conclusion

The above mentioned impacts of human activity on environment are just an introduction to the thinking that with the development in technology, the number of ways in which humans are affecting the environment is proliferating. The complexity, frequency and magnitude of impacts are increasing, partly because of rising population and also because of a general increase in per capita consumption. The US Council on Environmental Quality (1982) has remarked:

> If present trends continue, the world in 2000 will be more crowded, more polluted, less stable ecologically, and more vulnerable to disruption than the world we live in now. Serious stresses involving population, resources and environment are clearly visible ahead. Barring revolutionary advances in technology, life for most people on earth will be more precarious in 2000 than it is now unless the nations of the world act decisively to alter current trends.

In fact, nations of the world are coming closer at least on environmental issues and they are thinking in terms of sustainable development. The technology has produced several environmental problems and technology only, can solve these problems.

Reference

Haigh, M.J. (1978), *Evolution of Slopes on Artificial Landforms: Blaenavon UK*, Rresearch paper no. 183 (The university of Chicago, Department of Geography), Chicago, University of Chicago.

6

Degradation of Environment

In the present century, environmental degradation has emerged as a major global concern for human survival. In the opinion of the World Commission on Environment and Development, the future is to face an ever-increasing environmental decay, poverty, hardship and an ever more polluted world. The strains of the ecological crisis are so apparent that the task to preserve and protect the environment has become the primary requisite of the economics of development. Modernization is steadily transforming the nature into products. The resource squeeze has led to an ever-intensification of ecological damage. Land, water and air are being polluted every minute. Approximately 10,000 million tons of oxygen is being burned and about 24,000 million tons of carbon dioxide is being discharged every year. The green revolution has turned red. Deforestation is proceeding at an alarming rate depleting the genetic store. Mega technology has created problems of waste disposal and Industrial Revolution has now become a great hazard to the environment, as well as, to humanity.

Man's environment by now is sufficiently saturated by the complex chemical emissions, aerosols, toxic effluents, sewage, pesticides, solid wastes, polluted rains, dust and radiation. Global 2000 Report, 1980 reveals that the globe will become more crowded, more polluted, ecologically less stable and more vulnerable to disruption than the world we live in today. Degradation of environment has now become one of the most challenging problems not only of developing countries but also of the developed world. A large number of international and global organizations with the collective wisdom of scientists, planners and geographers have come up to settle the environmental issue confronting the nations and the physical world. The environmental crisis has convinced the world to use technology and resources to repair the damage already done to our environment and also to use substitutes for certain harmful chemicals in order to protect and preserve nature and natural resources.

Nature and Types of Environmental Degradation

The environment is an integrated system in which all its elements act and react in such a way that a balance is always maintained. All the physical elements; relief, structure, climate, natural vegetation, soil, water bodies, etc. are the determining factors of environment on the one hand, and man's activities on the other. Man is a user of the environment for his developmental activities and always disrupts this natural system and creates a background for environmental degradation. When this degradation is lesser, nature recovers it by its own system of recovery, but, whenever its quantity is more, it creates an imbalance in nature.

Environmental degradation is the deterioration of the environment through depletion of resources such as air, water and soil; the destruction of ecosystems and the extinction of wildlife.

According to Paul Harrison, 'rape of the earth has been central to the past progress of western civilization. The plunder and waste of irreparable resources, the poisoning and destruction of the fragile ecosystem, the belief that man can mould the nature with no need to take account of nature's reaction to such violation is a gross oversight'. Scientific and technological development has given several benefits to mankind but at the same time, it is also responsible for various types of environmental degradation. This problem is more in urban areas because of concentration of population, industries as well as of vehicles. Northam has observed two types of environmental degradation in urban areas – physical and social. 'Environmental degradation of physical nature involves contamination or impairment of use of naturally occurring elements of man's environment in which the natural element is the vehicle by which the pollutant is dispersed or disseminated.'[1] On the other hand, 'environmental degradation of a societal nature is somewhat different because it involves environmental contamination of urban places, more than of a natural element.'[2] The UN international strategy for Disaster Reduction defines environmental degradation as, 'the reduction of the capacity of the environment to meet social and ecological objectives and needs.'

In general, there are two main categories of environmental degradation: (i) physical or natural, and (ii) human. Although both these categories are interrelated with each other, for the convenience of study, their separate analysis is necessary.

Physical or Natural Environmental Degradation

There are certain natural phenomena which caused environmental degradation. Since these phenomena are natural, man's role is not significant, but at times man accelerates the process. Soil erosion, volcanism, floods, droughts, famines, desertification, thunder storms, cyclones, etc. are responsible for environmental degradation.

Environmental Degradation due to Soil Erosion

Soil is usually defined as 'any part of earth's crust in which plants root' and is formed as a result of long-term process of complex interactions between climatic, topographic and biological factors. Bare rock surfaces are exposed to various types of physical, chemical and biological processes which lead to physical and chemical disruption. Physical processes of weathering include action of water, temperature, glaciers and gravity which cause weathering of rocks through such processes as wetting drying, heating, cooling freezing, glaciation, solution and sand blast, etc. The chemical processes of weathering include hydration, hydrolysis, oxidation-reduction, carbonation, etc. All these processes disintegrate the soil but their impact on the environment is limited. On the other hand, soil erosion on a large scale leads to the degradation of the environment. It may be caused due to natural processes as well as by man. The main types of soil erosion are as follows:

1. normal or natural erosion,
2. fluvial erosion,
3. aeolian or wind erosion,
4. marine erosion, and
5. accelerated soil erosion by human actions.

Actually, normal soil erosion is a continuous process and its impact on environment is also limited, but when it occurs on a large scale due to water and wind erosion, it not only degrades the environment but is also responsible for the expansion of wasteland due to gullies and ravines, as has been in case of the Chambal region of Madhya Pradesh and Rajasthan. Four types of soil erosion due to water have been identified, these are:

1. sheet erosion,
2. rill erosion,
3. gully erosion, and
4. riparian erosion.

Soil erosion due to wind is a common phenomenon in arid and semi-arid regions and is often a cause of desertification or desert expansion. The process of soil erosion is always accelerated by man either by overgrazing, faulty land practices or by deforestation.

Figure 6.1 shows the areas of soil degradation in the world through deforestation, overgrazing and also due to over exploitation of vegetation.

There are certain phenomena associated with the process of soil erosion, such as soil creep, earthflow and mudflow, mass wasting, landslides, etc. The slow downslope movement of soil and regolith is known as soilcreep. It is a process in which particles slowly move down the hill slope. In humid climatic regions where slopes are steep, a mass of water-saturated soil may slide downslope during a period of a few hours in the form of an earthflow. When it takes place in the form of mass wasting, it becomes mudflow, a mud stream that pours down canyons in mountainous region and poses a serious environmental hazard. It also occurs in desert regions during a cloud burst. Earthflow and mudflow are the common causes of blockage of highways and railroad lines, supply of irrigation water and also damage property and cause loss of life.

The impact of soil erosion on the environment is both local as well as regional. The following are the general impacts:

1. formation of gullies and ravines and the whole region becomes an undulating area not suitable for human activities;
2. washout of the upper layer of the soil, specially along hillside slope;

Figure 6.1
Soil Degradation: The Hot Spots

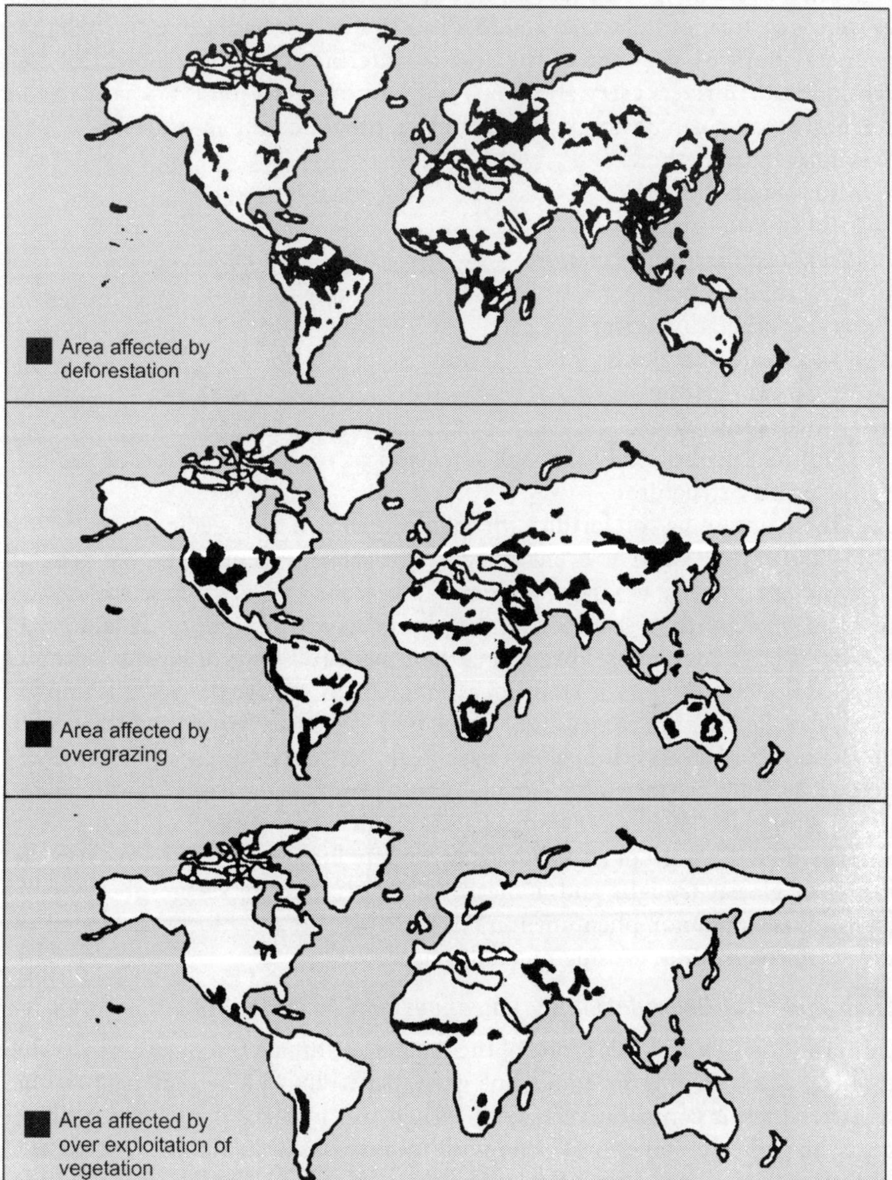

3. uprooting of plants due to soil cutting, thus responsible for deforestation;
4. loss of soil fertility;
5. loss of pastures;
6. expansion of deserts; and
7. increase in frequency of droughts and floods.

In fact, too much soil erosion is one of the causes of eco-imbalance. In India, the problem of soil erosion can be seen in Jammu and Kashmir, Himachal Pradesh, Uttar Pradesh, Bihar, Assam, Arunachal Pradesh, Sikkim as well as in areas of Rajasthan and Madhya Pradesh. It has been estimated that in India about 53 per cent of the area is suffering from the problem of soil erosion. Indian rivers carry about 600 million tonnes of soil every year to the sea. Soil erosion can be checked by adopting the following measures:

1. Biological methods
   - (i) Strip-cropping
   - (ii) Crop-rotation
   - (iii) Application of manure
   - (iv) Shelter belt
   - (v) Vegetation cover
2. Mechanical methods
   - (i) Contouring
   - (ii) Terracing
   - (iii) Control of gully through retention of run-off, diversion of run-off, new structures
3. By maintaining soil fertility through proper soil management system

Landslide is another event which may cause disruption in the general ecosystem. Landslide is a natural hazard and means any downslope movement of mass of regolith or bedrock under the influence of gravity. It is a rapid sliding of large rock masses beginning their descent as unit of blocks, without internal flowage. When the landslide is in the form of huge blocks, it is known as rock avalanche. In Switzerland, Norway or Canadian Rockies, villages built on the floors of steep-sided valleys have been destroyed by the sliding of rock masses. In India, reports of rockslide often appear in the news in hilly states like Jammu and Kashmir, Himachal Pradesh and North-East states. The environmental hazard of rockslides is extremely high for towns located on the floors of glacial troughs and at heads of fiords. The blockage of roads by landslide is a common phenomenon but sometimes it also blocks the course of the small rivers, thus, causing much damage.

Environmental Degradation by Volcanoes and Earthquakes

The eruption of volcanoes is one of the most dreadful natural phenomena. The molten rocks, boiling lava and hot gases penetrate out of that part of the surface which is comparatively weak. When this material comes out of these spots, an explosion takes place, and with thunderous noise, molten rocks, lava and gaseous clouds rise up towards the sky. All this results in a threat to the environment as well as to life.

Volcanism is not common throughout the world but is limited in the circum-pacific belt extending from Andes in South America, through the Cascades and the Aleutians into Japan and then south into the East Indies and New Zealand. There is also an important belt in the Mediterranean, which includes the active volcanoes of Italy.

A volcanic explosion poses a great threat to environment. On an average, a volcano, in an explosion ejects 150 cubic km of material. This material is of three types – solid material in the form of small and large pieces of rocks, ash particles and also pieces or volcanic bombs which shoot into the sky. The liquid material called magma comes out of volcano – also known as lava. A huge amount of gaseous material comes out of a volcano comprising of steam, carbon dioxide, sulphur in vapour form which produce poisonous gases when they combine with other gases. Gases like chlorine, fluorine, etc. are also emitted in large quantities.

The eruptions of volcanoes and lava flows are environmental hazards of the severest type, often taking a heavy toll on plants and animal life. Loss occurs from sweeping clouds of incandescent gases and suspended matter that descend the volcano slopes like great avalanches. When Mount Pelée erupted in 1902 in the West Indies, the burning material ejected out of it moved in the form of a cloud over the city called St. Pierre and only two people out of total 30,000 population survived. The explosions of Etna and Vesuvius in Italy destroyed many beautiful cities. In 1943, when Paricutin, 320 km west of Mexico City, exploded, created a mountain 167 metre high. In 1991, volcanic explosions in Japan and Philippines have damaged a lot of property and caused loss of human life and also polluted the environment. Similarly, volcanic activity in the Andaman & Nicobar Islands was responsible for the environmental degradation in this area.

Earthquakes are tremors which are produced by the passage of vibratory waves through the rocks of the earth. The devastation caused by earthquakes to the life of man and property, as well as to the ecosystem is very high. The earthquake belt of the world is also associated with the volcanic belt, but several other regions too have occurrence of earthquakes. Wherever earthquakes occur, apart from loss of life and property, landslides, change in course of rivers, loss of vegetation and animal life, destruction of dams and development of faults and fractures in earth's surface are common and all these events lead to environmental degradation. The list of devastating earthquakes in the world is too long to be enumerated here. In India, the occurrence of earthquakes in areas of Jammu and Kashmir, Himachal Pradesh, Uttar Pradesh, Bihar, West Bengal, Punjab, Delhi, Haryana, north-eastern states as well as in Maharashtra, are common. The most recent earthquake occurred on 30 September 1993 in Maharashtra and on 28 March 1999 in the hill district of Chamoli in Uttar Pradesh in which heavy loss of property and life was reported, apart from the whole regional ecosystem is getting disturbed.

Floods, Droughts and Famines

Floods and droughts both are natural hazards and are related with rainfall. If there is heavy rain in the catchment area of a river, floods will occur. On the other hand, if there is less rain or failure of monsoon, droughts will occur. In both the situations, their impact on environment creates problems and also affects the regional ecosystem.

Flood is a hydrological phenomenon that occurs in any drainage basin. Its magnitude and frequency varies primarily on the precipitation conditions in the catchment area. Most of the river valleys are subjected to varying degrees of flood incidences, either creating minor damages or devastating havoc at times. Such calamities of floods pose a serious problem to the whole environmental set-up of a region, bringing about some phenomenal changes in the physical environment ushering in accelerated erosional, transportational and depositional activities of the river with the consequent effects on soil. The devastating effects of flood can be seen on vegetation, agriculture, industry, settlements, sometimes causing loss of human and animal lives and of materials also.

The main cause of flood is excessive rainfall but its occurrence is also due to earthquakes, deforestation in the upper catchment areas, failure of dams, excessive snow melting, etc. The beneficial effects of floods have been established beyond doubt in the form of fertilization of soil. But its devastating effects far exceed the benefits.

In India, the problem of flood and challenge to the environmental setting is of gigantic magnitude, because of: (i) typical seasonal rainfall, and (ii) erratic nature of tropical cyclones and rainstorms. The following are the areas which may be termed as flood-prone areas:

1. Upper catchment and piedmont zone of the Himalayan river basins.
2. Ganga river valley – upper, middle and lower deltaic zone.
3. Brahmaputra valley and its tributaries.
4. Deltaic region of Mahanadi, Godavari, Cauvery and Krishna rivers.
5. Narmada and Tapti basins.

The common measures of the flood control are as follows:

- (i) Control of erosion in upper catchment areas.
- (ii) Construction of dams and reservoirs at suitable sites.
- (iii) Prevention of overflow in channels.
- (iv) Construction of multiple channels for distribution of the excess river water.
- (v) Development of shelters in flood-prone areas.

The flood problem should be taken up with more intensive preventive measures, in order to avoid environmental problems.

Drought is also a natural hazard which has its impact on the regional ecosystem. The areas which get less rainfall, i.e. amount of rain is too less to support agricultural production or even grasslands are usually called drought-prone areas. When droughts become severe, they cause famines, which are devastating for men, animals and also environment. Generally, droughts occur when rainfall is less than 25 to 50 per cent of the average. If it is below 50 per cent of the average, famines occur. In 1899, India faced the severest famine known as *Chappan-ka-Akal*. A similar famine also occurred in 1917. In both these famines several lakhs of people lost their lives. In 1992–93, Ethiopia also suffered from the grips of a bad famine, in which more than 30 lakh people died. Famine conditions are common in Rajasthan, Gujarat, Haryana, Madhya Pradesh, Orrisa (Odisha) and Andhra Pradesh in India. In Rajasthan, famine has now become a regular phenomenon.

During 1985–95, as many as 21 to 28 districts were under famine in Rajasthan. Similar situation remained during 1998 to 2002. In this period, whole of Rajasthan was under the grip of famine due to failure of monsoon in successive years.

The impact of droughts and famines on vegetation, soil, water resources, animals as well as on climatic conditions is far-reaching. Therefore, preventive measures should be taken.

Desertification

Desertification is a complex process affecting more than six hundred million people and approximately one-third of the earth's land surface. Desertification, to an ecologist or environmental scientist, is a systematic phenomenon involving deterioration of the ecosystem due to change in the pressure of human adaptation or use system. It is a process through which fertile or non-desertic land is converted into dry, infertile or desertic land.

According to Mann (1977), 'desertification can be defined as a package of processes which brings about certain basic changes in a particular ecosystem and converts it from a relatively non-desertic to desertic terrain. This involves an interplay of climatic, adaphic and biotic factors which may (a) work through time starting with a climatic accident and ultimately resulting in the formation of a desert, or (b) lead to accentuation of the arid conditions within an already desertic terrain, or (c) cause a creep of the desertic conditions within a desertic tract towards its adjoining areas'.[3]

In India, the arid zone covers about 12 per cent of the country's geographical area and occupies over 3,20,000 sq km of hot desert located in parts of Rajasthan (62%), Gujarat (19%), Punjab and Haryana (9%), and Andhra Pradesh and Karnataka (10%). In these regions as well as in other arid regions of the world, due to the high pressure on grazing lands and overexploitation of resources, degradation of environment is growing day by day. Increased desertification problems are also evident in the form of salinity hazard, seepage and waterlogging problems in irrigated areas of the desert. The increased pressure of livestock results in over-exploitation of resources leading to soil erosion and removal of surface vegetation. The monitoring of deserts is necessary and should be done by remote sensing methods, both orbital and aerial. In order to control desertification, the following steps should be taken:

1. Ban on cutting of desert vegetation,
2. Plantation of ecologically suitable plants,
3. Development of pasture lands,
4. Control overgrazing,
5. Stabilization of sand-dunes,
6. Proper use of available groundwater resources,
7. Management of land use and farming practices,
8. National watershed programme should be intensified, and intensifying
9. Public awareness should be generated.

Thunder Storms, Tornadoes and Tropical Cyclones

Due to abrupt changes in atmospheric conditions thunder storms, lightning, hail storms, tornadoes, and cyclonic storms occur and are the causes of environmental hazards. A thunder storm is an intense local storm caused by strong conventional activity. Thunder and lightning normally accompany the storm, and rainfall is heavy, sometimes due to cloudburst also. Violent surface winds may occur at the onset of the storm. The speed of the rising air in a thunder storm is about 112 km per hour. *Lightning* is also one of the environmental hazards that annually results in the death of many persons and livestock and causes forest fires and building fires. One of the most destructive products of the thunder storm is the hail storm. Hail consists of rounded lumps of ice ranging from 0.5 to 5 cm in diameter. Large hailstones are an environmental hazard as they can be extremely destructive to crops and flimsy buildings.

A tornado is a small, intense cyclone in which the air spirals at a tremendous speed. Associated with frontal air mass interactions in mid-latitude, tornadoes are the most violent storms. These storms are most frequent and violent in the United States, although these occur in Australia and in other places of mid-latitudes too. A tornado moves across the country with a wind speed of upto 400 km per hour in a funnel-like shape. The end of the funnel cloud may alternately sweep the ground, causing complete destruction of anything in its path.

Tropical cyclones are the most powerful and destructive types of cyclonic storms, which are also known as hurricane or typhoon. These storms develop over oceans in latitude 8° to 15° N and S, but not close to the equator. The tropical cyclone is an almost circular storm having low pressure in which winds spiral at high speed, accompanied by heavy rainfall. The environmental impact of tropical cyclones lies in the form of destruction of life, infrastructures as well as of forests and animals. In islands and coastal regions large scale destruction of cities and their inhabitants has been reported on several occasions. Bangladesh is a country which has suffered a lot by heavy destruction caused by tropical cyclones – the death toll by tropical cyclones in Bangladesh is almost unbelievable. In 1822, the toll was 40,000 while in 1876 and 1897 it was 1 lakh and 1.75 lakh, respectively. In 1970, 3 lakh lives were lost and in 1991 the death toll was 5 lakh. These storms are also responsible for changes in atmospheric and environmental conditions.

Environmental Degradation due to Human Actions

Man and his activities are the main causes of environmental degradation; even physical environmental degradation is accelerated by human actions. Right from the dawn of human civilization, man has used nature not only for his existence and fulfilment of his necessities but also exploited it and to a certain extent, destroyed the ecosystem. Although nature has its own system of recovery or maintenance of eco-balance, but when over-exploitation is done it fails to recover, thus, several environmental problems have emerged and now in the last decade of the 20th century, the whole world is facing the problem of environmental degradation due to technological, scientific development on the one hand, and overpopulation, urbanization and industrialization on the other. Environmental pollution is the result of man's unplanned and unwise use of technology. The various forms of pollution, viz. air, water, land, noise, radiation, etc. have been discussed in detail in Chapters 6, 7 and 8. In this chapter, the problem of pollution has been excluded and our discussion is limited to environmental degradation due to deforestation, construction of big dams, mining activities, use of pesticides, urbanization, industrialization, etc.

Deforestation

The herculean feats of economic gains and development have been at the cost of deforestation. Deforestation is one of the outcomes of various developmental activities having the character of extensional development and it is true that deforestation is a havoc wreaked by thoughtless destructive activities of development. Forest clearings were made not only for village settlements but also for cultivation and pastures. As the population increased, more forests were cleared for various uses. Apart from this, the commercial exploitation of forests is the main cause of deforestation. There was a time when 70 per cent of the land area was covered with forest and now the total forest cover has shrunk to 16 per cent only. Now-a-days, tropical deforestation has become a concern of all and its impact is not only limited to the region itself, but on climatic conditions of the whole world, which is alarming. The annual decline in forest cover in tropical countries has been depicted in Figure 6.2. It is clear from the diagram that in countries like Brazil, India, Indonesia, Thailand, etc. the rate of annual cutting of forests is very high and has become a great problem to the environment and ecosystem.

In India, deforestation is going on at a very fast rate. The story of overexploitation of forests started from World War II, when the British started

Figure 6.2

Area of Closed Forests Cleared Annually in Selected Tropical Countries

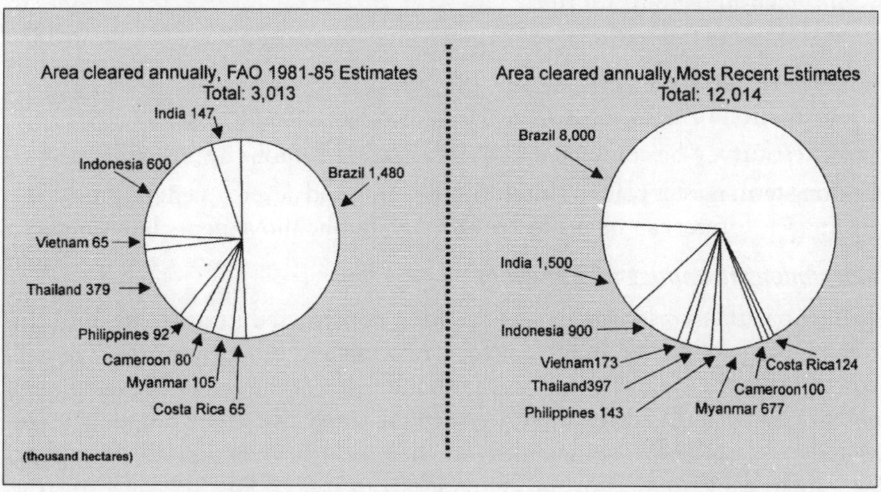

clearing forests to meet their needs. In the post-independence period, the process of exploitation has been accelerated to meet developmental projects of railways, industries, mining, river valley projects, defence, etc. The urbanization and expansion of agricultural activity has also contributed to the exploitation of forests. During the 1951–75 period, 43.28 lakh hectares of forest land has been lost to different developmental work. The commercial exploitation of forests is still going on in spite of several legal provisions to restrict it. Maximum deforestation has occurred in Madhya Pradesh, where loss of forest is about 2 million hectares, while in Maharashtra over a million hectares, Orissa, Andhra Pradesh, Jammu and Kashmir also over a million hectares and Rajasthan and Himachal Pradesh over half a million hectare has been cleared. But in terms of proportion, Rajasthan, Haryana and Punjab lost over half of the forest cover and states like Himachal Pradesh and Jammu and Kashmir lost over a third of their forest cover. In Tripura, Manipur, Meghalaya and Mizoram, the loss of forest cover was minimum during the 1970–90 period.

The impact of deforestation on the environment can be seen in the form of micro-climatic change, increase in temperature and decrease in humidity/rainfall, soil erosion in the form of sheet, rill and gully erosion, increase in frequency of floods and landslides, loss of soil fertility, etc. All these lead to unaccountable destruction and also pose serious ecological problems. Serious efforts will have to be made to protect and conserve forests both at government as well as public level. Due attention should be given to forest management. Some useful steps in this direction are:

1. a proper ratio between forest cutting and plantation of trees should be maintained;
2. actual growth of trees be supervised;
3. strict prohibition of the cutting of trees for fuel be imposed;
4. alternative media for wood such as moulded plastic, etc. be developed for furniture and other domestic articles;
5. protection from forest fires;
6. development of national parks and game sanctuaries;
7. social forestry be developed;
8. plantation of trees according to local ecological conditions;
9. forest survey be conducted and classification be done accordingly; and
10. long-term master plan for forest conservation and development be prepared. In fact, forests can be protected through public movements like *Chipko.*

Environmental Impact of Big Dams

Throughout the world, big dams have been constructed in order to fulfil the growing needs of power and water for irrigation and other purposes. There is no doubt that these dams have become an indicator of the level of development and have boosted the regional economy of the respective areas. But, during the last twenty years, environmentalists of the world have expressed the feeling that these dams may not be suitable from the point of view of the environment, or in other words, the ecosystem has been disturbed by these dams due to (i) submergence of large areas under water (ii) destruction of vegetation, wildlife and other animals, (iii) instability may provide sustainability to earthquakes, (iv) silting, and (v) large scale human migration, etc. In spite of these problems, dams have been constructed and this process will continue. But, at this point, it is necessary to examine big dam projects from the point of view of degradation of environment and then only implementation of the project be done. As far as possible emphasis should be given on small dams. Similarly, selection of dam site be done properly so that there is minimum impact on the environment.

In India, during post-independence period several dams have been constructed. Certainly, these dams have changed the entire scenario of power supply and agricultural development. Although their negative impact on environment is there, in comparison to the benefits provided, all these projects are considered successful.

The controversy over Tehri Dam and Sardar Sarovar projects is well known because of their adverse effects on environment.

The points raised by environment activists are:

1. The ecology will suffer, invaluable forests will be destroyed because of submergence of forest lands. The wildlife and birds, etc. living in these areas would face extinction.
2. The possibility of earthquakes would be more in view of the large size of the dam.
3. The deposition of silt in the base of dam may restrict flow of huge quantum of water.
4. The stagnant water of the reservoir would spread water borne diseases in adjoining areas.
5. The rehabilitation of tribals and other people may not be proper.

All these issues have been raised not only for above mentioned projects of big dams but for other projects also and experts are of the opinion that for sustainable development it is necessary to construct small dams. That will not only save cost and energy but also not be harmful for the environment. Whenever there is national necessity, the decision should be taken after considering the environmental issues also.

In the case of Tehri Dam, doubts about safety have been expressed by many on account of the fact that the dam is going to be located in a region of high seismocity and is, therefore, potentially unsafe. This doubt has however, been denied by experts. There is no doubt about the fact that Tehri Dam will change the entire ecosystem of the region, but looking at the benefits in the form of power generation and water supply, this project has not only been approved by the government but its work is near completion. In fact, issues raised regarding environmental degradation are present, but what is not generally pointed out is that environmental degradation is already there.

Similarly, the Sardar Sarovar project has been designed to provide irrigation water to drought hit areas of Gujarat, Madhya Pradesh, Rajasthan and Maharashtra. This project, which is estimated to cost over Rs.6,400 crores, will on completion, provide irrigation facilities to about 18 lakh hectares of land and create installed capacity of power of 1,450 MW. It will benefit about 25 lakh people in 3,340 villages. It will also provide drinking water to 295 lakh people residing in 131 towns and 4,720 villages in Gujarat. But it will create serious environmental problems, for which proper steps should be taken in advance in order to maintain the regional eco-balance.

Environmental Degradation due to Mining

Mining is one of the most important economic activities in the world. With growing knowledge of the mineral resources as well as due to technological and scientific developments, nowadays mining is performed on a very large scale. The extraction of minerals is done not only to fulfil regional or national needs but also for trade with the result that large scale mining operations have become a threat to the environment and are responsible for environmental degradation.

Mineral resources are the product of geological structure and their formation may take a very protracted period, extending over several thousands of years. Therefore, any mineral extracted cannot be formed again for use. There are three types of mineral resources, viz. non-metallic minerals, metallic minerals and mineral fuels. All these are useful to man in some way or the other. Therefore, mining operations are expanding day by day without taking into consideration the impact on the environment. Digging of land is carried out on a very large scale and its impacts on the environment are as follows:

1. In mining areas destruction of land due to deep and surface mining is a common feature. In USA alone, about 1,50,000 acres of land has become wasteland due to mining. In areas of coal, iron ore, building material and other ores, etc. one sees no signs of the original relief. In mining areas of Bihar, Orissa (Odisha), Madhya Pradesh and Rajasthan one can already see the destruction of land.
2. Heaps of unused material is also a great problem. Generally mine owners do not care for this waste material, therefore, it becomes the cause of environmental degradation.
3. Largescale destruction of natural vegetation and wildlife creates ecological problems.
4. Pollution due to dust spreading all over the area is not only a threat to the environment but also the cause of many diseases among the workers as well as among the residents of nearby areas.

   The other impacts of mining include:
   (i) expansion of barren land,
   (ii) creation of 'ghost town' after mining, and
   (iii) ecological crisis due to air, water, land and noise pollution.

Unplanned and unorganized mining is a danger to the environment and this should be checked. Other measures in this direction are:

1. levelling of mining areas,
2. re-use of waste material,

3. plantation in mining areas,
4. rehabilitation of people in mining,
5. measures to check air and land pollution,
6. improvement in mining areas,
7. health and safety measures should be taken,
8. policy of resource conservation should be adopted and alternative material should be developed, and
9. proper management and planned regulated mining can solve many problems.

Environmental Degradation due to Pesticides

The problem of contamination through pesticides in our food grains, dairy products, vegetables, fruits or in our living environment as a whole has been visualized very well. Pesticides have been used in farming: (i) to minimize the loss of food caused by pests and ensure a more bountiful harvest, and (ii) to decrease the extent of vector-borne diseases. Contrary to these benefits, it has seriously degraded environmental quality and altered the ecological balance, as shown in Figure 6.3.

The pesticides which have been widely applied throughout the world are carried forth by wind and water and distributed elsewhere. It was found that the same pesticides slowly decomposed into compounds that were toxic to fish and wildlife species in recent years. Worldwide, 70,000 chemicals are in use per day and 500 to 1,000 new ones are added each year. Between 4,00,000 to 2 million pesticide poisonings occur world-wide each year, of which an estimated 10,000 are fatal.

Figure 6.3

Pesticide Impact Cycle

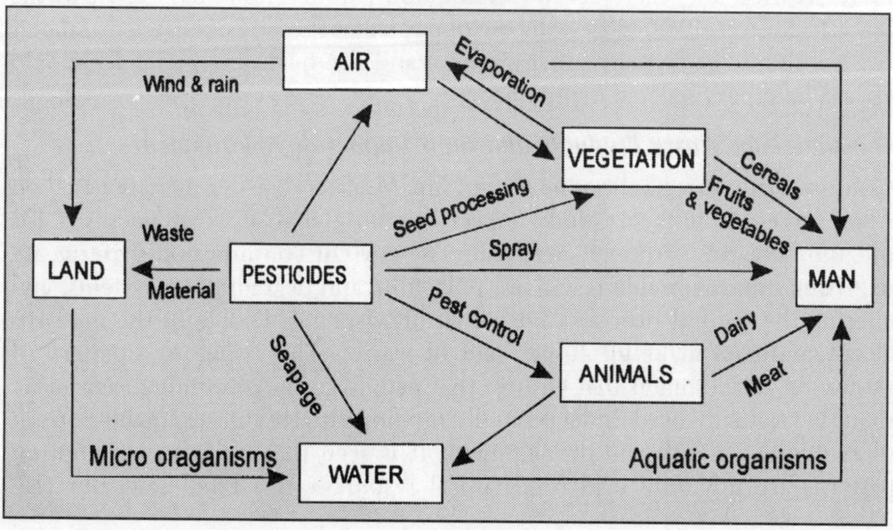

Research has proved that continuous use of pesticides has many negative effects on the environment, such as: (i) killing non-targeted organisms, (ii) accumulation in the food chain, (iii) building of immunities by the targeted pests, (iv) lower reproductive potential, (v) synergistic effects, (vi) toxic effects on water, (vii) pesticide residue in food and tissues, (viii) effects on vegetation and milk, and (ix) direct pesticide poisoning.

Pesticides have their impact on all three components of the earth, i.e. lithosphere, hydrosphere and atmosphere as depicted here:

Soil: Pesticides are toxic, hence it is assumed that their presence in soil will change or alter fundamental soil processes, such as: (i) organic matter decomposition, (ii) nitrogen transformation, (iii) sulphur transformation, (iv) phosphate availability, (v) trace element availability, and (vi) soil enzyme activity which influence soil fertility and productivity.

Water: The presence of toxic chemicals in water also has a significance as they are picked up by unicellular aquatic organisms like plankton and get accumulated in the body by a phenomenon called bioconcentration.

Air: The amount of pesticides released into the atmosphere, if larger than really required, pollutes the air.

Presently, in India, 123 pesticides are permitted and are used over a total area of about 142 million hectares. Among all pesticides, insecticides constitute about 80 per cent, followed by fungicides at 10 per cent, herbicides at 7 per cent, and others at 3 per cent. Pesticides like DDT, DDE, TDE, deidrin, endrin, heptachlor-epoxide and aldrin are harmful. Many developed countries have prohibited the use of those pesticides which have harmful effects on living organisms. Pesticides can even enter the body through the skin, lungs (inhaled) or the digestive tract. Nausea and vomiting, abdominal cramps, diarrhoea and involuntary defecation are the usual symptoms following gastro-intestinal poisoning.

Many countries have enacted laws for the safe use of pesticides. In India, there are two laws, i.e. Prevention of Food Adulteration Act, 1954 and the Insecticides Act, 1968. In fact, the use of pesticides should be restricted and pest control should be done by other methods and also quality of pesticides should always be supervised and maintained.

Consumerism, Waste Products and their Impact on Environment

Consumerism now has become way of life. Modern societies that are based on using large amounts of goods, especially those that are manufactured for one-time use, are extremely wasteful. The current consumption patterns are depleting non-renewable resources, poisoning and degrading ecosystems, and altering the natural processes on which life depends. People in the industrialized countries make up 20 per cent of wastes. This is due to a pattern of economic development that ensures that people go on consuming even more than they actually need. India is rapidly moving into this unsustainable pattern of economic growth and development. It is seen that today's consumption patterns are not only depleting natural resources at a rapid rate, but also widening the inequalities in consumption in different societies. Money is not the only way to measure the cost of an item that we use. When one adds up all the raw materials and energy that goes into the manufacture of goods or the services provided by nature that one uses during a day's activities, the toll on the environment is large. When this cost is multiplied over a lifespan, the amount is staggering. If one considers the over-utilization in each family, city or a country, the impacts are incredibly high. For example: two hundred billion cans, bottles, plastic cartons and paper cups, are thrown away each year in the 'developed' world. 'Disposable' items greatly increase this waste. Rather than compete on quality or reliability, many industrial consumer products are made for one-time use. Buying quality products that are warrantied against failure or wearing out, learning about the raw materials that things are made of, and an appreciation of their origin from nature's storehouse, as well as knowing the conditions of the workers that make them, are some ways of resisting consumerism and decreasing waste.

While there may be some new appliances and cars that are more productive and energy-efficient, discarding the old often leads to an almost total waste of the energy and material already invested in these products. This alone may more than nullify the energy savings of the new product. This is a tricky problem. Figure 6.4 indicates the waste and pollution in the manufacturing-consumer society.

Consumerism is related to the constant purchasing of new goods, with little attention to their true need, durability, product origin, or the environmental consequences of their manufacture and disposal. It is driven by huge

Figure 6.4

Waste and Pollution in the Manufacturing Consumer Society

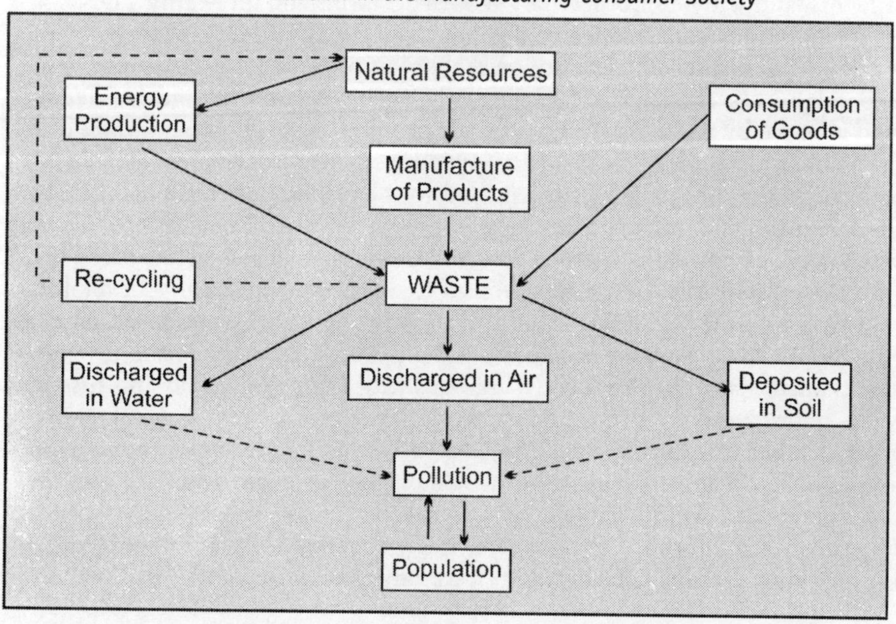

sums spent on advertising designed to create both a desire to follow trends, and a personal feeling of satisfaction based on acquisition.

Consumerism interferes with the sustainable use of resources in a society by replacing the normal common sense of desire for an adequate supply of life's necessities, with an insatiable quest for things that are purchased by larger and larger incomes to buy them. There is little regard for the true utility of what is bought. An intended consequence of this strategy, which is promoted by those who profit from consumerism, is to accelerate the discarding of the old, either because of lack of durability or a change in fashion. Especially in developed countries, landfills are being rapidly filled with cheap discarded products that fail to work within a short time and cannot be repaired. In many cases, consumer products are made psychologically obsolete by the advertising industry long before they actually wear out.

The inordinate amount of waste that is generated by consumer-oriented societies around the world is now a serious environmental issue. Most human activities are related to production and consumption cycles which produce excessive amounts of waste in the form of solid, liquid and gaseous waste products. The problems of waste management in the urban and rural sectors are different. Rural communities used them effectively. With the advent of an industrial civilization, the highly complex technological processes for production of goods have rapidly increased problems due to inadequate waste disposal. This creates a heavy burden on natural resources, degrades the environment and creates health hazards. With the rapid increase in population, the amount of waste in terms of quantity and quality has increased waste management pressures many-fold in recent years. If the high quantities of waste generated continues, mankind will be drowned under heaps of garbage and streams of sewage. Our health will be affected by dangerous industrial effluents, and will be smothered by clouds of smoke and unhealthy gases.

The increasing demands of consumption on the finite resources of the planet, increasing level of environmental pollution, and the problems of waste disposal must be changed to the careful utilization of resources and recovery of used material by waste recycling. Therefore, the reuse of goods and waste utilization should become a part of the production-consumption cycle. Current patterns in the industrial sector have led to the disposal of waste in a careless and uneconomical manner. For example, it is estimated that the per capita production of domestic waste is many times higher in a developed country when compared to a developing country. Unfortunately, many developing countries are now working out similar wasteful trends through development, but do not have the same economic potential to handle the waste this new unsustainable strategy produces. Large quantities of solid, liquid and gaseous waste is produced by urban industrial communities in the form of plastic, paper, leather, tin cans, bottles, mineral refuse, and pathological waste from hospitals. Dead animals, agricultural wastes, fertilizer and pesticide overuse, and human and animal excreta are essentially rural concerns. The waste is either discharged into the atmosphere, water sources, or buried underground. These wastes are not considered to have any economic value. This attitude towards waste has led to disastrous effects on environment besides the over exploitation of natural resources.

Environmental Degradation due to Technological Development

With the help of scientific and technological development, man has achieved new dimensions in his development and has improved his quality of life by adding several amenities and facilities. The industrial development is proceeding at a very fast rate; similarly, revolution in the transportation sector has also occurred not only in developed countries but in developing countries also. In the fields of agriculture, resource utilization, power consumption, engineering and in other scientific fields, man has achieved success of high level. But the current strong feeling is that all these developments have been responsible for the degradation of the environment and there is an urgent need to evolve an environmental friendly technology. It is said that 'technology produced the crisis and technology can solve it'. Some of the problems produced by technological and scientific developments are as follows:

1. Degradation of environment and expansion of pollution through industrialization.
2. Unorganized mining.
3. Rapid deforestation.
4. Pollution due to transportation.
5. Danger of leakage of radioactivity from atomic reactors.
6. Disturbances in eco-sensitive areas like ozone layer, tropical forests, Antarctica, etc.
7. Population growth, urbanization, decline in death rate and other demographic changes are responsible for more and more exploitation of resources.
8. Extinction of several species of wildlife, birds and plants.
9. Too much stress and strain due to machines has created several health problems.

Many international organizations recognize environmental degradation as one of the major threats facing the planet, since humans have only given one earth to work with, and if environment becomes irreparably compromised, it could mean the end of human existence.

In brief, the quality of environment is decreasing day by day. It should be maintained through environment planning, people's participation, environmental status evaluation, and environmental legislation and administration.

Notes

1. Northam, R.M. (1995), *Urban Geography*, John Wiley, New York, p. 76.
2. Ibid., p. 76.
3. Mann, H.S. (1977), Desertification: Overview, *Arid Zone* 16(3), pp. 1–4.

7

Air Pollution

Air, which is a mixture of gases, moisture and some inert material, controls life on earth. It is a reservoir of oxygen needed by man and other animals, and of carbon dioxide which in essential for plants. Any contamination in air may disturb the whole atmospheric system which is an insulating blanket around the earth. Without air, there would be no clouds, no winds, no rain, no snow and no fire. In other words, there would be no life on earth. The atmosphere, which shields the earth from ultraviolet radiations, becomes progressively tenuous with increasing distance from the earth. About 75 per cent of earth's atmosphere lies within 16 kilometres from the surface, and 99 per cent of it lies below an altitude of 30 kilometres.

The structure of the atmosphere is controlled by various gases present in the air. The clear, dry air at sea level contains 78 per cent nitrogen, 21 per cent oxygen, 1 per cent argon, and 0.03 per cent carbon dioxide and other gases as shown in Table 7.1.

All the major as well as minor components of the air/atmosphere are biologically important and the participation of each in the living processes are in some cases critically sensitive to slight changes in concentration. Whenever the proportion of the components is disturbed by man, that becomes a cause of air pollution, which nowadays has become a major global problem.

Table 7.1

Composition of Clear, Dry Air at Sea Level

Components	*Per cent by Volume*
Nitrogen	78.084
Oxygen	20.9476
Argon	0.934
Carbon dioxide	0.0314
Neon	0.001818

Cont'd...

...*Cont'd*

Helium	0.000524
Methane	0.0002
Krypton	0.000114
Sulphur dioxide	0 to 0.0001
Hydrogen	0.00005
Nitrous oxide	0.00005
Xenon	0.0000087
Ozone	Summer 0.000007 Winter 0.000002
Nitrogen dioxide	0 to 0.000002
Iodine	0 to 0.000001
Ammonia	0 to trace
Carbon monoxide	0.1 ppm–0.2 ppm

Source: H.J. Sanders (ed.), *Chemistry and the Atmosphere*, The American Chemical Society.

Nature of Air Pollution

In an ecosystem, all the elements act and react in an integrated form, therefore, this life supporting system continues in a systematic way. But when any disturbance is created by man even in a micro form, there is a case of pollution. Growing industrialization and transportation and the increasing use of pesticides and unwanted chemicals in the air has rendered the whole atmosphere polluted and its impact is very dangerous not only to man and other living organisms but also to the environment itself.

Air pollution may be defined as an imbalance in the quality of air so as to cause ill effects. According to Maxwell, 'our enormously accelerated abuse of the atmosphere has become a health hazard and a threat to life, damaging both plants and animals in areas polluted with poisonous fumes, dust, and smoke'.[1] The industrial, automotive and domestic activities have resulted in increasingly outrageous insults to the atmosphere. 'In air pollution, the pollutants includes gases, solids and liquid particles of both organic and inorganic chemical classification.'[2] The different types of pollutants are continuously introduced into the atmosphere and are removed by natural processes of cleansing. But when pollution exceeds the atmosphere's self-purifying capacity, accumulation of pollutants occurs causing serious hazards for environment and other organisms including humans. The following are the main reasons for the increasing rate of air pollution:

1. Poisonous gases and other particles emitted from industries without any treatment.
2. Heavy increase in the number of automobiles and their emission (smoke and other dangerous gases).
3. Increased use of chemicals and petrochemicals.
4. Population concentration in cities.
5. Fast rate of deforestation.

6. Tests of experiments of atomic weapons.
7. Tests of chemicals and biochemical weapons.
8. Space research and satellite wastes.
9. Unorganized mining and traditional practices of the use of fuelwood, etc.

Certain natural events such as volcanic eruption, dust storms, etc. are also causes of air pollution. In order to understand the nature of air pollution, it is necessary to know the various sources of pollution.

Sources of Air Pollution

Air pollution is the result of the combined effects of several pollutants. These pollutants are associated with each other and also react with other elements, therefore, it is difficult to categorize them, but for proper understanding they can be divided into the following categories on the basis of their origin, nature, size, impact, etc.

1. Particulate matter can be divided into two types, viz. natural and man-made. The natural form of particulate matter is the result of volcanic dust and gases, mineral dust, sea-salt crystals. Another class of natural particulate matter is smoke from forest fires and grass fires. Living plants release pollens and spores into the air, these are organic compounds. From forest trees, certain hydrocarbons called terpenes are also released into the atmosphere.

   Man-made particulate matter comes from many sources, but the major source is from the combustion of hydrocarbon fuels - petroleum products, coal, peat and wood. Combustion of solid wastes is another source. Other kinds of matter introduced into the atmosphere are industrial chemicals, fly ash, refining fossil fuels, mining and smelting ores, as well as pollutants discharged through quarrying, farming activities, etc. Use of various types of solvents and also radioactive elements are the causes of air pollution created by man.
2. Another classification is: (i) primary, and (ii) secondary pollutants. The primary pollutants are the gaseous and other solid micro-particles inducted to the atmosphere. These pollutants are emitted and as such are not found in the air. The most important gaseous pollutants are carbon monoxide, oxides of sulphur, hydrogen sulphide, hydrocarbons, oxides of nitrogen, ozone and other oxidants. Secondary pollutants are the result of chemical reactions. Evidences and experiments indicate that exhaust gases of automobiles contribute more in the formation of secondary pollutants. For example, oxides of nitrogen produced in the combustion of petroleum and other fuels emitted into the atmosphere, yield ozone in the presence of sunlight. It is to be noted that ozone is not emitted as such into the atmosphere but formed only from primary pollutants.
3. According to the chemical composition, air pollutants can be divided into organic pollutants and inorganic pollutants. Others have divided them into solid, liquid and gaseous pollutants. The gaseous pollutants are

carbon monoxide, sulphur dioxide, hydrogen sulphide and organic sulphide, hydrogen fluoride, hydrogen chloride, oxides of nitrogen, aldehydes and organic acids, etc. Particulate pollutants consist of both solid and liquid particles. They vary in sizes from 0.01 micron to 20 microns. Dust, fume, mist, spray, smoke are included in this category.

4. According to the source type, pollutants can be classified as being produced from:
   (i) combustion,
   (ii) transportation emissions,
   (iii) industrial processes,
   (iv) use of solvents, and
   (v) radioactivity.

Combustion processes yield particulates such as fly ash and smoke and oxides of sulphur and nitrogen. The amount of sulphur dioxide emitted depends upon the contents in the fuel. High temperature processes such as thermal fixation of atmospheric nitrogen yield larger quantities of oxides of nitrogen. Carbon Monoxide is also emitted from combustion. The other contaminants that are produced by combustion includes acids and aldehydes.

The simplest form of combustion is the use of fuels in domestic use. In India as well as in many developing countries, wood, coal, cow-dung and kerosene oil are commonly used. All these materials yield carbon dioxide, carbon monoxide, sulphur dioxide, etc.

Automobiles may be considered as the main source of air pollution, specially in urban areas. Automobile exhausts release smoke and to a certain extent lead particles also. Smoke contains the gaseous pollutants carbon monoxide, hydrocarbons and oxides of nitrogen. The pollutants emitted by petrol and diesel driven vehicles are as given in Table 7.2.

Table 7.2

Pollutants Emitted by Petrol and Diesel Driven Vehicles

Emissions	*Petrol Engine*	*Diesel Engine*
Carbon monoxide	3	–
Sulphur dioxide	40 ppm	400 ppm
Nitrogen oxides	1100 ppm	300 ppm
Unburnt hydrocarbons	150 ppm	20 ppm
Particulate matter	0.1 g/m^3	0.01 g/m^3

The rapid rate of industrialization has resulted in more and more air pollution. Various industrial processes release almost all types of pollutants into the air. Some industries like cement, iron and steel, fertilizer, petro-chemical, etc. are of great concern because of the difficulty in controlling the emission of pollutants from them. Acid rain has become a great threat to the environment. The use of solvents is increasing with the growing use of paints, spray, polish, etc. Due to the presence of hydrocarbons in these materials, air pollution is caused, which is dangerous for health. Similarly, spray of pesticides in agriculture is responsible for air pollution even in rural areas.

Nuclear material, when released into the air, is hazardous for all living organisms. Nuclear weapon testings, nuclear reactors, chemical processing of plants, research institutes and hospitals contribute many radio nuclides to the atmosphere. Table 7.3 indicates the main sources of air pollution and their impact.

Table 7.3

Sources and Effects of Common Air Pollutants

Pollutant	*Anthropogenic Sources*	*Health Effects*	*Environmental*
Carbon Monoxide (CO)	Secondary pollutant formed by chemical reaction of VOCs and NOx in the presence of sunlight	Breathing problems, reduced lung function, asthma, irritates eyes, stuffy nose reduces resistance to colds and infections, premature ageing of lung tissue.	Damages crops, forests and other vegetation, Damages rubber , fabric and other materials; smog reduces visibility
Nitrogen Oxides (NOx)	Burning of gasoline natural gas, coal, (cars are a major source of NOx)	Lung damage, respiratory illnesses, ozone (smog) effects	Ozone (smog) effects, precursor of acid rain which damage trees, and soil. Acid rain also causes buildings, statues and monuments to deteriorate.
Ozone (O_3)	Burning of gasoline, natural gas, coal & oil	Reduces ability of blood to bring oxygen to body cells and tissues	
Volatile Organic Compounds (VOCs)	Full combustion, solvents, points	Ozone (smog) effects, cancer and other serious health problems.	Ozone (smog) effects vegetation damage
Particulate Matter	Emitted as particles or formed through chemical reactions, burning of wood, diesel and other fuels, industrial processes, mining activities.	Eye, nose and throat irritation, lung damage, bronchitis, cancer, easy death.	Reduces visibility, Ashes, smoke , soot and dust can dirty and discolour structures
Sulphur Dioxide (SO_2)	Burning of coal and oil, industrial processing	Respiratory illness, breathing problems, damage of lungs.	Acid rain which can damage trees and soil. Acid rain also causes deterioration to monuments
Lead	Combustion of fossil fuels and leaded gasoline, paint, smelters, battery manufacturing	Brain and nervous system damage, digestive and other problems. Some lead containing chemicals cause cancer in animals	Harm to wildlife and livestocks

Among above mentioned sources of air pollution, Carbon Monoxide (CO) accounts for more than 50 per cent of air pollution nationwide and worldwide. CO is a colourless, odourless, flammable gas, it is a product of incomplete combustion of carbon-containing material. The other common sources of air pollutants are Ozone (O_3), Nitrogen Oxides (NO), Volatile Organic Compounds (VOCs), Particulate Matter, Sulphur Dioxide (SO_2) and Lead. Apart from above mentioned air pollutants, there are also 'hazardous air pollutants' (HAPs) also commonly known as 'toxic air pollutants', which includes 188 chemicals specified as hazardous in the 1990 US Clean Air Act Amendment. Some of the hazardous air pollutants are, benzene, toluene, ethylene glycol, chloroform, methyl bromide, formaldehyde, parathion, styrene, vinyl chloride, PAMs, asbestos, metals – arsenic, cadmium, mercury, nickel, etc.

Specific Phenomena Related with Air Pollution

Ozone is normally present in the atmosphere at about 0.05 ppm at sea level. It is produced naturally in the atmosphere by the action of electric discharges on oxygen. Ultraviolet radiation in the wave length of 260 mm is absorbed by oxygen molecules, causing them to react in this manner. Radiation from the sun probably causes high concentration of ozone in the outer portions of the earth's atmosphere. Ozone is an important component of smog (smoke + fog). Ozone plays a key role in the photo-chemical formation of atmospheric pollutants as well as being end product in the intricate complex of reactions that take place. The ozone layer provides a cover to the earth's surface and protects the earth from ultraviolet rays.

Until 1974, atmospheric scientists were proceeding in their research on possible impacts of nitrogen oxide jet engine exhausts on the ozone layer. But later it was found that a new threat to the ozone layer has emerged from synthetic chemicals called chlorofluorocarbons. These are simple compounds of the elements chlorine, fluorine and carbon. The chlorine can destroy ozone. Recent scientific studies indicate that if chlorofluorocarbon production continues to grow at the present rate, the compound will enter the stratosphere in quantities capable of seriously depleting the ozone layer, normally termed as 'holes in ozone layer'.

The use of chlorofluorocarbon is increasing because of the demand of 'personal care products' such as deodorants, hair sprays, shaving creams and countless other consumer cosmetic products as well as in refrigeration. The ozone layer serves as a shield, protecting the troposphere and earth's surface from most of the ultra violet radiation found in the sun's rays. If these ultra-violet rays reach the earth's surface in full intensity, all exposed bacteria would be destroyed; plants and animal tissues would be severely damaged. In this protective role, the presence of the ozone layer is an essential factor in man's environment.

Greenhouse Effect

The temperature at the surface of the earth is maintained by the energy balance between the sun rays that strike the planet and the heat that is radiated back into space. Some of the heat is absorbed and retained by the earth or objects on the surface. Much of this does not pass through the air envelope to outer space but is absorbed by the carbon dioxide and water vapour in the atmosphere and adds to the heat that is already present. Thus, carbon dioxide acts like the glass of a greenhouse, and on a global scale, tends to warm the air in the lower levels of the atmosphere. This is called the greenhouse effect which is also responsible for the increase in temperature over the earth's surface.

Scientists are of the opinion that carbon dioxide will increase global temperature significantly. There is no second opinion about the fact that due to industrial and other forms of air pollution, the carbon dioxide content is increasing. Volcanic eruptions are also responsible for the increase of carbon dioxide. Volcanic activity during historical period as well as in present times has thrown out dust and ash which spread throughout the entire atmosphere and last for years.

There is evidence that the temperature of the entire earth has risen slightly during recent decades. Between 1885 and 1940, atmospheric warming was a world-wide phenomenon. There was an increase in mean annual air temperature of about 0.5°C. The greatest increase in mean annual temperature was in the northern hemisphere between 40°N and 70°N latitude where there was an increase in the average winter temperature of 1.6°C. After 1940, this process of global warming has become much slower, but with the increase in carbon dioxide content of the atmosphere from combustion, the prediction of global warming is there. This will result in the melting of glaciers and melting of snow and increase in the sea level, which may destroy not only the coastal regions but also other parts of the earth.

With the increase in temperature, scientists have predicted that the process of melting of ice caps of Antarctic, Greenland, etc. may lead to a rise in the sea level. Some of them have pointed out that there is a rise of four feet in every 10 years. There is a crucial relationship between the presence of the polar ice cap and global temperatures and if it happens, its impact on climatic conditions of the world will be drastic and this will change the entire pattern of our ecosystem.

Acid Rain

Acid rain is the outcome of sulphur dioxide gas (SO_2), released into the air by the combustion of fossil fuels. This readily forms sulphuric acid (H_2SO_4) in the minute water films of suspended droplets in the air over cities. The washout of sulphuric acid by precipitation results in rainwater with an abnormally high content of the sulphate ion, a condition known as acid rain. Nitric acid, formed by reactions involving the pollutant nitrogen oxide, may also contribute to the acidity of rainwater.

The degree of acidity is measured in terms of pH value of rainwater. The pH of rainwater as measured in the laboratory is 5.65 owing to the dissolution of atmospheric CO_2 in it. Rain, having a pH value of less than 5.65, the value resulting from the equilibrium between atmospheric CO_2 and pure distilled rainwater at 25°C, can be termed acid rain.

In recent years, it has been pointed out by water chemists that pH value of rain in north-western Europe and USA is going down to 3 and 5 respectively, which is an indication of acidity in rainwater. The pH numbers are on a logarithmic scale, these values mean that rain in these areas has now often 100 to 1,000 times more acid than before. After studying rainwater for 20 years US scientists have reported average pH value of about 4. They have also observed values as low as 2.1 in rainwater at certain localities. The acid rain is now not only the problem of industrial developed world but is also seen in countries like India. Khemani et al. (1989), in an article 'Spread of Acid Rain Over India', have summarized the observations as shown in Table 7.4.

Table 7.4

Average pH Values of Rainwater over Cities in India

High pH Values (> 7.0)	*Medium pH Values (7.0–6.0)*	*Low pH Values (6.0–4.5)*	*Very low pH Values (< 4.5)*
Ahmedabad	Agra	Trivandrum	Chembur(Bombay)
Allahabad	Bhopal	—	—
Amritsar	Calcutta	—	—
Bikaner	Chandigarh	—	—
Jaisalmer	Delhi	—	—
Jodhpur	Nagpur	—	—
Lucknow	Visakhapatnam	—	—
Srinagar	—	—	—

Source: Khemani, L.T. et al., (1989) Spread of Acid Rain Over India, *Atmospheric Environment*, Vol. 23, pp. 757–762.

Some of the adverse environmental effects of acid rain are: excessive leaching of nutrients from plants, foliage and the soil, disturbance of balance of predators and prey in aquatic ecosystems, various metabolic disturbances to organisms; acidification of lakes and streams and the corrosion of structures.

Smog

'Smog' is the name given for 'smoke fog'. The term was first used by H.A. Des Voeux in 1911 in a report on smoke fog deaths. During the autumn of 1909 in Glasgow, Scotland, 1,063 deaths were attributed to smoke and fog. In 1930, in Meuse Valley of Belgium, a large number of people got ill and many were killed by smog. The 'black fog' that settled on London in December 1952 caused 3,500 to 4,000 deaths. It was fog accompanied by thermal inversion over most of the British Isles. In Pennsylvania, smog killed 20 people and about 40 per cent population became ill. The residents of the Kanto Plains area of Yokohama, Japan, were afflicted with an accumulation of smog during winters of 1945 and 1946. Similar type of attack was reported in New Orleans, Louisiana, in 1959. In recent years, with the increase in atmospheric pollution, the instances of smog resulting in serious injury or deaths have also become common.

The smog is of two types (i) industrial smog, and (ii) photochemical smog. The industrial smog is a combination of smoke and fog. This is a greyish mixture of soot, sulphurous compounds, and water vapour. This kind of smog continues to be found wherever industries are concentrated and where coal is the primary energy source. On the other hand, photochemical smog is produced when several pollutants from automobile exhausts-nitrogen oxides and volatile organic carbon compound are acted on by sunlight. (Figure 7.1)

Figure 7.1

Industrial and Photochemical Smog. (a) Industrial smog, or grey smog, occurs when coal is burned and the atmosphere is humid. (b) Photochemical smog, or brown haze, occurs when sunlight acts on vehicle pollutants.

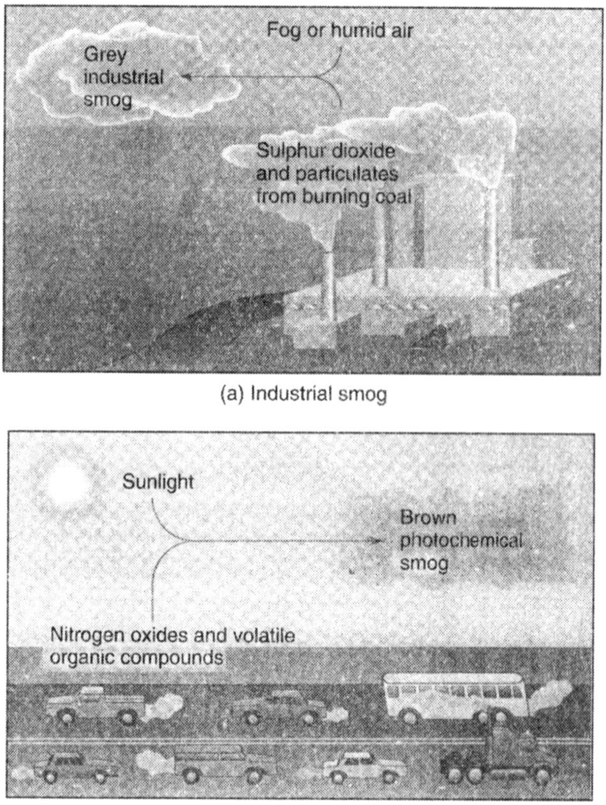

(a) Industrial smog

(b) Photochemical smog

The most irritating and injurious components of smog are the products of reactions in the atmosphere between oxygen, ozone and emission pollutants. The mixture of these products is called photochemical smog. These are reactions in which oxygen, ozone, nitrogen oxides and hydrocarbons produce those compounds which are toxic and irritating.

The constituents of smog are quite toxic and are responsible for respiratory and cardiac difficulties. Eye irritation is the most common symptom of smog injury. Similarly, bronchial asthma, chronic bronchitis and pulmonary emphysema are associated with smog. It is also injurious to animals and plants and one of the main causes of environmental degradation.

Pollution by Vehicular Exhaust

The automobile is man's greatest achievement in minimizing distances. The number of automobiles is increasing day by day and has become a cause of air pollution and degradation of the environment. The automobile, with its internal combustion engine emits poisonous gases which are harmful to human health and is the most serious pollution problem of the technological age.

Exhaust emissions from diesel engines include carbon monoxide, hydrocarbons, sulphur oxides, organic acids, etc. The two primary pollutants are carbon monoxide and nitrogen dioxide, both of which are extremely poisonous gases and are present in the atmosphere of urban areas, specially in metropolitan areas. A third form of pollutant is constituted by the unburned hydrocarbons of gasoline. Lead is also a very toxic compound and its main source in the environment is believed to be from leaded gasoline used as fuel for internal combustion engines. The lead usually added to gasoline is the organic compound tetraethyl lead (TEL), which is extremely poisonous. The presence of lead in the atmosphere is a threat to the environment as well as for all living organisms.

It becomes clear from the above analysis that the problem of air pollution is increasing with the growth and expansion of industries and automobiles. It is high time that all of us should know the harmful effects of air pollution and also evolve technology to control it.

Effects of Air Pollution

Air pollution has now become a widespread problem and every individual in one way or the other is facing problems caused by air pollution. Its impact can be seen locally, at regional level as well as at global level. At local and regional levels its manifestations are in the form of alterations in: (i) visibility, (ii) intensity of sunshine, (iii) precipitation amount, and (iv) acid rain. Its global effects are: (i) change in natural climate by rise of temperature, melting of snow, (ii) increase in carbon dioxide, (iii) increase in particulates, (iv) holes in ozone layer, etc. Several aspects of air pollution, such as, effects on the ozone layer, greenhouse effect, smog and acid rain have already been discussed. The effects of air pollution can be grouped under the following heads:

1. effects on human health,
2. effects on animals and plants,
3. effects on atmosphere, and
4. other effects.

Effects on Human Health

Some environmental poisons can cause acute illness and even death. Others may be harmful, but the disease may take years or even decades to appear. Air pollution mainly affects the respiratory system. Bronchitis, emphysema, asthma and lung cancer are some of the chronic diseases caused due to exposure to polluted air. It is feared that lung cancer is caused mainly due to polluted air because carcinogens are found in the polluted air. Its mortality rate is higher in urban areas. Figure 7.2 shows the various effects of air pollution on the human body. Sulphur dioxide is the most serious and widespread air pollutant. Its lower concentration is a cause of spasms in the smooth muscle of bronchioles and its higher concentration induces increased mucus production. Sulphur dioxide is also considered to cause cough, shortness of breath, spasm of the larynx and acute irritation to the membranes of the eyes. SO_2 also acts as an allergenic agent. When it reacts with some compounds, sulphuric acid is formed which may damage lungs.

Figure 7.2

Effects of Air Pollution on the Human Body

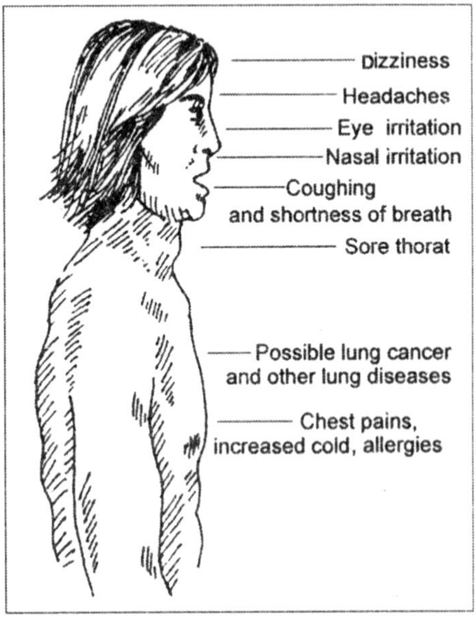

Carbon Monoxide often affects the oxygen carrying capacity of blood. Nitric oxide is reported to be a pulmonary irritant and its excess concentration may cause pulmonary haemorrhage. Hydrogen sulphide is also toxic. Lead emitted from automobile exhausts is a cumulative poison and is dangerous particularly to children and may cause brain damage.

The particulate pollutants such as asbestos, silica, carbon, beryllium, lead, etc. are capable of exerting a noxious (fibrotic) local action in the interstitial areas of the lungs. Radioactive elements are also harmful to man and other living organisms. As described earlier, smog has a killer effect, which is the result of air pollution. The death toll by smog varies from few persons to thousands. In December 1952, about 4,000 persons died in London due to smog. Similar cases have been reported in London itself in 1956, 1957 and 1962 in which the death toll was between 700 and 1,000 persons. In other countries also smog deaths have been reported. In fact, the growing air pollution has now become a health hazard for man.

Effects on Animals and Plants

The impact of air pollution on animals is more or less similar to that of effects on man. Chronic poisoning results from the ingestion of forage contaminated with atmospheric pollutants. Among the metallic contaminants, arsenic, lead and molybdenum are important. Fluoride is another pollutant which causes fluorosis among animals. A number of livestock have been poisoned by fluorides and arsenic in North America. Bone lesions in animals due to excessive fluorides have also been reported.

Air pollution has caused widespread damage to trees, fruits, vegetables, flowers and in general, vegetation as a whole. The total annual cost of plant damage caused by air pollution in USA alone has been estimated to be in the range of 1 to 2 billion dollars. The most dramatic early instances of plant damage were seen in the total destruction of vegetation by sulphur dioxide in the areas surrounding smelters. When the absorption of sulphur dioxide exceeds a particular level, the cells become inactive and are killed, resulting in tissue collapse and drying of leaves. Cotton, wheat, barley and apple are more sensitive to this pollutant.

Fluorides are responsible for various types of injuries to plants. The leaves of apple, apricot, fig, peach, etc. are more susceptible to air borne fluorides. Fluorides seem to interfere with the photosynthesis and respiration of plants. Smog also causes injury to plants. Similar impact of ozone can be seen in the form of lesions on plants. Chlorine, ammonia, hydrogen sulphide, etc. are also harmful to vegetation.

Effects on Atmosphere

Some of the effects of air pollution on atmospheric conditions have already been discussed such as effect on ozone layer, greenhouse effect, etc. There is an increase in the carbon dioxide concentration in the air due to increased combustion of fossil fuels. Carbon dioxide absorbs heat strongly and the radiative cooling effect of the earth is thus decreased. The rising of temperatures and ozone holes are some of the problems which have attracted the attention of the scientists of the world. These problems are not related to any region or a country but are the global problems and their impact on world climate may be hazardous to the whole world. The local weather conditions are highly susceptible to air pollution. Its impact on temperature, humidity, rainfall and clouds is apparent. The 'smog dome' on large urban centres is the result of air pollution. Due to air pollution, visibility also reduces.

Other Effects

Air pollution can also cause damage to property and materials. The smoke, grit, dust and oxides of sulphur have harmful effects on structures. In 1972, when an oil refinery at Mathura was opened, its impact on Taj Mahal became a major issue. Sulphur dioxide is the most damaging of gaseous pollutants. Aluminum alloys, copper and copper alloys, iron and steel are corroded when exposed to contaminated air. Hydrogen sulphide reacts with lead paints to form lead sulphide thereby producing a brown to black discolouration. The damage caused by air pollution to structures is not serious but from an aesthetic point of view, it is not desirable.

Indoor Air Pollution

Indoor Air Pollution means pollution inside the home or in a covered area. A study conducted at Harvard University indicate a startling observation: 'regardless of which community they studied – its location, whether rural or urban; lightly or highly industrialized; and regardless of sex, age, smoking habits and occupation – indoor air pollution was the major source of exposure to many air pollutants.' (K. Hill, 2007) This fact is true for whole of the world that indoor air pollution is more harmful to the people because we spend most of the time inside the house.

The main sources of indoor air pollution and pollutants released by them are given in Table 7.5 and also shown Figure 7.3.

Table 7.5

Indoor Air Pollutants

Sources	*Pollutants released*
Household products	
• Paints, stains, thinners, strippers, polishes, cleaners, solvents, air fresheners • Cosmetics, perfumes, and colognes • Hobby supplies (paints, glues, metals, wood, glass, others) • Propellants in aerosol sprays • Pesticides and their solvents	These various products can emit VOCs such as formaldehyde, benzene, toluene, xylene, and hexane. Many other VOCs are also possible, depending on the products use. Some also emit particulates and other pollutants.
Household furnishings	
• Drapes, upholstered furniture, pressed-wood in cabinets, walls, sub-flooring	New Furnishings. Formaldehyde and other VOCs
• Carpets, shelving, other surfaces, 'dust catchers' (also, see biological hazards)	Older furnishings. Dust, biological pollutants
• Carpets, furnishings, bedding	Dust mites and dust
Dust and dirt getting into the home	These can contain pesticide residues, metals, and biological pollutants
Combustion sources	
• Oil or gas furnace, unvented kerosene or gas stoves, wood stoves, fireplaces • Gas water heater or clothing dryer • Environmental tobacco smoke (ETS)	Without good controls, combustion appliances release carbon monoxide, nitrogen oxides, sulphur dioxide, particulates (sometimes formaldehyde, benzene, and other VOCs) ETS emits these and many others too.
Biological hazards	
• Moist areas (basement, bathroom, etc.)	Mold VOCs bacteria, bacterial toxins, viruses If not kept clean, these can emit all of the above
• Humidifiers and dehumidifiers	Animal dander, dried saliva, urine, Pollen and dust
• Pets, intruding rodents and insects • Outdoor air	
Hazards old and new	
• Old paint	Lead
• Old insulation (furnace pipes, old tiles)	Asbestos
• Construction materials, wood panelling	VOCs including formaldehyde
• New carpets, padding, adhesives	VOCs
• Painting, stripping old paint, etc.	VOCs
Basement floor (cracks, sump openings)	Radon
Household water	
• Municipal chlorinated drinking water]	Radon
• Well water	Chloroform and other VOCs Radon, sometimes VOCs

Figure 7.3
Sources of Indoor Air Pollutants

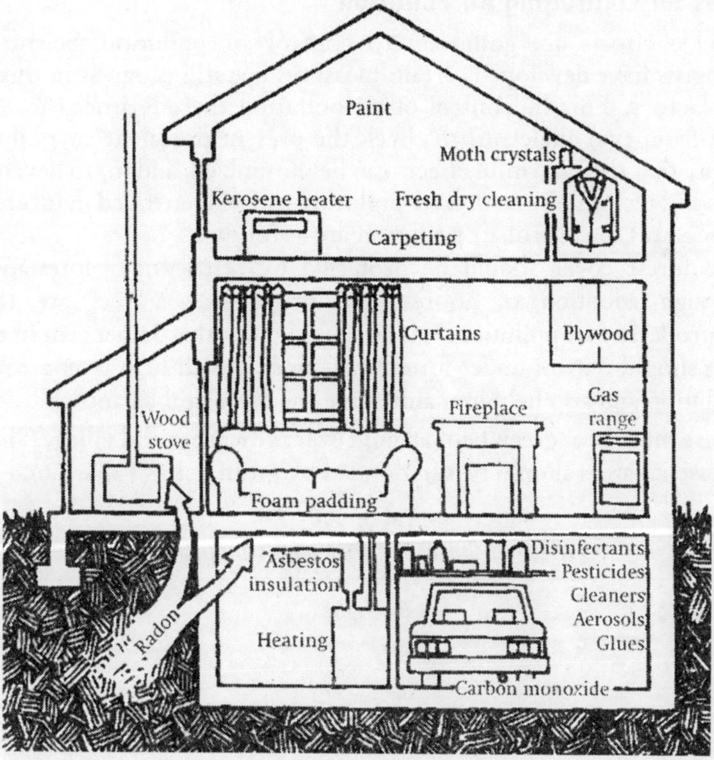

The contaminants affecting indoor air quality spring from many sources (Table 7.4) can be listed as:

- Particulates, which are emitted by combustion appliances including wood stoves, fire places, gas stoves, and kerosene heaters.
- VOCs are emitted from construction materials (such as particle board) and many consumer products including cleansers. polishes, paints, moth balls, glue, rugs, furniture, drapes, clothing, and hobby materials. Chlorinated water emits VOCs too.
- Biological contaminants grow in moist areas in homes and become airborne.
- The radioactive gas radon seeps up from rocks and soil under homes into indoor air, and also enters air from well water pumped into the home.
- Older homes may have special problems such as flaking lead paint or asbestos. Some may have accumulated more biological contaminants from persistent wet places or old rugs.

The effects of indoor air pollution are manifold, which may be minor or major depending on the pollutant, its concentration, how one is exposed and the length of exposure. Chronic effects can result from ongoing exposure, such as irritation of eyes, nose and throat, allergies from airborne microorganism, flu like symptoms or headaches from carbon monoxide, etc.

Measures for Controlling Air Pollution

World-wide efforts are going on to control air pollution. Scientists and technologists have developed certain measures but still progress in this field is not satisfactory. For the control of air pollution there is a need to deal this problem from two angles: (i) to check the present prevalent air pollution in such a way that their harmful effects can be minimized, and (ii) to develop such methods and technology so that air pollution can be controlled in future. Some of the measures to control air pollution are as follows:

1. The forest cover should be protected by restricting deforestation and through adoption of afforestation programmes. Trees are the best controllers of air pollution. It is an accepted fact that 33 per cent of the land area should remain under forest/tree cover. It will help in controlling air pollution and also help in maintaining the ecological balance.
2. There must be a 'green belt' around every township and village. Similarly, industrial areas should be surrounded by green belts (Table 7.6).

Table 7.6

Width of Green Belt for Various Industries

Industry	*Examples*	*Area of Works (ha)*	*Width of Green Belt*
Heavy industry	Oil refineries, chemical works, metallurgical and seaport industries, nuclear reactors	>500	<2 km
Heavy industry	Machine manufacture, ship-building, big harbour industries, power stations	200–500	> 1 km
Medium-heavy industry with much air pollution	Manufacture of straw boards, artificial fibres, and ceramic products, cement works	100–200	500 m or more
Medium-heavy industry with little air pollution	Manufacture of cars, foods, textiles		200 m or more
Light industry with some air pollution	Tanneries, textiles and food industries	50–100	50–100 m
Light industry with some air pollution	Manufacture of electronic apparatus and domestic machines		
Service industry	Printing works, bakeries, laboratories	10–50	>100m
Workshops, handicrafts, etc.	Fashion studios, photo printing shops, potteries	1–10	>50 m

Adapted from *Manual on Urban Air Quality*, WHO Regional Publications, European Series No. 1, Copenhagen, 1976.

3. The main source of air pollution is the automobiles, therefore, their engines should be redesigned in such a way that their emissions cause minimum pollution. Several steps have been taken and some technology has also been developed, but still it is in a very preliminary stage. Apart

from technological changes, some steps on individual level should be taken by auto owners, such as:
(i) regular service of the vehicle,
(ii) change or replacement of old engines,
(iii) use of filters and after burner,
(iv) the engine should be well tuned, and
(v) mixing of anti-tibs in diesel and petrol, etc.

4. The use of steam engine by the railways should be stopped or minimized. As far as possible, electrically operated rail engines should be used, which are pollution free.
5. In countries like India, traditional use of fuelwood, coal, etc. should be checked and newly devised smoke free furnaces be used.
6. The industrial areas should be located at a certain distance from residential areas. The selection of the site should be done with consideration of air direction.
7. Steps should be taken for immediate check on forest fires. For this purpose, proper arrangement and sound infrastructure is necessary.
8. In industries, arrangements for pollution control should be done. Only after full arrangements of effluent treatment is done, permission for production should be given.
9. With the help of chemical reactions, harmful effects of pollution can be controlled. Measures for such methods should be taken and the methodology and material should be available to the concerned persons.
10. Some methods of controlling air pollution are filtering, settling, dissolving, absorption, etc. For these methods, cheap devices should be developed.

Actual prevention of air pollution is not so simple as explained above. However, much of the pollution could be minimized or controlled without undue cost by careful planning. For this, the following steps will be helpful:

Control of Gaseous Pollutants

The equipments which can be used to control gaseous pollutants are classified as combustion, absorption and adsorption equipments. Combustion is applicable to pollutant gases which are oxidizable. In petrochemical, fertilizer, paint and varnish industries, combustion control equipments are useful. Absorption is a diffusional process in which the transfer of gas molecules into a liquid phase takes place. Such equipments include spray chambers, packed towers and sieve plate contractors. Absorption is the means of controlling air pollution which occurs due to some gases, vapours and inflammable compounds which cannot be treated by other means.

Control of Emission from Motor Vehicles

The emission control techniques include tune-ups, catalytic reactors and engine modifications. A high air-fuel ratio will reduce the concentrations of both carbon monoxide and hydrocarbons. The modified engines have an efficient system of burning of the fuel. At present, there is no system of reduction of sulphur dioxide although, researches in this direction are in progress throughout the world.

Control of Aerosol Emissions

This can be controlled by arresters and scrubbers. Arresters include inertial separators, filters and precipitators. The electrostatic precipitator is considered to be the most effective device for preventing the emission of dust from fuel gases and is the standard equipment for large power stations.

Control by Fuel Selection and Utilization

Coal and oil are the primary fuels in which smoke, grit and sulphur dioxide are the major pollutants. Coal pollutes more because it releases more smoke and carbon in the air. Instead of coal, oil can be used. But in oil, the amount of sulphur dioxide emission is higher. Therefore, fuel selection should be done properly and the harmful impact be restricted by inducing other chemicals. In order to control smoke, the coal is pulverized before being used. When oil is used for fuel, it is essential that the ratio of air should be maintained constant in order to prevent smoke emissions.

Control of Air Pollution by Site Selection and Zoning

Selection of an industrial site is the most important factor through which impact of air pollution can be minimized. The industrial site should be selected considering: (i) residential areas, (ii) nature of industries, and (iii) climatic conditions, specially the direction of winds. The site of each and every industry should have an approval from the department of environment. It should be approved in taking into consideration the long term perspective rather than immediate effects. Within the city the location of industrial zone is a critical aspect and town planners should have a check over its location as well as unorganized growth.

Apart from these controls legal controls are also there. The UNO and WHO have issued guidelines to its members for the formulation of laws regarding prevention of air pollution and accordingly all the countries of the world have enacted certain laws for its control. In India, the Prevention of Air and Water Pollution Acts, 1974, 1981; the Air Prevention and Control of Pollution Act, 1981 and the Environmental Protection Act, 1986, have been enacted for air pollution control. These laws should be imposed properly, Apart from this, the most important aspect is public awareness about air pollution because the government alone cannot prevent pollution. It can be checked only through the combined efforts of the government, NGOs and the public.

Notes

1. Maxwell, K.E. (1973), *Environment of Life*, Dickenson Pub. Com., California, p. 128.
2. Strahler A.N. and Strahler A.H. (1977), *Geography of Man's Environment*, John Wiley, New York, p. 54.
3. Hill, Marquita K. (2007), *Understanding Environmental Pollution*, Cambridge University Press, Cambridge, UK, p. 401.

8

Water Pollution

Water is an essential resource for life on earth. We drink it, bathe in it, relax in it, fish in it, keep cool with it, irrigate the plants, produce energy with it and also use it for transportation and recreation. Many ancient civilizations have developed in river valleys of the Nile, Indus, Tigris and Euphrates, etc. Even now also, water resources provide a base for social, cultural and economic development. It seems that water is abundant, but usable water is very limited and creates a serious conservation problem in many places where it is needed. Apart from scarcity of water, we are also facing a problem of water pollution not only of rivers and lakes but also of underground water. Most of the rivers and lakes of the world are now facing the problem of water pollution. In developed country like United States of America 40 per cent of rivers and 46 per cent of lakes are polluted for fishing, swimming, or aquatic life. Polluted drinking waters are a problem for about half of the world's population. Each year, there are about 250 million cases of water-based diseases. In developing countries, 70 per cent of industrial wastes are dumped untreated into waters where they pollute the usable water supply. The intolerable burden of chemical and human waste products have become a threat to aquatic life as well as to human health. Therefore, much attention has been given to study the various facets of water pollution, not only to understand the nature and effects of water pollution but also for its control.

Sources of Water and its Quality

The planet earth has 71 per cent of its area as oceans and the oceans contain 97 per cent of all water in the form not suitable for human consumption. Much of the remainder is frozen in the ice caps and glaciers. The water in rivers and lakes is very small. Less than 1 per cent is in the form of ice free fresh water yet this negligible portion of the planet's water is crucially important to all forms of terrestrial and aquatic life. Apart from surface water available in rivers and lakes, underground water resources are important. The percentage of land water distribution over earth's surface (other than oceans) is as follows:

Rivers and streams	00.014%
Fresh water lakes	01.4%
Salt lakes and inland seas	01.2%
Soil moisture and seepage	00.7%
Ground water (one km deep)	48.3%
Ground water (below one km deep)	48.3%

Fresh water is a scarce commodity with greatest amount locked in glaciers and ice caps. But it is constantly replenished in a cyclic way known as hydrological cycle. The hydrological cycle is a simplified description of water movement from place to place. This water cycle pattern has been depicted in Figure 8.1.

Figure 8.1

The Hydrological Cycle

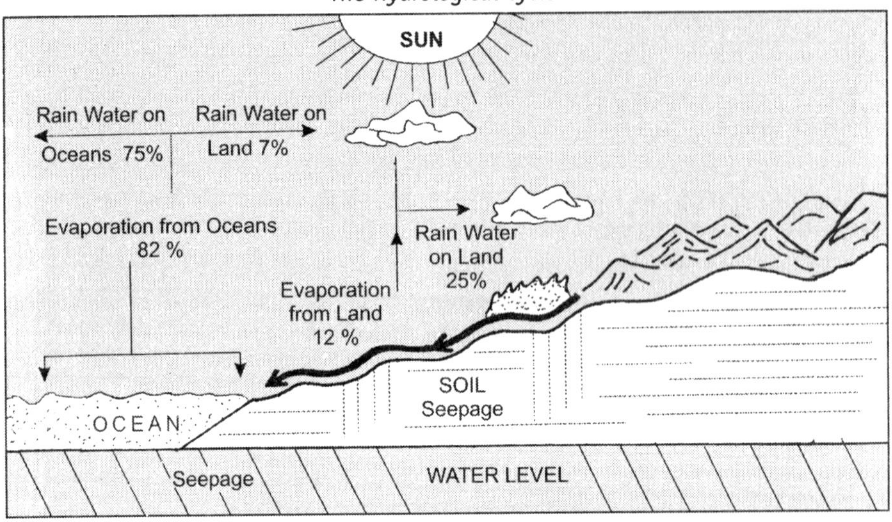

The cycle begins with evaporation by solar energy. The formation of clouds (condensation) leads to subsequent precipitation which results in the conversion of salt water to fresh water. The precipitation on the surface may run-off as rivers or may infiltrate into the soil and rock and form ground water. Surface and sub-surface water returns to the ocean unless evaporation interrupts the cycle.

The quality of water is the most important factor. From chemical point of view, water is the monoxide of hydrogen (H_2O) in which two parts of hydrogen and one part oxygen are present. But this form of pure water is not available in nature. As soon as water vapour enters into the atmosphere after condensation, impurities like dust particles, gases, etc. are mixed in it and when it reaches the surface, several mineral and chemicals are also mixed in it. The quality of drinking water depends on the quantity of harmful elements present in it. Drinking water should be clear, odourless and tasteless and its pH value should be between 7 and 8.5. In 1971, WHO has given the limits of the various impurities in drinking water as shown in Table 8.1.

Table 8.1

Permissible Limits of Impurities in Water

	Impurities	*Max. Permissible Limit (mg/litre)*
1.	Total solid matters	500
2.	Hardness	2 mEq/Litre
3.	Calcium (Ca)	75.00
4.	Magnesium (Mg)	30.00
5.	Sulfate (SO_4)	200.00
6.	Chlorides	200.00
7.	Iron (Fe)	0.10
8.	Manganese (Mn)	0.05
9.	Copper (Cu)	0.05
10.	Zinc (Zn)	5.00
11.	Arsenic (As)	0.05
12.	Cadmium (Cd)	0.005
13.	Cyanide (Cn)	0.05
14.	Lead (Pb)	0.05
15.	Mercury (Hg)	0.001
16.	Phenolic elements	0.001

The permissible organic impurities include Bacillus coli not more than one in 100 m/litre and other coli bacteria not more than 10. In natural water impurities are found in three forms:

1. suspended impurities,
2. colloidal impurities, and
3. dissolved impurities.

Suspended impurities float on the water surface as dust particles, particles of minerals, paper, cloth, foam, leaves and floating vegetation as well as other organic and inorganic materials. Colloidal impurities are those which are mixed with water and difficult to separate, such as silica glass, oxides of various minerals, micro bacteria, etc. On the other hand, dissolved impurities become part of the water during its flow. These include carbon dioxide (CO_2), sulphur dioxide (SO_2), methane (CH_4), sodium (Na^+), potassium (K^+), calcium (Ca^{++}), magnesium (mg^{++}), iron (Fe^{++}), ammonium (nh4+), etc. Apart from these, man has induced several impurities in the water, which are responsible for water pollution.

Water Pollution

Water pollution simply means contamination of water due to any external material, or in other words, introduction of something to natural water which makes it unsuitable for human consumption. WHO has defined water pollution as 'any foreign material either from natural or other sources that may contaminate the water supply and makes it harmful to life, cause of its toxicity, leads to reduction of normal oxygen level of water, causes aesthetically unpalatable effects and spread of epidemic diseases'. Owen has defined it as 'any unreasonable contamination of water which lessens its value to man'.[1] In general, 'water pollution may be defined as the adverse change in composition or condition of the water such that it becomes less suitable for the purposes for which it would be suitable in its natural state. The changes include physical changes, chemical changes and biological changes'.[2] According to Gilpil, 'the deterioration in chemical, physical and biological properties of water brought about mainly by human activities',[3] is the water pollution. Whatever additions are done to the water either by natural processes or by human activities invariably, they change the natural qualities. The problem of water pollution was recognized by Hippocrates (450 BC), who suggested filtration and boiling as remedial measures. With the fast increase in our industrial civilization, the demand for water is also increasing day by day. At the same time, population increase, urbanization, improper sewage disposal, unsafe industrial wastes, radioactive wastes, oil, etc. have polluted our water resources so much so that 70 per cent rivers and streams not only of India but of all the countries contain polluted water.

Sources of Water Pollution

Water pollution is caused by several sources, which are not independent in nature but interact with one another. Generally, one or two factors become prominent and may be considered the primary source of water pollution. The various sources of water pollution can be classified as follows:

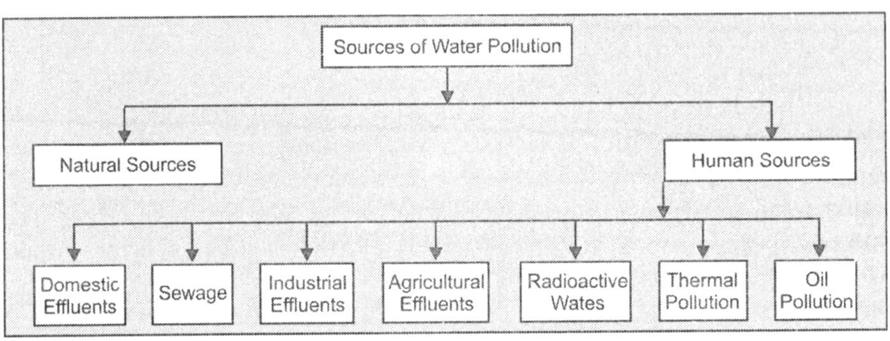

Natural Sources

There are some natural elements which create water pollution. These are gases, soil, minerals, humus material, waste created by animals and other living organisms present in the water. During rains, one can see the direct impact of soil on the water of rivers, tanks and other water bodies. The brown and dirty water is the result of mud mixed in the water, which becomes clear after some time. Similarly, several types of natural suspended and colloidal impurities are also there in the water. These elements may be organic or inorganic and may even be harmful. The minerals like sodium, potassium, calcium, magnesium, iron, etc. are also present in the water, and if their quantity is more than the permissible limit, these are harmful. Some poisonous minerals such as nickel, beryllium, cobalt, lead, mercury, cadmium, etc. are very harmful and are responsible for water pollution. Sometimes water-borne vegetation also becomes a cause of pollution.

Human Sources

There is no doubt that water pollution is the result of human activities. The rapid growth of population, urbanization, industrialization and increasing use of chemicals have resulted in water pollution and this problem is increasing day by day in spite of several measures taken in this direction. This is not only a problem of developed countries and urban areas but has also become an uncontrollable problem of developing countries as well as of rural areas. A misconcept developed right from the dawn of human civilization is that water bodies can be used as dumping grounds for all types of wastes and this is responsible for water pollution throughout the world. This is not only hazardous to human health and other organisms but also disturbs the delicate ecosystem. Table 8.2 indicates sources or type of contaminates responsible for water pollution and their impact.

For detailed analysis, the human sources of water pollution can be classified into the following categories:

1. domestic effluents and sewage,
2. industrial effluents,
3. agricultural effluents,
4. radioactive wastes,
5. thermal pollution, and
6. oil pollution.

Domestic Effluents and Sewage

Man, for his various domestic purposes such as drinking, preparation of food, bathing, for cleaning the house, cooling etc. uses on an average 135 litres of water per day. About 70 to 80 per cent of this is discharged and drained out, which through municipal drains poured into, in many cases, a river, tank or lake. This water is known as domestic waste water in which when other waste material such as paper, plastic, detergents, cloth, other waste materials are mixed, is known as municipal waste or sewage. The domestic waste water and sewage is the main source of the water pollution. This is the inevitable and unfortunate fallout of urbanization. As it decays, this organic waste depletes the oxygen from water and upsets the natural balance of the aquatic ecosystem.

Table 8.2

Water Pollution Contaminants and their Impact

Source or Type of Contaminant	*Impact*
1. Degradable wastes	
Domestic and municipal sewage and other oxygen demanding industrial wastes - they are acted upon by the bacteria and oxygen dissolved in water and reduced to inorganic form, as also in quantity.	Depletion of dissolved oxygen in water - harmful for fish and other aquatic life.
2. Plant Nutrients	
Phosphatic and nitrogenous compounds dissolved in water from industrial wastes, or due to washing away of fertilizers due to heavy rains from fields.	Stimulate growth of algae/plankton and other plants resulting in clogging of waterways and rendering the water unfit for human use.
3. Infectious Agents	
Polluted, stagnant water breeding grounds for parasites, bacteria, viruses of all kinds.	Cause water-borne diseases and outbreak of epidemics such as ameobiosis, dysentery, cholera, typhoid, etc.
4. Synthetic Organic Compounds	
Such as pesticides, agricultural chemicals, detergents, industrial wastes, DDT, etc.	Instance of cadmium, lead and mercury poisoning in human beings. Cause neurological impairment and even death.
5. Inorganic Minerals and Other Chemicals	
Asbestos and acids, etc.	Asbestos produces lung cancer. Acids cause allergies, ulcers, skin diseases. Also have catastrophic impact on fish and aquatic life.
6. Radioactive Elements	
Wastes from nuclear power plants and nuclear fuel reprocessing plants are packaged and buried under water/oceans.	Potentially hazardous in the event of leakages and induce radiation-related illness.
7. Sediments	
Soil and mineral particles washed into streams or water.	Reduce amount of sun-light available for marine plants, cause clogging of filter plants.
8. Thermal Pollution	
Water used for cooling in electric, thermal, nuclear power plants, when recirculated in water bodies.	Can raise lake water temperature by 70–100°C aggravate diminution of dissolved oxygen, kills marine fish and plant life.

Municipal sewage is considered to be the main pollutant of water. Most of the sewage receives no treatment before discharge, specially in developing countries like India. In Delhi alone, 120 crore litres of water is consumed, per day out of which 96 crore litres waste water is drained into the Yamuna river through 17 big drains. All the 47 towns located on the bank of river Ganga drain their sewage into it. With the growth of population the quantity of waste water is also increasing in addition to the production of large quantities of sewage. Sewage contains decomposable organic matter and exert an oxygen demand on the receiving waters. The common organic materials found in a sewage are soaps, synthetic detergents, fatty acids, proteinaceous matters such as amines, amino acids, amides and amino sugars. In addition to the above components it also contains numerous microorganisms. Sewage also contains pathogenic bacteria and viruses derived from human faces. Untreated waste water is often the carrier of viruses and bacteria and, along with household sanitation practices, has been linked to high infant mortality rates in developing countries. Even where most sewage is treated, as in the developed world, recent measurements of faecal coliforms in some countries indicate increasing pollution. Sewage supports the growth of other forms of life that consume oxygen, it is measured in terms of Biochemical Oxygen Demand (BOD). It is the lack of oxygen that kills fish and other aquatic life.

In recent years, there has been considerable growth in the use of detergents, which causes severe water pollution. Many modern detergents contain phosphates, which are essential components of agricultural fertilizers. When phosphate detergents are discharged into waterways, they supply a needed nutrient and promote rapid growth of algae. This enrichment process is known as 'eutrophication'. In many areas of the world, aquatic weeds have multiplied explosively. They have interfered with fishing, navigation, irrigation and even production of hydroelectricity. In developing countries, human population and settlements are growing fast, often faster than waste water treatment facilities can be provided. Thus, much of the untreated waste water and sewage is discharged into rivers and other water bodies, making the water unsuitable for drinking and is a main cause of water pollution.

Industrial Effluents

Industrial activities generate a wide variety of waste products which are generally discharged into water courses. Major contributors are the pulp and paper, chemicals, petro-chemicals and refining, metal-working, food processing, textile, distillery, etc. The wastes, broadly categorized as heavy metals or synthetic organic compounds, reach bodies of water either through direct discharge or by leaching from waste dumps. In developed countries of the world, many industrial discharges are strictly controlled. Yet, water pollution continues from accumulations of wastes discharged over the past 100 years. But, in developing countries, industrial discharges are largely uncontrolled, thus, a main cause of water pollution.

All the Indian rivers have been polluted by industrial effluents. The holy river Ganga has become a highly polluted river due to various types of industrial discharges. Along the Ganga, several chemical, textile, tanning, pulp and paper, petrochemical, rubber, fertilizer and other heavy industries are located and all of them discharge their waste water and other effluent, directly or indirectly, into the river, resulting in the pollution to such an extent that even the Ganga Action Plan, to control water pollution, has failed. From Delhi Industrial area alone, more than 8 lakh tonnes of industrial waste is discharged into the river Yamuna.

Damodar river of Bihar is a highly polluted river due to industrial wastes discharged from Bokaro, Rourkela, Indian Iron and Steel Company (CISCO), Bengal Paper Mills, Sindhri Fertilizer Factory, etc. A study reveals that from Durgapur Plant, 1,800 cu/m washed coal has been discharged into the river.

Similarly, from CISCO, 15,000 cu/m and from Bengal Paper Mills, 12,000 cu/m per day, industrial waste is discharged into river water. Likewise, about 10 to 15 tonnes of sulphuric acid from Sindri and 5 to 10 tons toxic chemicals are discharged into the river water. The story of the Hooghly river of West Bengal is also similar. Its water has been polluted to such an extent that even fish fertilization becomes difficult. The paper mills located along the river discharge about 11.4 tonnes liquid wastes into the river. Same is the story of other rivers of India excluding rivers of North-East region. Chambal, Narmada, Kaveri, Godavari, Mahanadi and other small rivers have been polluted and this has not yet been stopped, rather more and more pollutants have been discharged, resulting in greater water pollution.

The nature and effects of the pollutants from the effluents of paper and pulp, textile, food processing, chemicals, metal, petroleum are as follows:

Paper and pulp industries' effluents from paper and pulp industry include wood chips, bits of bark, cellulose fibres and dissolved lignin, in addition to a mixture of chemicals. All these produce a sludge which blankets fish spawning grounds and destroys certain types of aquatic life. The effluents contains chlorine, sulphur dioxide, methyl mercaptans, etc. which are considered to be highly poisonous to fish.

Textile industries' effluents are alkaline in nature and have a higher demand for oxygen.

Food processing industries include dairies, breweries, distilleries, meat-packing, etc. where the waste products include fats, proteins and organic wastes. These industries discharge wastes containing nitrogen, sugar, proteins, etc. All these wastes have a higher BOD and are responsible for water pollution.

Chemical industries include acid manufacturing industries, alkali manufacturing, fertilizer, pesticides and several other industries. The effluents from these industries contain acids which have corrosive effects. The effluents from fertilizer industries contain phosphorus, fluorine, silica, and large amounts of suspended solids.

Metal industries usually discharge effluents containing copper, lead, chromium, cadmium, zinc, etc. which are toxic to man as well to aquatic life. These wastes also contain acids, oils, greases and cleansing agents.

Petroleum industries include oil refineries and petro-chemical plants. The effluents include hydrocarbons, phenolic compounds and other organic and inorganic sulphur compounds.

Other industries which pollute water are tanneries, soaps and detergent industries, glass, electroplating, bleaching, atomic plants, explosive factories, etc.

Mining operations can result in metals leaching into the acidic effluents, thus adding to the metal load in rivers, lakes and ground water. Discharge of mercury from gold mining activities has polluted some streams in Brazil and Ecuador and created serious health problems. With reference to water pollution through mercury, mention of Minimata Gulf incident must be made. In 1950, near the Japanese coast, in Minimata gulf, fishermen suffered from blindness, weakness, mental illness, paralysis, etc. It was found that effluents discharged from a plastic factory contained mercury which entered the fish and by eating those fish, all the fishermen suffered from effects of mercury poisoning. The problem of water pollution through industrial effluents has become a major environmental problem and sufficient measures should be taken to control it.

Agricultural Effluents

Agricultural water pollution is caused by fertilizers, insecticides and pesticides, farm animal wastes and sediments. In recent years, use of chemical fertilizers has increased manifold. The green revolution of India is a reflection of the increased use of fertilizers. The chemicals used in fertilizers enter the ground water by leaching and the surface waters by run-off. The nitrates, when mixed with water may cause methemoglobinemia in infants. Incidences of nitrate poisoning are also there in livestock. The plant nutrients, nitrogen and phosphorus are reported to stimulate the growth of algae and other aquatic plants.

The use of various types of pesticides and insecticides in agriculture is also one of the causes of water pollution. Their presence in water is highly toxic to man and animals, because all these have a high persistence capacity, i.e. their residues remain for long periods.

The farm animal wastes often pose serious problems of odour and water pollution. These wastes also contain pathogenic organisms which get transmitted to humans. Sediments of soil and mineral particles washed out, from fields also cause water pollution. They fill stream channels and reservoirs and reduce the sunlight available to aquatic plants.

Radioactive Wastes

Radioactive elements such as uranium and radium possess highly unstable atomic nuclei. This disintegration results in radiation emission which may be highly injurious. During nuclear tests, radioactive dust may encircle the globe at altitudes of 3,000 metres or more, the same often comes down to the earth as rain. Eventually, some of the radioactive material, such as Strontium 90 (which can cause bone cancer), percolates down through the soil into ground water reservoirs or is carried out into streams and rivers. In both cases, public water supplies may be contaminated. The construction of more nuclear reactors and the increasing use of radioactive materials in medical research represent other potential contamination sources.

Thermal Pollution

Most of the thermal and electric power plants also discharge considerable quantities (about 66%) of hot effluent/water into nearby streams or rivers. This has resulted in thermal pollution of our water courses. Thermal pollution is undesirable for several reasons. Warm water does not have the same oxygen holding capacity as cold water. Therefore, fishes like black bass, trout and walleyes, etc. which require a minimal oxygen concentration of about 4 ppm, would either have to emigrate from the polluted area or die in large numbers. When the temperature of the receiving water is raised, the dissolved oxygen level decreases and the demand for oxygen increases hence, anaerobic conditions will set in resulting in the release of foul gases. Thermal pollution is considered hazardous for the whole aquatic ecosystem. Several industries have installed cooling towers, where the heated water is cooled. But even so, thermal pollution has become a serious problem for water bodies located near thermal plants.

Oil Pollution

The spread of oil in the sea has become a common feature nowadays. Oil is transported across oceans through tankers and either due to some accident or leakage, oil spills onto the water and causes the degradation of aquatic and marine environment. Between 1968 and 1983, there were more than 500 tanker accidents that involved oil spills. Altogether, more than one million tons of oil was released. A dramatic incident was that of the tanker 'Torrey Canyon', when it struck off the southern tip of the British Isles in March 1967. The Torrey Canyon was the largest oil spill up to that time. The pollution caused widespread destruction of many forms of marine life despite strenuous efforts to clean up the spill. Similarly, on 16 March 1978, the oil tanker Amoco Cadiz, lost its steering off the coast of Brittanny in France and the total spilling of oil was 1.6 million barrels. Such accidents have become very common due to technical problems or heavy marine traffic. During the 1991 Gulf War, there was heavy bombing on oil tanks which resulted in the spilling of oil. The impact of this oil spill on the marine ecosystem in this area has not yet been remedied. Offshore drilling operations also contribute their share of oil to the sea. The total quantity of oil that finds its way into sea each year is very large. It has been estimated that about one million tonnes of oil spills into the ocean each year from tankers and oil drilling operations.

In 2010, there was a huge oil spill near Louisiana coast (America) in which more than 150 km have been polluted by the oil spill. Similarly in April, 2010, the Transocean Oil Rig exploded. The disaster damaged the Gulf of Mexico coast causing one of the biggest environmental disasters in US history.

Harmful Effects of Water Pollution

Since use of water is universal to all living organisms, the impact of its pollution is also widespread. The effects of water pollution become severe because water is the controlling factor in the 'food chain', therefore, its impact is not only widespread but intensive also. The harmful effects of water pollution can be divided into three groups, viz. (i) effect on man, (ii) effect on aquatic life, and (iii) other effects.

On Man

On a worldwide scale, the pollution of water supplies is probably responsible for more human illness than any other environmental influence. The diseases so transmitted are chiefly caused by microorganisms and parasites. Sewage and polluted water are responsible for several water-borne diseases. Some of the diseases caused by polluted water are cholera, typhoid, infantile diarrhoea, dysentery, infectious hepatitis, polio, giardiasis, amoebic dysentery, jaundice, etc. Cholerais, an illness caused by ingestion of the bacterium *Vibriocholeral,* results rapidly in massive fluid depletion and death in a very large percentage of patients. It is transmitted by drinking water contaminated with the faeces of infected individuals. In developing countries, deaths by cholera are still common. Guineaworm is another disease caused by polluted drinking water. In India alone about 18 lakh people suffer from this disease. *Schistosomiasis* is a group of diseases caused by infection through worms in water. There are certain chemicals present in polluted water which may cause diseases like nepurities, wrist drop, foot drop, etc. In spite of full measures taken for the supply of purified drinking water in USA, there are incidents of diseases caused by water contamination in one way or other. The developing countries of Africa, Asia and Latin America, where arrangements of water treatment are limited or negligible, a large section of the population is susceptible to water-borne diseases.

Table 8.3 indicates the injections organisms commonly present in polluted water and diseases caused by them :

Table 8.3

Infectious Organisms in Polluted Water

Organism	*Diseases*	*Symptoms*
Bacteria	Typhoid, paratyphoid, bacillary dysentery, gastroenteritis, cholera	Headache, diarrhea, abdominal cramps, fever, nausea, vomiting occurring several hours to days after ingestion; severity varies depending on organism
Viruses	Viral gastroenteritis (up to 100 types in raw sewage[a]); also polio and hepatitis A	Symptoms seen depend on the specific virus
Parasites	Diseases from protozoans including: • *Giardia labia and Cryptosporidium* Diseases from parasitic worms: • Tapeworms, roundworms, hookworms	Protozoa: mild to severe diarrhea, including bloody diarrhea Worms: symptoms vary depending on specific worm – abdominal pain, anaemia, fatigue, weight loss

On Animals, Marine Life and Vegetation

The effect of water pollution is more on aquatic life, because their existence depends on water and when there is any disturbance in their ecosystem, the impact is maximum on them. In polluted water, due to abundant growth of algae, the oxygen content becomes lesser, causing the death of fishes and other organisms. It is estimated that during the last 20 years, there is a decrease of about 40 per cent in aquatic life. Algae are responsible for imparting a peculiar taste and odour to many of the water supplies and also are a cause of gastroenteritis. A report by Wheler indicates that algal poisons usually act on the central nervous system and skin and are reported to be capable of producing cirrhosis of the liver.

There are many cases on record of the destruction of marine life by polluted waters. Mass killing of fish was among the earliest and most dramatic result of indiscriminate pollution of water. Another type of fish killer is pollution from municipal sewage. Industrial, thermal, and radioactive pollution, etc. are also threats to marine life. Because of water pollution large number of ducks also die every year. For example, 1,40,000 ducks died of botulism in California during 1970. Occasional deaths of animals have also been reported by drinking of polluted water. If water pollution is more severe, the process of photosynthesis is also obstructed which affects the growth of aquatic vegetation. All the chemicals that are drained into the water have harmful effects on every organism that survives there.

Other Effects

Due to water pollution, the physical and physiological nature of water has also changed. The colour of water changes due to pollutants, mainly due to organic dyes. The colour in itself is not harmful, however, aesthetic considerations make coloured water unsuitable for any purpose. The turbidity caused by the presence of colloidal or suspended matter is the characteristic trait of sewage and industrial effluents. Generally, the more the turbidity, the stronger is the sewage or effluent and the effects are worse. Foam is the product of water mixed with soaps and detergents. It consists of a suspension of air bubbles in water medium. Industrial effluents containing chemical substances as iron, chlorine, phenols, etc. affect the taste of the water. The decomposition of organic matter and algae, fungi and filamentous bacteria also imparts peculiar tastes. Similarly, the odour of water also changes due to the presence of pollutants. Even small amount of sulphides cause bad odour in water. The pollutants also reduce the transparency of water. Oil spills have their effect over evaporation. The industrial effluents which contain several types of chemicals are also responsible for the loss of soil fertility. In fact, the effects of water pollution are multidimensional in nature. Apart from above mentioned direct effects there are several indirect effects also. In general, water pollution has now become a threat to the ecosystem and an important cause of environmental degradation.

Measures for Controlling Water Pollution

As described earlier, water pollution now has become a worldwide problem. Not only developing countries but developed countries are also facing this problem. However, developed countries have initiated several methods not only to control water pollution but also for the purification of polluted water. But most of these techniques are expensive and even those countries are also facing difficulties in controlling water pollution. Apart from technological methods there are other measures which are useful in minimizing water pollution. Some of them are as follows:

1. All domestic and municipal effluents be drained in water bodies only after proper treatment.
2. As far as possible domestic waste water be used for irrigation, it is very useful for horticulture.
3. Similarly, sewage can be mixed with soil, it will act as a fertilizer.
4. The drinking water sources and its nearby area be kept clean and regular cleaning be done.

5. There must be strict regulations for industrial effluents, that they must not discharge highly toxic water into water bodies. Every industry should develop its own effluent treatment plant.
6. Use of pesticides in agriculture should be limited and only standard quality pesticides be used.
7. There is a need to conserve several water purifying organism.
8. In the Indian context, there should be a complete ban on the disposal of dead bodies in rivers. The bank of water bodies are often used as public latrines, this should be strictly checked.
9. Algae and other water borne vegetation should be cleaned regularly.
10. Chemicals such as potassium permanganate be spread regularly in order to protect water from microorganisms.
11. Under present practices, pure, clean, drinking water is used to flush toilets and irrigate lawns. There are two possibilities to check this loss – first, waste water from bath room and sinks be recycled and be used for irrigation and flushing of toilets and second, separate water supply system be provided for drinking water and water for other uses.
12. The most important aspect would be to educate people not to pollute water. This awareness will help in solving this problem to a great extent.

Treatment of Sewage

The treatment of sewage is one of the important measures, which aims in the removal of BOD, phosphorous, nitrogen, solids and bacteria. The composition of sewage is complex, and it differs depending upon the sources, the type of treatment or lack of it. The process of treating sewage is broadly classified as *primary, secondary* and *tertiary* (Figure 8.2).

Primary Treatment

Primary treatment consists of removing floating and suspended solids by mechanical means. More than one half of the suspended solids can be removed by primary treatment as shown in Figure 8.3.

In this process, first the large solids are screened out and grease and scum are removed. Screening is done through metal bars first spaced 25 to 50 mm apart and may range down to 0.8 mm openings. Sand and other coarse material is removed by grit chambers. After screening and removal of grit, the waste water is run directly into settling or sedimentation tanks, the process by which the suspended solids are removed by gravitational setting. Primary sludge is a problem because it is bulky and must be removed. It also contains 94 to 99 per cent water. In some cases, sludge is dried in beds with some water removed by filtration. A report of the American Chemical Society indicates that primary treatment reduces 60 per cent suspended solids, 35 per cent BOD, 30 per cent COD (Chemical Oxygen Demand), 20 per cent nitrogen and 10 per cent phosphorus.

Figure 8.2
Sewage Treatment Process

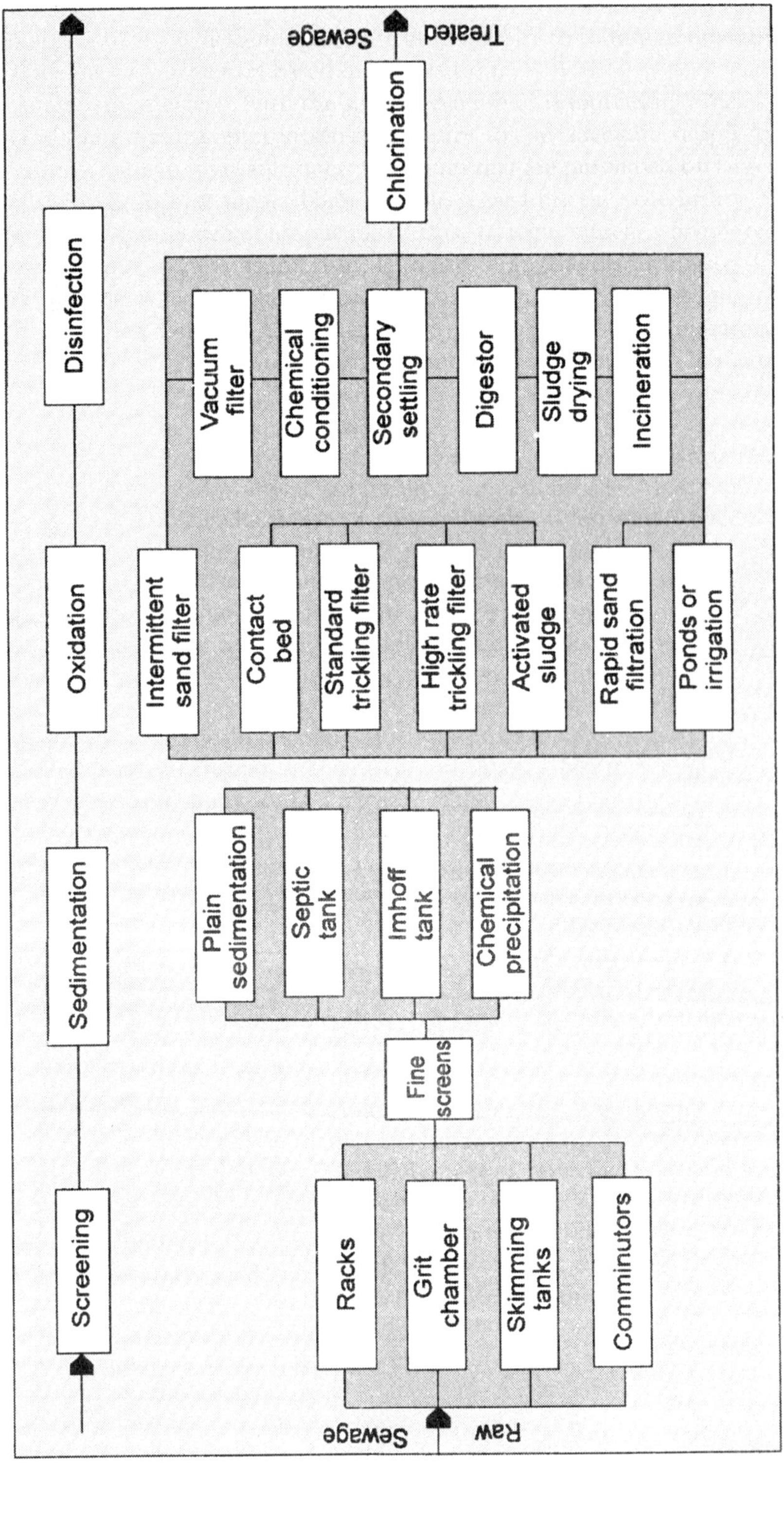

Figure 8.3
Primary Sewage Treatment

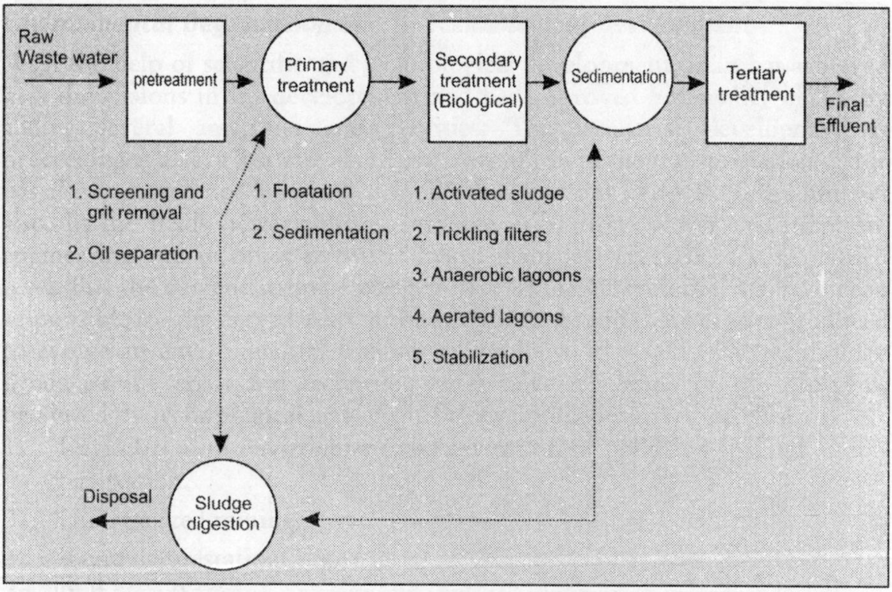

Secondary Treatment

Secondary treatment of waste involves the biological degradation of organic material by microorganisms under controlled conditions.

The usual method is to bring about the biological oxidation of the organic material under aerobic conditions, in which the waste is aerated to supply oxygen for the microorganisms. The degraded material settles out in secondary settling tanks and afterwards removed by sedimentation. The clarified waste water is discharged in the outfalls to rivers, lakes or oceans (Figure 8.4).

Tertiary Treatment

Tertiary treatment aims at further purification of waste water and also for its recycling. The number of methods used for tertiary treatment are:

1. chemical coagulation and filtration,
2. adsorption,
3. chemical oxidation,
4. desalination, and
5. oxidation ponds.

However, there is a growing need for advanced procedures that will provide a product capable of being re-used for various purposes. The recycled waste water can be used for irrigation, mainly for non-food crops such as grasslands, lawns, play grounds, etc. The renovation of waste water to a quality that would permit its re-use for a variety of purposes is a major objective of current research in this field.

Figure 8.4
Secondary Sewage Treatment

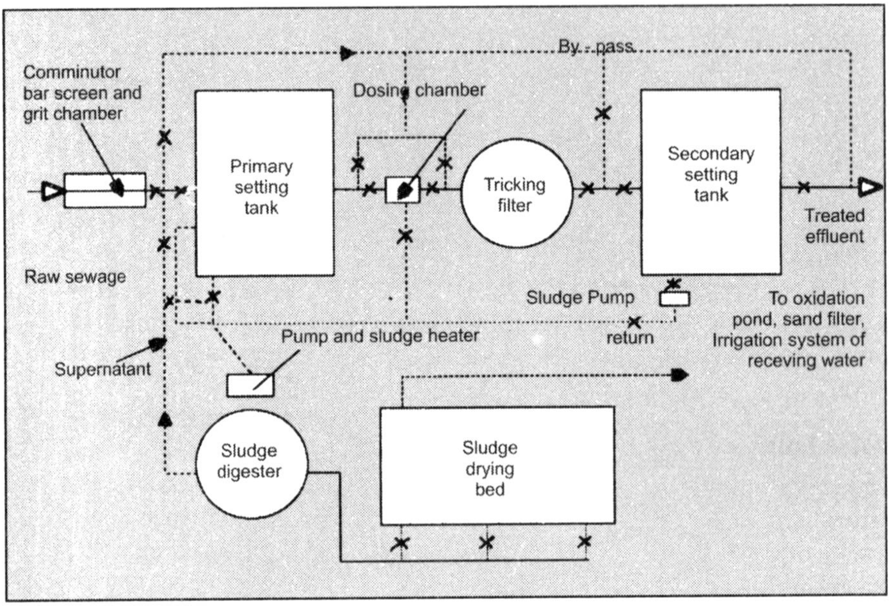

Notes

1. Owen, S. Oliver (1971), *Natural Resource Conservation: An Ecological Approach*, Macmillan, New York, p. 141.
2. Manivasakam, N. (1992), *Environmental Pollution*, NBT, New Delhi, p. 34.
3. Gilpil, A. (1978), *Dictionary of Environmental Terms*, Routledge, Kegan and Paul, London, p. 171.

9

Noise Pollution, Land Pollution and Pollution by Radioactivity

Noise Pollution

Sound is a normal feature of our life and a medium through which communication is possible. It may be pleasant, like the sound of running water of a stream or the sweet sound of birds in the forest or the pleasant sound produced by musical instruments. On the other hand, it may be the noisy and harsh sound of lightning and thunder or noise produced by machines, automobiles, railways, aeroplanes or even the blaring sound of loudspeakers or some musical instruments, but all sounds are not noise. 'Noise is any sound that is not wanted. It is one of the more common forms of atmospheric pollution'.[1] Sound is caused by the vibration of molecules, while noise radiates from vibrating surfaces. More and more noise is the creation of modern civilization and has now become a major environmental pollutant specially in urban areas. In 1972, the UN Environment Conference at Stockholm, noise pollution has been accepted as a problem, which needs proper control.

Noise pollution is an environmental problem that has only recently received much attention, yet it has been decades in the making. The perception of noise as a problem is different for different people. Noise at certain levels might be an aggravation or nuisance for some, while at higher levels, noise might become a more widespread or general aggravation. At still higher levels, noise might fall within 'discomfort range' and at even higher levels, it may lead to physical and psychological damage.

Measurement of Sound/Noise and its Intensity

The most popular measure of noise level is the 'decibel' measured by an instrument known as decibel metre. A sound between 0 and 1 decibel is about the weakest that the average human can hear. For testing purposes, 0 decibel is considered to be the threshold of hearing. A whisper is about 20 decibels and an average speaking voice is about 60 decibels. The loudest sound that a person can stand without discomfort is about 80 decibels. Automobile horns may reach 90 decibels and a jet airplane may have an intensity of about 140 decibels. The levels of common noise is given in Table 9.1 and has also been depicted in Figure 9.1.

Figure 9.1

Levels of Common Noise

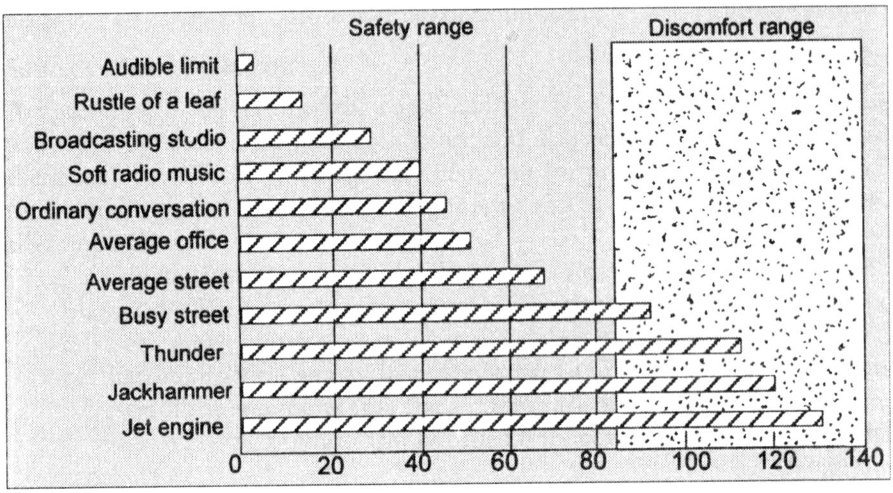

(After t. Berland)

Table 9.1

Intensity of Various Sounds

Sound Source	*Intensity (in decibels)*	*Response Criteria*
	0–10	just audible
Broadcasting studio	20	—
Soft whisper (15 feet)	30	very quiet
Library, bedroom slow radio	30–50	quiet
Light auto traffic, air conditioning unit	50–70	intrusive
Freeway traffic, freight tram	70–80	annoying
Motor cycle, heavy traffic	80–100	very annoying; hearing damage
Rail noise	110–130	intolerable
Lighting-thunder	120	intolerable
Carrier deck jet operation	140–150	painfully loud

In most of the countries, maximum limit fixed for noise is 75 to 85 decibels. According to scientists noise above 90 decibels for a long time may result in deafness. In most of the cities in India, the noise level is 75 to 90 decibels but is more in industrial areas, near aerodromes, railway stations and highways, etc. But since noise pollution is increasing day by day, planning for its control should be done. Actually, loudness alone is not the sole cause of noise problem, although this is the most apparent characteristic of noise that creates environmental problems. Pitch of the noise also influences the degree of annoyance – the higher the pitch, the greater the annoyance. Similarly, duration of the noise also determines how annoying or irritating noise will be. The degree of noise in an enclosed space or indoors is different from the level of outdoor noise.

Sources of Noise Pollution

The sources of noise can be divided into two categories, viz. natural and artificial.

Natural sources are associated with natural phenomena like lightning, thunder, volcanic eruption, earthquakes, sound of the ocean waves, etc. These reasons are local or regional and generally their impact is very limited and not very harmful. On the contrary, noise produced by man is much greater and harmful.

Artificial sources are responsible for noise pollution. The sources of noise pollution in an industrial society are manifold. The main sources in cities and other highly developed countries are noise associated with transportation, automobiles, motorcycles, rail engines, buses, fire brigade, police cars, ambulances, air planes, industries, high pitched musical instruments, etc. During religious or domestic functions and in political meetings during elections, noise pollution often increases due to use of electronic medium of sound.

The main source of noise, however, comes from transport. The number of road vehicles, particularly diesel engine vehicles, has increased the level of noise enormously. This problem is more in metropolitan cities where a number of vehicles ply simultaneously and their noise is unbearable. The horns and sirens used by these vehicles, specially the pressure horns are very noisy. The value of horns and sirens except in emergencies, is questionable. The whistle of rail engines is very harmful to the ears. Busy traffic streets, highways, bus stands and railway stations have problems of noise pollution. Larger and faster jet aircrafts have been built over the years and have become one of the most troublesome sources of noise pollution. For example, O'Hare International Airport, in Chicago, the busiest airport in the world, has more than a thousand operational jet flights on an average day. This means a flight in or out every 40 seconds. The residents within 20 km radius from the airport are facing great problems not only in their conversations, television viewing, sleep and relaxation but it has also begun to take toll on their health. The air traffic has increased to such an extent that nobody likes a new air field in his neighbourhood.

Noise associated with manufacturing, building, construction, mining, road building, etc. are second in intensity to transportation noises. But their impact is felt more by factory and other workers. It is reported that noise inside factories can become a health hazard causing deafness. Industrial noise, particularly from mechanical saws and pneumatic drills, is unbearable and a nuisance to the ears. The impact of noise in a factory varies with age groups.

For workers below 30 years of age, the 95 decibel noise is not harmful, but for 40 to 50 years age group this limit is 84 decibels and for 50 to 60 years age the bearable limit is 80 decibels. In mining, noise produced by blasting and digging is also a source of uncomfortable noise.

Nowadays, music has also become noisy and the electronic medium and instruments have further developed its intensity. In a closed space, high volume music is harmful and is responsible for noise pollution. Similarly, during social and religious functions, marriages and other gatherings, including political meetings, the use of high volume loudspeakers has become a common feature and an irritating cause of noise pollution. The popular 'pop music' may be entertaining but its ultimate impact is not healthy on man.

Effects of Noise Pollution

The most notable effect of noise pollution is on hearing. Violent noise can cause temporary or permanent impairment of hearing, thus a cause of deafness. Continual noise can lead to gradual decline in auditory acuity and eventual deafness. Acute damage occurs to the ear drum when exposed to very loud sudden noises.

Noise causes several undesirable effects also. Damage to hearing and loss of sleep are only two of the obvious results. Noises produce irritability and a feeling of fatigue and may reduce a worker's efficiency. There is evidence that noise is one of the major cause of stress and anxiety. It also has its impact on blood pressure and is a causes of other cardiovascular diseases. The rate of heart beat may be affected by noise. A recent report indicates that blood is also thickened by excessive noises. Eosinophilia, hyperglycemia, hypokalaemia and hypoglycemia are also caused by a change in blood and other body fluids due to noise. Noise also causes headaches and irritability. The effects of noise on the foetus are not fully known. Medical scientists have noted that an unborn child will move and kick when there is a loud noise. Some of studies indicate that children born in areas of high noise have some defects in comparison to the children born in peaceful areas.

Control of Noise Pollution

Since noise pollution is created by man, it can be controlled by adopting certain measures. Some of the measures are as follows:

1. Noise producing industries should be located away from residential areas.
2. Inside industries proper arrangements to minimize noise be done by constructing sound proof walls and also to provide such instruments to workers which can protect their ears from noise.
3. Old machines often create more noise, therefore, all the machines should be well maintained and replaced if necessary.
4. The automobile horn should be designed in such a way that the noise it produces may not be harmful.
5. Use of horn should be minimum and pressure horn should be banned as has been done in many countries.

6. The noise created by musical instruments, and other indoor equipments can be checked by individuals in their own interest.
7. The noise created by railways can be checked by construction of ballastless rail tracks.
8. Special arrangements be done to check noise near air bases. Control of aircraft noise requires several changes, which should be done.
9. Dense tree plantations around or nearby areas of noise can also reduce its intensity.
10. Every government has enacted certain laws to control noise pollution. The basic need is the proper implementation of these laws and regular supervision. In countries like India, the problem of noise pollution is still in its primary stage, therefore, measures to control this problem are necessary.

Soil and Land Pollution

The Problem of Solid Wastes Disposal

Although one-fourth of the surface of the earth is land surface, half of this surface is not useful to man due to permanent snow cover, deserts, mountains, etc. and only about 448 lakh sq km of land can be used by man for his multifarious activities. Any misuse or wastage of land may create problems not only for the present generation but for the future ones as well. The problem of land and soil pollution is growing day by day with the rapid growth of population, urbanization, industrialization, agricultural and other developmental activities. Land pollution is the result not only of man's misuse of land but more due to solid waste disposal. The disposal of solid waste is more a problem of developed nations but now it has also become a grave problem of all the countries of the world. Solid wastes are dumped in the ocean or open grounds; mining wastes are produced at the rate of millions of tons a day; slag heaps and mill tilings accumulate near processing operations; and industrial refuse contaminates streams and lakes. In USA alone, solid waste produced annually includes 30 million tons of paper and paper products, 4 million tons of plastics, 30 billion bottles, 60 billion cans, 100 million tyres, 8 million junk and abandoned automobiles and millions of major appliances of innumerable makes, sizes and kinds. No one knows the total amount of the world's waste, but it is certain that the problem of solid waste disposal now has become a threat to our environment, specially the environment of our cities.

Land pollution can be defined as the 'changes in physical, chemical and biological conditions of the soil through man's intervention or misuse of land and resulted into degradation in quality and productivity of the soil'. The problem of land pollution differs from water and air pollutions in the respect that the pollutants remain in place for relatively longer periods. Rapid urbanization with the consequent increase in buildings has resulted in the reduction of lands for the wastes to be disposed. Dumping of industrial and municipal wastes causes leaching by toxic materials which seep into the soil and affect the ground course. Agricultural practices introduce pesticides, fertilizers and manures to the land resulting in both biological and chemical contamination. Thus, land and soil pollution is increasing day by day by hazardous materials and microorganisms enter the food chain or water and are consequently ingested by man. Therefore, there is an urgent need not only to control soil pollution but also for its proper management.

Sources of Land Pollution

The soil and land pollution is the result of several sources, which can be categorized under following heads:

1. domestic and municipal wastes,
2. industrial and mining wastes,
3. agricultural wastes,
4. radioactive materials, and
5. biological agents.

Domestic and Municipal Wastes

One of the main causes of land and soil pollution is the growing quantity of domestic and municipal wastes. Household garbage includes food scraps, old newspapers, a variety of plastic items, bottles, discarded papers, wood, lawn trimmings, glass, canes, old appliances, tyres, worn-out furniture, broken toys and a host of other items. The total quantity of solid waste is large and increasing. In United States, municipal solid waste averaged 1.2 kg per person per day in 1920. The quantity rose to 2.3 kg in 1970 and 3.6 kg in 1980 and now its quantity is more than 4.5 kg. Urban wastes comprise both commercial and domestic wastes including dried sludge of sewage. In general, all the urban solid wastes are referred to as 'refuse'.

The amount of solid waste generated is directly related with prosperity. In contrast to the 3.6 kg per person per day in the United States, residents of Australia produce 0.8 kg per person per day. The average person in India produces only about 0.2 kg per day. The general composition of municipal wastes is as given in Table 9.2.

Table 9.2

Average Composition of Municipal Refuse in Developed Countries (% by weight)

Paper wastes	—	42.0
Wood and bark	—	2.4
Grass	—	4.0
Brush	—	1.5
Cuttings green	—	1.5
Leaves dry	—	5.0
Leather goods	—	0.3

Cont'd...

...Cont'd

Rubber	—	0.6
Plastic	—	0.7
Oil, paints	—	0.8
Linoleum	—	0.1
Rags	—	0.6
Street refuse	—	3.0
Household dirt	—	1.0
Unclassified	—	0.5
Total Rubbish	—	64.0
Garbage	—	10.0
Fats	—	2.0
Total Food Wastes	—	12.0
Metals	—	8.0
Glass and ceramics	—	6.0
Ashes	—	10.0
Total Non-Combustibles	—	24.0
Grand Total	—	100.0

It has been estimated that in 45 major cities of India, the quantity of average per day municipal waste is about 50,000 tonnes. The National Environmental Engineering Institute (NEERI), Nagpur, has given details of the municipal waste in Indian towns as shown in Table 9.3.

Table 9.3

Nature of Solid Waste in Indian Towns

(in percentage)

Material Composition	*Towns According to Population Size*			
	Upto 2 Lakh	*2 to 5 Lakh*	*5 to 20 Lakh*	*Above 20 Lakh*
Paper	3.09	4.74	3.80	7.07
Plastic	0.57	0.59	0.81	0.86
Metals	0.51	0.39	0.64	1.03
Glass	0.29	0.34	0.44	0.76
Ash and dust	46.60	39.97	41.81	31.74
Mixed material	33.41	39.76	40.05	41.74
Carbon	12.36	12.51	11.95	15.92
Nitrogen	0.60	0.61	0.50	0.51
Phosphorus, Phosphate	0.70	0.71	0.67	0.59
Potassium	0.70	0.73	0.72	0.67

It becomes clear from the above table that in our cities 90 per cent wastes include ash, dust, mixed material and carbon, while in developed countries paper, plastic, glass, metal, etc. dominate the waste content.

The dumping of domestic and municipal wastes is a serious problem in cities because of its impact on environment and public health. Solid wastes may or may not cause diseases in man but are hazardous to health. Diseases such as dysentery, diarrhoea, plague, malaria and numerous others are the result of the indiscriminate dumping of wastes.

Industrial and Mining Wastes

The disposal of industrial solid wastes is the major source of soil pollution by toxic chemicals. The industrial wastes are mainly discharged from coal and mineral mining industries, metal processing industries and engineering industries. They contain toxic metals such as lead, copper and chemicals having acids and are responsible for soil pollution. It has been reported that about 50 per cent of raw materials ultimately become waste products in industry and about 15 per cent of it are toxic. The chemicals discharged from the industries often enter the surface or ground water or poison the soil or crops. The production of consumer goods also involved environmental problems including land and soil pollution. Table 9.4 lists some of the effects of the production of consumer goods.

Table 9.4

Production of Consumer Goods and its Effects on Land/Soil

Process	*Activity*	*Effects*
Extraction	Mining and oil drilling	Surface land disruption, acid mine drainage, mine failings, sludge ponds, oil spills.
	Agriculture	Disruption of natural habitats, soil erosion, fertilizer run-off, poisonous pesticides.
	Forestry	Habitat disruption, soil erosion, pesticides.
Processing	Manufacturing	Soil and underground water pollution, loss of soil productivity, air and water pollution.
Energy	Energy conversion and transmission	Depletion of resources, thermal and radioactive pollution, disruption of land for transmission rights. Litter, land-use disruption, release of hazardous wastes, dust, etc.

The expansion of mining activities in many countries of the world has now become a main cause of land pollution due to loss of soil and destruction of land. Mining has become a threat to the environment because it leads to the huge quantity of waste generally not useful to man and its reprocess is not economical. Mining operations produce about 1.35 billion tonnes of debris each year.

Agricultural Wastes

Agricultural practices also pollute the soil. According to an estimate, agricultural activities produce more than 1.8 billion tonnes of waste each year. About three-quarters of this is manure. Much of this manure is piled in dumps where it pollutes streams and waterways. Yet, at the same time, farmers across the continent are suffering from worn-out and depleted soils. Other agricultural wastes include branches and slash left over from logging apart from animal wastes. In addition to fertilizers and pesticides, soil conditioners and fumigants are used in agriculture. Organic compounds containing lead, mercury and arsenic, when applied to a land will accumulate on the soil permanently and introduce these toxic metals into plant products.

Radioactive Materials

The radioactive wastes produced by nuclear testing laboratories and industries reach the soil and accumulate there. Wastes from nuclear reactors contain ruthenium–106 and rhodium–106, iodine–131, barium–140, lanthanum–140, cerium–144, etc. All the radionuclides deposited on the soil emit gamma radiations, and are harmful to soil as well as for plant growth.

Biological Agents

The excreta of humans, animals and birds is also a source of soil pollution by biological agents. Digested sewage sludge which is used as a manure also causes soil pollution. In the developing countries, intestinal parasites constitute the most serious soil pollution problems. Faulty sanitation, waste water and wrong methods of agriculture also induce soil pollution. There are three groups of pathogenic organisms that pollute the soil, viz. (i) pathogenic organisms excreted by man, (ii) pathogenic organisms excreted by animals, and (iii) pathogenic organisms found in contaminated soil.

Harmful Effects of Land/Soil Pollution

The following are the harmful effects of the land and soil pollution:

1. Land and soil pollution is responsible for loss of fertility and productivity of soil.
2. The decomposition of the various types of waste material causes harmful gases and bad smell, which not only pollute environment but is also harmful if mixed with chemicals.
3. The municipal and domestic waste is often discharged in water bodies, thus responsible for water pollution.
4. The dumping of waste in oceans is a common practice and a main cause of imbalance of the marine ecosystem.
5. Particles in the sewage may clog the micro holes of the soil and also destroy the microorganisms necessary for the soil.
6. Industrial and chemical wastes are also responsible for the pollution of underground water.
7. Those bacteria which are transmitted from man to soil infect man causing dysentery, cholera, tuberculosis, typhoid, paratyphoid fever, etc.

In fact, land and soil pollution is one of the main causes of air and water pollution, therefore its control and proper management is necessary not only in developed countries but also in developing countries.

Control of Land and Soil Pollution: Methods of Disposal of Wastes

Land and soil pollution can be controlled by adopting measures for the disposal of solid wastes, which unfortunately involves large finances and technology. Prior to disposal, the wastes have to be collected from the places by trucks to the disposal area. It is estimated that about 80 per cent of the total cost of solid waste management is spent for collection only. The methods used for the disposal of waste are open dumps, ocean dumping and sanitary landfills.

Dumping is a popular and an inexpensive means of solid waste disposal. *Open dump* is the least acceptable method but prevalent in most of the developing countries, including India and a cause of soil as well as general environmental pollution. In this method, waste is collected and hauled to a site where it is dumped on the ground. Organic matter rots or is consumed by insects, rats or hogs. In some communities, the pile is set afire in the evening to reduce the total volume and the odour. All this creates serious ecological problems. The dump itself is a potential source of disease. Rain erodes the dump, and the polluted water flows into nearby rivers and ground water reserves.

Sanitary landfill is commonly used for final disposal of solid wastes. This is cheaper than other disposal methods and have the advantage of avoiding acute pollution problems associated with discharging wastes into waterways or polluting air from incineration. But the main disadvantage over open dumps is in the aspect of public health. Leaching by surface water or ground water may contaminate streams and other water supplies. The availability of land for the dumping of the growing volume of wastes is also a problem.

Ocean dumping of wastes is practised by many coastal cities. It was considered to be the favourite disposal method under the prevalent notion that the ocean is an inexhaustible sink. We are now learning that the ocean does not have an infinite capacity for covering, holding, absorbing or decomposing materials, nor the capacity for anywhere near the load of pollution that would result if the rate of ocean disposal continues, as in the past. There were seven major categories of waste - harbour dredgings, industrial wastes, municipal sewage sludge, refuse and garbage, construction and demolition debris, military explosives and chemicals and miscellaneous wastes like agricultural land run-off, pesticides, fertilizers, sewage and garbage from vessels, oil spills, etc. Now the environmentalists have warned that ocean dumping is neither a safe nor a desirable method of the disposal of waste. In 1970, the Council on Environmental Quality recommended a series of drastic actions to control ocean dumping because it upsets the ecological balance of regions of the sea.

Incineration is a method through which the volume of waste can be reduced by 60 per cent and it also reduces the public health problems. This method is adaptable over a wide range of capacities from small domestic incinerators to large centralized municipal plants. It can handle a mixture of garbage and rubbish, and the physical nature of the residue - called *clinker* - aids in the disposal of the waste, although it aggravates the problem of air pollution.

Chemical processing of solid wastes is also applied but it is costly and technically not feasible for poor countries. But due to prospect of energy production and recovery of certain usable materials, the salvage of some wastes by chemical methods is now considered economically feasible. The reclamation of non-ferrous scrap metals has long been an established industry.

Composting municipal refuse to convert it into a fertilizer and soil conditioner is appealing not only because it appears to be a good way to recycle the resources in solid wastes, but also because of the beneficial nature of the compost. This may be considered as useful method for the disposal of municipal refuse, specially in developing countries.

New methods of waste disposal that are under trial in various parts of the world are:

- The transformation of the organic content into sugar or proteins.
- The heating of organic refuse under anaerobic conditions to convert it into useful gases such as methane, which is usable as fuel, or into a liquid product.
- The compression of refuse into building blocks that can be sheathed with more durable materials. One such plant is in operation in Japan.
- The compression of refuse into briquettes that can be used as fill. A mixture of fly ash, dried sewage sludge, incinerator residue and river and lake dredgings is under study.
- The use of degradable plastic bags instead of trash cans in the collection and transport of municipal refuse.
- The transport of refuse as a liquid slurry in pipelines, a method now under study.

Apart from refuse and recycling of waste, there are methods through which waste can be utilized properly.

Energy from Refuse

In many metropolitan areas of developed countries the garbage is burnt in a well-engineered furnace. The heat from the fire is used to boil water and produce steam. Then, the steam is sold for industrial use. But there are problems in this type of burning also such as – wet garbage and food scraps – are difficult to burn and burying PVC products produces hydrogen chloride gas, which reacts with water to produce a strongly corrosive liquid, hydrochloric acid. In spite of these difficulties, incineration may become profitable in the future because:

1. Increasingly large quantities of dry paper and cardboard have appeared in refuse, thereby increasing the fuel content of the trash.
2. The price of fuel has skyrocketed and therefore the value of steam has also gone up, thus it will be more economical day by day.
3. The rising cost of land has made it harder to find adequate sites for landfills, thus demanding alternative solutions, and, burying of refuse is the most suitable alternative.

Recycling of Waste Material

Most refuse contains a wealth of raw materials that can be easily reused or recycled. A large quantity of metals can be obtained from:

Table 9.5

Various Recycling Routes for Some Common Wastes

Waste	*Recycling Possibilities*
Paper	Use the backs of business letters for scrap paper or personal stationery, lend magazines and newspapers to friends
	Repulp to reclaim fibre
	Compost
	Incinerate heat
Glass	Purchase drinks in deposit bottles and return them, use other bottles as storage bins in the home
	Crush and remelt for glass manufacture
	Crush and use as aggregate for building material or antiskid additive for road surface
Tyre	Recap usable casings
	Use of swings, crash guards, boat bumpers, etc.
	Shred and use of manufacture of new tyres
	Grind and use as additive in road construction
Manure	Compost or spread directly on fields
	Ferment to yield methane; use residue as compost
	Convert to oil by chemical treatment
	Treat chemically and re-use as animal feed
Food scraps	Save for meals of leftovers
	Sterilize and use as hog food
	Compost
	Use as culture for yeast for food production
Slaughter house and butcher shop-wastes	Sterilize and use as animal feed Render compost

waste dumps, as also paper, glass, tyres, manure, etc. Table 9.5 indicates the possibilities of recycling the various waste materials.

It becomes clear from the above table that if the recycling process is adopted properly, not only, can the problem of waste disposal be minimized but it will save financial resources also. An estimate indicates that about 21 per cent of the paper in United States was recycled, compared with about 50 per cent in Japan. If the rate of recycling were increased to 50 per cent in United States, approximately 100 million trees would be saved annually, and the energy saved would be enough to supply 7,50,000 homes with electricity. If people would separate their household trash into piles of aluminum, steel, paper, plastic and food scraps, then perhaps all these materials could be recycled.

New Approach in Waste Management

The above mentioned methods have proved quite useful for waste management. But, now-a-days, there is more emphasis on people's participation in waste management. There is a growing adoption of the polluter-pays principle, i.e. NIMBY and NIABY, which is as follows:

The public are becoming better informed on environmental issues, and may object to real or imagined pollution and waste hazard or nuisance. Special-interest groups or the public can adopt a 'not-in-my-backyard' (NIMBY) attitude. NIMBY is often invoked when authorities are proposing to site a hazardous or 'nuisance' waste or pollution treatment plant. Some proposals are opposed by environmental activists and NGOs whenever they are to be sited (for example, deep underground storage of high-level nuclear waste) – a 'not-in-anybody's-backyard' (NIABY) response.

Apart from the polluter-pays principle, over last few decades, pollution and waste management have been aided by:

— a trend towards more proactive planning and management;
— increased release of information on pollution and waste risks, as a result of government, NGOs, media and international agency activity;
— improved legislation;
— better policing;
— spread of better environmental quality standards more widely applied;
— better methods and equipment for monitoring and assessment; etc.

Radioactive Pollution

Nature and Sources

Radioactive substances are among the most toxic materials known. The magnitude of their injurious effects is in a range that might be said figuratively to represent a 'quantum jump'. H.P. Jammet, Chief of Atomic Hygiene and Radio Pathology, French Atomic Energy Commission, France, defines radioactive pollution of the environment as 'the increase in natural background radiation, emerging from the activities of man involving the use of naturally occurring or artificially produced radioactive materials. The atmosphere screens out much of the sun's radiation, including most of that which would be lethal to life'.

Naturally occurring, radioactive elements are present in the rocks, water, and air and in all living organisms. The composite of all forms of natural radiation that we are exposed to, is called background radiation. For millions of years, background radiation remained at a relatively constant level. Recently, however, the radiation that we are normally exposed to has increased and can be expected to increase further. Man-made radiation includes that from X-ray machines, radioactive fallout, nuclear reactors, research laboratories, industrial and medical uses of radioactive materials and radioactive wastes from the production of radioactive materials.

Environmental radiations may be divided into two types: (i) natural, and (ii) man-made radiation.

Naturally occurring radiations are: (a) cosmic radiations from the outer space reaching the earth's surface, and (ii) terrestrial radiation from natural radio-isotopes present in the earth's crust. Cosmic ray bombardment of the atmosphere continuously produces a few radioactive materials of short half-life. The half-life of a radionuclide refers to its period of radioactivity and has a bearing on the length of time that a radioactive substance might remain active as an environmental pollutant. Table 9.6 indicates the half-life and radiation of the important radioactive elements.

Table 9.6

Commonly Used Radionuclides

Material	*Half-life*	*Radiation*
Carbon–14	5, 568 yrs	beta
Cobalt–60	5.24 yrs	beta, gamma
Hydrogen–3(tritium)	12.26 yrs	beta
Iodine–131	8.08 days	beta, gamma
Phosphorus–32	14.3 days	beta
Plutonium–239	24, 360 yrs	alpha, gamma
Radium–226	1622 yrs	alpha, beta, gamma
Strontium–90	27.7 yrs	beta
Uranium–235	710 million yrs	alpha, gamma
Uranium–238	4.51 billion yrs	alpha, beta, gamma

Source: Maxwell, K.E., *Environment of Life*, 1973, p. 230.

Man-made radiations originate from the activities of man involving the use of radioactive materials. They are used for the production of nuclear weapons, nuclear fuel and for the production of electric power. One of the most harmful uses of radioactive material is in the nuclear weaponses, specially in atom bombs. The world has already seen the nature of destruction as well as impact of radioactivity on man and environment during the second world war when atom bombs were dropped over Hiroshima and Nagasaki, the two Japanese towns. During the last forty years, there have been a number of nuclear weapon testings either in sea or underground, resulting in the increase in the level of radiation of the environment. A study revealed that the radioactive materials which cause such explosions are fission products as Strontium–90, Caesium–137, Iodine–131 and unused explosives. These elements either settle down in the soil or water or remain suspended in the atmosphere, thus becoming a major cause of radioactive pollution.

Nuclear fuels used in operation of reactors also contribute to pollution. The heat liberated during fission and activation process is converted into electricity. During this process, two types of wastes are formed: (i) the fission products remaining in both the primary and secondary fuels, and (ii) extraneous activation products in the coolant. The disposal of radioactive or nuclear wastes is a great problem and a cause of environmental pollution. It includes remaining metal products at the site of mines, fission products and activation products. These are of three categories, viz. low level, intermediate level and high level. Among these, high level products remain in the environment for several hundred years. Similarly, leakage in nuclear reactors often became the cause not only of environmental pollution but also of the death of hundreds of people as happened in case of leakage at Chernobyl Atomic Reactor in erstwhile USSR on 26 April 1986.

Effects of Radioactive Pollution

Different forms of radiation have different effects. Visible light and infrared heat rays and certain kinds of radiation are generally beneficial. Some of the other forms of radiation, for example X-rays, are more energetic and are therefore, biologically injurious because of their destructive action on cells and tissues. Because these highly energetic forms of radiation tend to split substances, including living matter, into ions, they are called ionizing radiations. The other forms include electromagnetic radiation and particulate radiation. Ultraviolet radiation is highly injurious to the tissues of some organisms. It is well recognized that sun-burn is caused largely by the ultraviolet content of the sun's radiation and that excessive exposure can cause serious injury to both the superficial and deep layers of the skin.

The environmental hazards from fission reactors include: (i) the possibility of a serious accident that releases a large amount of radioactivity, (ii) the hazard of even the usual small radioactive emissions with normal operations, and (iii) the pollution caused by radioactive wastes. The fate of a radioactive contaminant is also dependent on ecological conditions. Radioactive pollution causes deposition of the contaminants on ground and surface waters. The water consumed by plants thus acts as a medium for radioactivity. This contamination is passed on to vegetation and animals through the nourishment from soil and water.

Man is considered to be the ultimate sufferer through reactions and interactions. Radioactive particles present in air have a direct impact on man's respiratory system, while indirect contamination occurs by consumption through the food chain. The ill-effects of radioactive pollution are numerous. Radiation can cause cancers, abnormal births, skin diseases, mutations in man. Chronic exposure to radiations leads to leukemia in an individual and even affects an unborn child in the womb. The radiations from medical X-rays, colour TV, luminous dials of clocks and watches, X-ray fluoroscopes are also contributing to the risks. As described earlier, any leakage of radioactivity from reactors is not only a health hazard but a cause of death and other disorders to thousands of people. Similarly, nuclear testing in sea waters not only creates disorder in the oceanic ecosystem but becomes a cause of the destruction of marine life.

Monitoring, Prevention and Control of Radioactive Pollution

In spite of certain harmful effects of radioactivity, nuclear power and other uses of radioactive elements have become an integral part of modern development. Many of the environmentally conscious countries in the world depend on nuclear power significantly for their electricity generation. At present, 17 per cent of the world's electricity is generated through nuclear sources. Nuclear source is considered clean, compact and concentrated. It is clean because its impact on the environment is minimal compared to other energy generating sources. To make it environmentally safer, there is a need for monitoring and prevention measures.

Monitoring of radiation is very necessary for the control of radiation. For this, sampling as well as its proper analysis is necessary. The samples of air, water and soil have to be collected regularly and their analysis both manual and instrumental be done, not only to know the actual level of radioactive pollution but also for its control.

Prevention and control measures are to be adopted to control radioactive pollutants. All measures should so aim that the level of radioactive pollution should not exceed the maximum permissible limit. The industrial wastes having radioactive elements should only be discharged after proper treatment. The high activity wastes, should not be discharged. Now, a few techniques have been developed through which these wastes can be used if treated properly. The disposal of the wastes into sea is also not a safe method but if nothing is possible, these wastes should be sealed in concrete-filled steel drums and discharged into depths of 1,000 fathoms or more.

Control of the emission of radioactive pollutants is very essential. In uranium mines, wet drilling should be employed. Closed-cycle systems with gaseous coolants of very high purity may be used in the case of nuclear reactors to prevent extraneous activation products. There is a need to control radiation hazard for workers and officials working in reactors or other places having danger of radioactivity. There is a need for regular monitoring and clean-up of surface, careful washing is necessary, apart from this proper kit to protect from radiation effects and health care of workers is also necessary. What is more, personnel at power stations must be trained in radiation protection and radiation hazard control measures. The UN and other agencies of the world are in favour of total ban on nuclear weapons, but, very little progress has been made in this field. Let us hope that the time will soon come when our world will be free from the dangers of atom bombs and other nuclear weapons.

Note

1. Maxwell, K.E. (1973), *Environment of Life*, Dickenson Publishing Company, California, p. 175.

10

Environmental Hazards and Disasters

Natural hazards originate in highly dynamic processes whose elemental nature expresses itself in their unpredictable and uncontrollable character. Those natural hazards that cause losses of human life and great material damage to economics are called disasters and catastrophes. Two aspects of such natural hazards are noted: their potential for disaster and their actual catastrophic results. Not all potentially dangerous hazards are disastrous. In contrast to their potential danger which remains the same, the intensity of disasters is governed by five major factors:

1. Historical and social conditions, as well as the level of economic development of a country or a region, which affect their vulnerability.
2. The scope of the damage that natural hazards may cause derives from forms of land use. It matters who and what are exposed to natural hazards.
3. The scope of disasters through natural hazards depends on a region's geographic position within a sequence of processes.
4. Natural hazards are often not so dangerous when taken separately but cause predicaments when combined with other processes.
5. In most cases, the scope of disasters produced by natural hazards is governed not by a single characteristic but by the scale, duration and intensity of corresponding phenomena. While natural hazards are always potentially disastrous, actual resulting damage varies in accordance with social, economic and geographic conditions.

In studying the problem of natural hazards, the relevance of modern scientific and technical advances as well as the growing influence of human activities on natural environment should be recognized. These activities perturb and modify nature's original dynamic state. As a result, the self-regulating properties of natural processes either vanish or are weakened. It would be logical that having disturbed the environment's natural state, man should also assume corresponding regulation functions.

The scientific interest in environmental hazards is of fairly recent origin. Attempts to defend individual buildings against earthquakes date back at least

2000 years and a concern for risk can be traced back to the earliest recorded times. But systematic research into natural hazards really begun with the work of Gilbest White (1936, 1945), who was the first to recognise that engineering schemes were not the only way to tackle flood problems in the USA. Over next 20–30 years, the activities of White and his colleagues were extended to other geophysical hazards. As a result, hazard-related research became established in a wide variety of academic disciplines ranging from geology to sociology not only in USA but in Europe, Russia, Japan, China, etc.

The pace quickened in the 1970s for several reasons. First, extreme natural events suddenly became more prominent. In large part this was due to the climate fluctuations of that decade which produced the Sahelian drought, the failure of the Peruvian anchovy harvest, the 1975/6 drought in North-West Europe and the severe North American winters of 1976/7 and 1977/8. These events exposed the vulnerability of many countries, including advanced nations, to climatic variability. Second, neo-catastrophism became fashionable amongst physical geographers, who also saw a need to make their work more relevant to human affairs and natural hazards became a strong focus for the study of the relationships between nature and society. Third, there was growing belief that some of the apparent inability to cope with hazards lay in differences between the 'real' world and how it was viewed in practice by managers and decision-makers. In turn, this gave momentum to the research on environmental hazard perception already begun by some human geographers.

Thus, the 1970s saw the publications of several important books. These works represent a plateau of achievement which, in the 1980s, was modified by three further changes in perspective. First, more emphasis was given to the relationships between under-development and vulnerability to natural hazards in the Third World, in particular the extent to which economic dependency exacerbates the effects of geophysical events. Second, following important signals of technological hazards in the 1970s – such as the Flixborough (UK) explosion in 1974, the release of dioxin at Seveso (Italy) in1976 and the nuclear incident at Three Mile Island (USA) in 1979 – so-called 'man-made' hazards began to receive more attention. The year 1984 was a clear turning point with several major industrial accidents, including the release of methyl isocyanate at Bhopal (India) which alone may have caused as many as 6,000 deaths. Third, the multidisciplinary nature of hazard research widened to the point where earlier distinctions between 'natural' and 'man-made' hazards were ever more difficult to sustain.

In terms of policy development, an international programme to reduce the losses from natural hazards – first proposed in 1984 – was finally adopted in December 1989 by the United National General Assembly proclaiming the 1990s as the International Decade for Natural Disaster Reduction (IDNDR). Its objective, as stated in the annex to resolution 44/236, is to reduce through concerted international action, especially in developing countries, the loss of life, property damage, and social and economic disruption caused by natural disaster, such as earthquakes, windstorms, tsunamis, floods, landslides, volcanic eruptions, wildfires, grasshopper and locust infestations, drought and desertification and other calamities of natural origin.

Nature and Definition of Hazards and Disaster

The terms 'hazard', 'risk' and 'disaster' are sometimes taken as synonymous but have some difference in nature and characteristics. *Hazard* is best viewed as a naturally, occurring or human induced process or event with the potential to create loss. *Risk* is the actual exposure of something of human value to a hazard and is often regarded as the combination of probability and loss. On the other hand when large members of people exposed to hazard are killed, injured or damaged in some way, the event is termed a *disaster*.

Small and Witheric (1995) has defined environmental hazard as 'a natural occurrence (event) which threatens, or actually causes damage and destruction to human settlements, man-made structures such as roads and dams, economic activities such as agriculture, mining, etc. Among the most serious environmental hazards are river floods, earthquakes, tsunamis, volcanic eruptions, landslides and glacial outbursts.' On the other hand, disasters are the 'geophysical event of an extreme nature which exerted a heavy toll upon life and property. They are usually called "natural disasters" or natural hazards' and refer to extreme events whose energy levels are such that human societies have no control over them' (Goudie, ed. 1990). A disaster can be ostensively defined as any tragic event stemming from events such as earthquakes, floods, catastrophic accidents, fires or explosions. It is a phenomenon that can cause damage to life, property and destroy the economic, social and cultural life of the people. 'They involve widespread human, material, economic or environmental impacts, which exceed the ability of the affected community or society to cope using its own resources' (UNISDR, 2009). In fact, there is very thin line between hazard and disaster , to some extent they are similar but different in nature and characteristics.

Geographers have defined environmental hazards in terms of physical environment or in other words, considered them as natural hazards. Accordingly, environmental hazards are those elements of the physical environment harmful to man and caused by forces extraneous to him. But, most of the environmental hazards have both natural and human components. For example, flood problems may be exacerbated by fluctuations in climate, such as increased storm frequency, and by human activities, such as land drainage and deforestation. The loss of life caused by a tropical cyclone can by greatly reduced by means of a warning message derived from monitoring by satellites and weather radar. Traditionally, natural hazards have been viewed in an ecological framework (Figure 10.1). This distinguishes between natural events and their interpretation as natural hazards (or resource). Since the earth is a highly dynamic planet, most natural events show a wide range of variation through time in the use of energy and materials for environmental processes. The outer limits of this behaviour we call 'extremes' and certain statistical measures, notably magnitude-frequency relationships, are used to describe such extremes, but, extreme natural events are not considered disasters unless they cause large-scale deaths or damage to humans.

Figure 10.1
Environmental Hazards

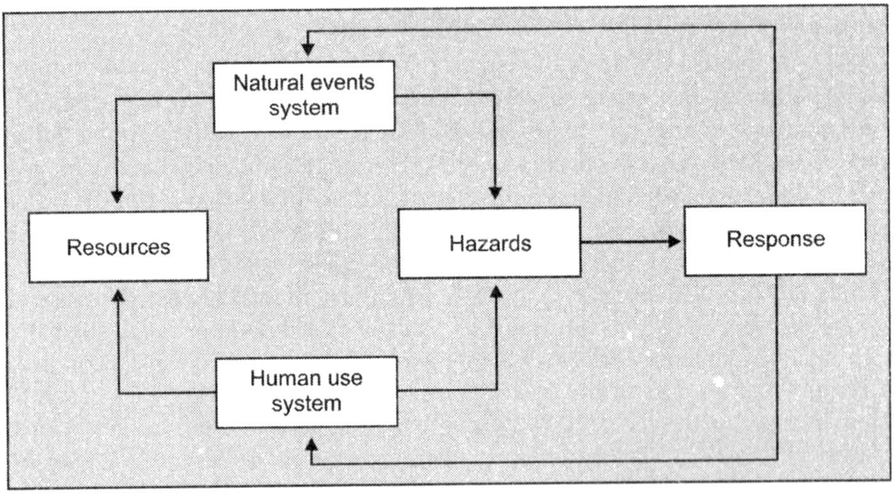

The term 'environmental hazard' has the advantage of including both natural and human dimensions. It also implies a spectrum of hazard types. Figure 10.2 shows that hazards range from 'natural' (geophysical) events, through 'technological' (man-made) events to 'social' (human behaviour) events. Some attempts can also be made to scale the range of hazards according to whether the impacts are intense and local, or diffused and widespread within a society. The extent to which hazards are voluntary or involuntary is particularly important. The degree of individual human responsibility for disaster increases greatly from the essentially accidental geophysical hazards (earthquake, tsunami) to the large self-induced social hazards (smoking, mountaineering).

Figure 10.2 represents a more wide definition of environmental hazard, e.g. 'the threat potential posed to man or nature by events originating in, or transmitted by, the natural or built environment' (Kates, 1978). The common features of the environmental hazards/disaster are:

1. The origin of the damaging process or event is clear and produces characteristic of threats to human life or well-being.
2. The warning time is normally short.
3. Most of the direct losses, whether to life or property, are suffered fairly shortly after the event, i.e. within days or weeks.
4. The exposure to hazard, or risk, is largely involuntary, normally due to the location of people in a hazardous area.
5. The resulting disaster occurs with an intensity that justifies an emergency response.

Thus, the working definition of environmental hazard is:

Figure 10.2

A General Spectrum of Environmental Hazards from Geophysical Events to Human Activities

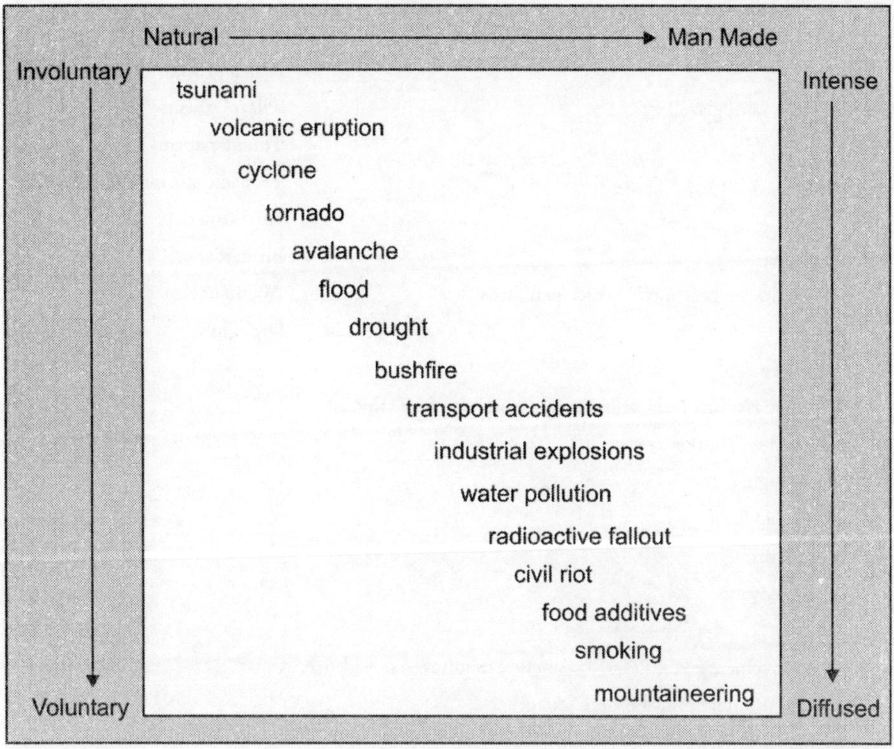

extreme geophysical events, biological processes and major technological accidents, characterized by concentrated releases of energy or materials, which pose a largely unexpected threat to human life and can cause significant damage to goods and the environment.

Classification of Hazards/Disasters

Several types of environmental hazards have been mentioned in the above discussion. But, a general classification is necessary for proper understanding and analysis. While classifying the environmental hazards and/or disasters, the following points should be taken into consideration:

1. aerial extent of damage zone,
2. intensity of impact at a point,
3. duration of impact at a point,
4. rate of onset of the event, and
5. predictability of the event.

Hewitt and Burton (1971) have listed potentially hazardous environmental agencies as given in Table 10.1.

Table 10.1
Potentially Hazardous Environmental Agencies or Processes

1.	*Atmospheric*	
	Single element	*Compound hazards*
	Rain	Rain and wind storms
	Freezing rain('glaze')	'Glaze' storms
	Hail	Thunderstorms
	Snow	Tornadic storms and tornadoes
	Wind	Hurricanes
	Lightning	Blizzards
	Temperature: 'heat wave', 'cold spell', frost	'White-out'
	Fog	Drought
2.	*Hydrologic*	
	Flooding: riverine (rain, snowmelt, natural damburst floods)	
	Lake and sea-shore wave action	
	Waterlogging	
	Sea-ice and icebergs	
	Run-off drought	
	Glacier advance	
3.	*Geologic*	
	Mass-movements: landslides, avalanches, mudflows, subsidence, etc.	
	Erosion (foundations, soils, etc.)	
	Earthquakes	
	Volcanic eruptions	
	Shifting sands	
4.	*Biologic*	
	Severe epidemics in humans	
	Severe epidemics in plants	
	Severe epidemics in domestic and wild animals	
	Animal and plant invasions (e.g. locusts)	
	Forest and grassland fires	
5.	*Technologic*	
	Transport accidents	
	Industrial explosions and fires	
	Accidental releases of toxic gas	
	Nuclear power plant failures	
	Failures of public buildings or other structures	
	Germ or nuclear warfare	

Traditionally, hazards have been classified as 'natural' and 'anthropogenic or man-made' . The natural hazards are those which arise from geo-tectonic forces or weather-related occurrences; where as anthropogenic hazards are the product of human actions. Although both these are often inter-related. Figure 10.3 depicts the types of environmental hazards:

Figure 10.3

Classification of Hazards

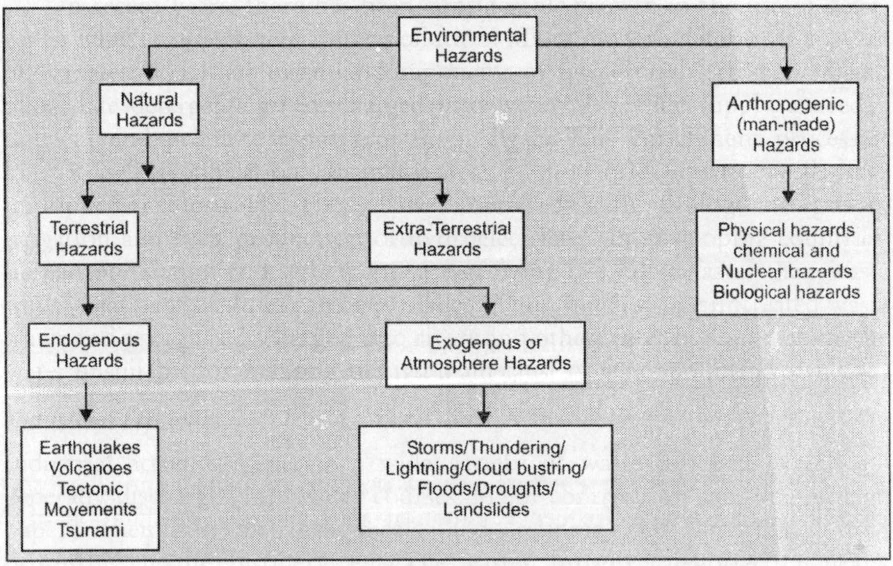

Basic Classification of Environmental Hazards

Hazards can also be divided according to the damage caused by them.

Although, such division is not effective in all parts of the world because of the difference in development, technology, etc. For example an ordinary disaster may include:

— at least 100 people dead, or
— at least 100 people injured, or
— at least US $ 1 million damage.

But such occurrences have great spatial differences in less developed countries and more developed countries, therefore, classification based on losses in not effective.

Development Chain and Human Response to Natural Hazards

Hohenemser et al. (1983) has viewed technological hazards as a sequence of events leading from human needs and wants to the selection of a particular technology, though to harmful consequences.

A schematic illustration of the causal chain of hazard development with an example of drought disaster has bcen depicted in Figure 10.4.

Figure 10.4
Causal Chain of Hazard Development

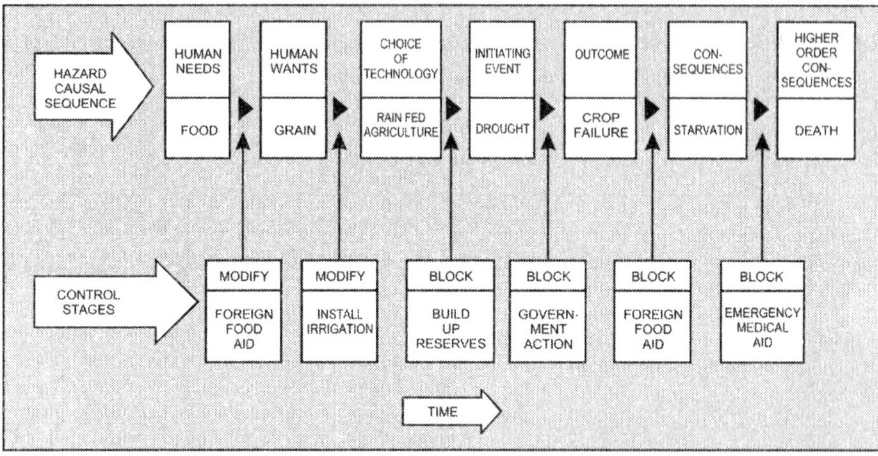

In Figure 10.4, the top line indicates seven stages of hazard development. The stages are linked by causal pathways denoted by triangles. Six control stages are linked to pathways between hazard states by vertical arrows. Each is described generally as well as by specific control actions designed to eliminate or reduce the evolving hazard.

The human response to natural hazards depends upon the nature of natural hazard, i.e. magnitude, frequency, duration and spacing. The duration of a

Figure 10.5
Human Response to Natural Hazards

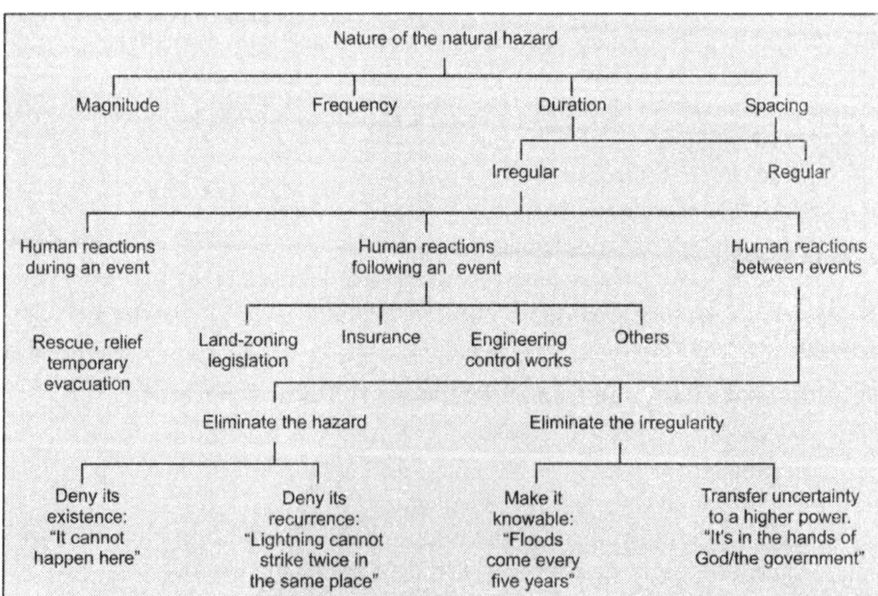

hazard is either regular or irregular, but mostly it is irregular. In this case human reactions are of three types viz., during the event, between the event and after the event, and, response to these three stages are different. The overall picture of these events and human response to natural hazards is depicted in Figure 10.5.

Common Environmental Hazards

The polygenic character of natural hazards of the world derives from its diversity of natural conditions. Potentially, disastrous natural phenomena are also closely associated with certain regions and geographic zones. Such phenomena often develop at the interface of physically different natural environments: sea and land, atmosphere and ionosphere, etc. In the world, most common and disastrous hazards are volcanism, earthquakes, landslides, floods, drought and famines, thunder storms, tornadoes and tropical cyclones.

Volcanism

The eruption of volcanoes is one of the most dreadful natural phenomena. The molten rocks, boiling lava and hot gases penetrate out of that part of the surface which is comparatively weak. When this material comes out of these spots, an explosion takes place, and, with thunderous noise, molten rocks, lava and gaseous clouds rise up towards the sky. All this results in a threat to the environment as well as to life.

Volcanism is not common throughout the world but is limited in the Circum-Pacific belt extending from Andes in South America, through the Cascades and the Aleutians into Japan and then south into the East Indies and New Zealand. There is also an important belt in the Mediterranean, which includes the active volcanoes of Italy.

A volcanic explosion poses a great threat to environment. On an average, a volcano, in an explosion ejects 150 cubic km of material. This material is of three types – solid material in the form of small and large pieces of rocks, ash particles and also pieces or volcanic bombs which shoot into the sky. The liquid material called *magma* comes out of volcano – also known as *lava*. A huge amount of gaseous material comes out of volcano-steam, carbon dioxide, sulphur in vapour form that which produce poisonous gases when they combine with other gases. Gases like chlorine, fluorine, etc. are also emitted in large quantities.

The eruptions of volcanoes and lava flows are environmental hazards of the severest type, often taking a heavy toll on plants and animal life. Loss occurs from sweeping clouds of incandescent gases and suspended matter that descend the volcano slopes like great avalanches. When Mount Pelée erupted in 1902 in the martinique in the Caribbean sea, the burning material ejected out of it moved in the form of a cloud over the city called St. Pierre and only two people out of total 30,000 population survived. The explosions of Etna and Vesuvius in Italy destroyed many beautiful cities. In 1943, when Paricutin, 320 km west of Mexico city, exploded, created a mountain 167 metre high. In 1991, volcanic explosions in Japan and Philippines have damaged a lot of property and caused loss of human life and also polluted the environment. Similarly, volcanic activity in the Andaman & Nicobar Islands was responsible for the environmental degradation in this area.

Earthquakes

Earthquakes are tremors which are produced by the passage of vibratory waves through the rocks of the earth. The devastation caused by earthquakes to the life of man and property, as well as to the ecosystem is very high. The earthquake belt of the world is also associated with the volcanic belt, but several other regions too have occurrence of earthquakes. Wherever earthquakes occur, apart from loss of life and property, landslides, change in course of rivers, loss of vegetation and animal life, destruction of dams and development of faults and fractures in earth's surface are common and all these events lead to environmental degradation. The list of devastating earthquakes in the world is too long to be enumerated here. In India, the occurrence of earthquakes in areas of Jammu and Kashmir, Himachal Pradesh, Uttar Pradesh, Bihar, West Bengal, Punjab, Delhi, Haryana, North-Eastern states as well as in Maharashtra, are common. The most recent earthquake occurred on 30 September 1993 in Maharashtra and on 28 March 1999 in the hill district of Chamoli in Uttar Pradesh in which heavy loss of property and life was reported, apart from the whole regional ecosystem being disturbed. Major earthquakes incidences that have occurred in the world are as follows.

Date	*Place*	*Losses*
18th April 1906	San Francisco	More than 3000 people dead
1st September 1923	Tokyo	Died one lakh 42 thousand people
2nd May 1960	Chile	Villages destroyed
26th July 1963	Macedonia	One thousand died
31st May 1970	Peru	66 thousand died
28th December 1972	Nicaragua	10 thousand died
19th September 1985	Mexico	10 thousand died
7th December 1988	N.W. Armenia	25 thousand died
21st June 1990	Gilan, Iran	40 thousand died
30th September 1993	Western & Southern India	10 thousand died
17th January 1995	Kobe (Japan)	6,430 died
27th May 1995	Sakhalin Island	1,989 died
May 1997	East Iran	1,600 died
30th May 1998	North Afghanistan	4,000 died
17th August 1999	Turkey	4,000 died
21st September 1999	Taiwan	2,500 died
12th November 1999	N.W. Trukey	400 died

Cont'd...

...Cont'd

26th January 2001	Bhuj (India)	20,000 died
31st October 2002	Italy	Building destroyed
24th February 2003	W. China	260 died
1st May 2003	S.E. Turkey	160 died
21st May 2003	Algeria	2,000 died
26th December 2003	Bam city (Iran)	41,000 died
25th February 2004	Morocco	400 died
8th October 2005	Muzaffarbad (POK)	50,000 died
27th May 2006	Java	5,700 died
30th September 2009	Penang (Sumatra)	a few thousand
12th January 2010	Port-au-Prince (Haiti)	2,30,000 died

In fact, earthquakes are most devastating and disastrous among natural hazards.

Floods, Droughts and Famines

Floods and droughts both are natural hazards and are related with rainfall. If there is heavy rain in the catchment area of a river, floods will occur. On the other hand, if there is less rain or failure of monsoon, droughts will occur. In both the situations, their impact on environment creates problems and also affects the regional ecosystem.

Flood is a hydrological phenomenon that occurs in any drainage basin. Its magnitude and frequency varies primarily on the precipitation conditions in the catchment area. Most of the river valleys are subjected to varying degrees of flood incidences, either creating minor damages or devastating havoc at times. Such calamities of floods pose a serious problem to the whole environmental set-up of a region, bringing about some phenomenal changes in the physical environment ushering in accelerated erosional, transportional and depositional activities of the river with the consequent effects on soil. The devastating effects of flood can be seen on vegetation, agricultural, industry settlement, sometimes causing loss of human and animal lives and of material also.

The main cause of flood is excessive rainfall but its occurrence is also due to earthquakes, deforestation in the upper catchment areas, failure of dams, excessive snow melting, etc. The beneficial effects of floods have been established beyond doubt in the form of fertilization of soil. But its devastating effects far exceed the benefits.

In India, the problem of flood and challenge to the environmental setting is of gigantic magnitude, because of: (i) typical seasonal rainfall, and (ii) erratic nature of tropical cyclones and rainstorms. The following are the areas which may be termed as flood-prone areas:

1. Upper catchment and piedmont zone of the Himalayan river basins.
2. Ganga river valley – upper, middle and lower deltaic zone.

3. Brahmaputra valley and its tributaries.
4. Deltaic region of Mahanadi, Godavari, Cauvery and Krishna rivers.
5. Narmada and Tapti basins.

The common measures of the flood control are as follows:

(i) Control of erosion in upper catchment areas.
(ii) Construction of dams and reservoirs at suitable sites.
(iii) Prevention of overflow in channels.
(iv) Construction of multiple channels for distribution of the excess river water.
(v) Development of shelters in flood-prone areas.

The flood problem should be taken up with more intensive preventive measures, in order to avoid environmental problems.

Drought is also a natural hazard which has its impact on the regional ecosystems. The areas which get less rainfall, i.e. amount of rain is too less to support agricultural production or even grasslands are usually called drought-prone areas. When droughts become severe, they cause famines, which are devastating for man, animals and also environment. Generally, droughts occur when rainfall is less then 25 to 50 per cent of the average. If it is below 50 per cent of the average, famines occur. In 1899, India faced the severest famine known as Chappan-ka Akal. A similar famine also occurred in 1977. In both these famines several lakhs of people lost their lives. In 1992–93, Ethiopia also suffered from the grips of a bad famine, in which more then 30 lakh people died. Famine conditions are common in Rajasthan, Gujarat, Haryana, Madhya Pradesh, Orissa and Andhra Pradesh in India, In Rajasthan, famine has now become a regular phenomenon.

Thunder Storms, Tornadoes and Tropical Cyclones

Due to abrupt changes in atmospheric conditions thunder storms, lightning, hail storm, tornadoes, and cyclonic storms occur and are the cause of environmental hazard. A thunder storm is an intense local storm caused by strong conventional activity. Thunder and lighting normally accompany the storm, and rainfall is heavy, sometime due to cloudburst also. Violent surface wind may occur at the onset of the storm. The speed of the rising air in a thunder storm is about 112 km per hour. Lightning is also one of the environmental hazards that results annually in the death of many persons and livestock and causes forest fires and building fires. One of the most destructive products of the thunder storm is the hail storm. Hail consists of rounded lumps of ice ranging from 0.5 to 5 cm in diameter. Large hailstones are an environmental hazard as they can be extremely destructive to crops and flimsy buildings.

A tornado is a small, intense cyclone in which the air spirals at a tremendous speed. Associated with frontal air mass interactions in mid-latitude, tornadoes are the most violent storms. These storms are most frequent and violent in the United States although occurring in Australia and in other places of mid-latitudes. A tornado moves across the country with a wind speed of upto 400 km per hour in a funnel-like shape. The end of the funnel cloud may alternately sweep the ground, causing complete destruction of anything in its path.

Tropical cyclones are the most powerful and destructive types of cyclonic storms, which are also known as hurricane or typhoon. These storms develop over oceans in latitude 8° to 15° N and S, but not close to the equator. The tropical cyclone is an almost circular storm having low pressure in which winds spiral at high speed accompanied by heavy rainfall. The environmental impact of tropical cyclones lies in the form of destruction of life, infrastructures as well as of forest and animals. In islands and coastal regions, large scale destruction of cities and their inhabitants has been reported on several occasions. Bangladesh is a country which has suffered a lot by heavy destruction caused by tropical cyclones – the death toll by tropical cyclones in Bangladesh is almost unbelievable. In 1822, the toll was 40,000 while in 1876 and 1897 it was 1 lakh and 1.75 lakh respectively. In 1970, 3 lakh lives were lost and in 1991 the death toll was 5 lakh. These storms are also responsible for change in atmospheric and environmental conditions.

The Hurricanes are chronic disasters in the Gulf coastal and Atlantic coastal areas of USA. There is a practice in USA to name the hurricane such as *Katrina*, *Walima*, *Andrew*, *Hugo*, *Agnes*, *Betsy*, *Camille*, *Floyd*, etc. This practice now has also been followed by India, China (typhoon *Carry*, typhoon *Betty*) and Japan also. In Myanmar, cyclone *Nargis* name was given to the most disastrous cyclone occurred on 2 May, 2008.

Landslide is another event which may cause disruption in the general ecosystem. Landslide is a natural hazard and means any downslope movement of mass of regolith or bedrock under the influence of gravity. It is a rapid sliding of large rock masses beginning their descent as unit of blocks, without internal flowage. When the landslide is in the form of huge bocks, it is known as rock avalanche. In Switzerland, Norway or Canadian Rockies, villages built on the floors of steep-sided valleys have been destroyed by the sliding of rock masses. In India, reports of rockslide often appear in the news in hilly states like Jammu and Kashmir, Himachal Pradesh and north-east states. The environmental hazard of rockslides is extremely high for towns located on the floors of glacial troughs and at heads of fiords. The blockage of roads by landslide is a common phenomenon but sometimes it also blocks the course of the small rivers, thus, causing much damage.

Tsunamis

Tsunamis are high energy waves in the oceans generated by high magnitude earthquakes in the ocean floors (exceeding 7.5 on Richter scale), or by violent central volcanic eruptions or by massive landslides of coastal lands or of submerged continental shelves and slopes or in deep oceanic tranches. Tsunami is a Japanese word meaning thereby *harbour waves*. Tsunami consists of Japanese words tsu + nah + me = *tsu* (means harbour) and *nami* (means waves). Tsunamis are called as *seismic sea waves* because most of the tsunamis are generated by undersea seismic events (earthquakes). These are also called high energy tidal waves. Tsunami moves away from the centre of origin with high speed and low crests across the ocean and is usually not noticed as the massive ocean waves move silently but assume destructive form as these travel through shallow waters of continental shelves and approach coastal waters and cause widespread devastation along the coastlines of low height and gentle slope.

Tsunamis are generally divided into the following two types:

1. distant tsunami or deep sea tsunami; and
2. local tsunami.

After being originated in the deep waters, initial tsunamis are split in two e.g. distant tsunami and local tsunami. Distant tsunami moves out to the deep ocean (open ocean) while local tsunami travels towards the coasts. Thus, two tsunamis move in opposite directions. Distant or deep tsunamis travel much faster than local tsunamis but it is the local tsunami that causes destruction in the coastal zones. Tsunamis are produced due to large-scale displacement of immense volume of sea water due to sudden tectonic disturbances in the sea floor. The changes and disturbances in sea floor are produced by a host of causative factors such as faulting in the sea floor, slumping and massive undersea landslides, sliding of large blocks of ice near the fiord coasts, avalanches submarine volcanic eruptions, undersea seismic events etc.

Tsunamis occur mostly in the Pacific Ocean, where in 86 per cent of the total tsunami occurrences are the products of tsunami genic undersea earthquakes. The well known tsunami in India has been occurred on December 26, 2004 when Sumatra tsunami hit the Indian coasts of Tamil Nadu, Andhra Pradesh, Orissa and Kerala.

Adjustment and Management

Adjustment to Hazard

Most environmental hazards which pose a recognizable threat to human populations produce some response designed to reduce the level of risk. Theoretically, the actions taken can attempt to manage the full range of hazards which affect an area but the information and resources necessary for such an approach are rarely available. At the other extreme, sometimes nothing can be done but simply accept recurrent losses. Within this range, all intentional hazard adjustments can be organized into three groups. These are:

- *Modifying the loss burden*: The most limited responses seek to modify losses by spreading the financial burden as widely as possible beyond the immediate victims through relief and insurance programmes. These schemes are essentially loss-sharing, rather than loss-reducing measures. But in some circumstances, both disaster aid and insurance can also be used to a limited extent to encourage future loss reduction.
- *Modifying hazard event*: This involves reducing losses by adjusting damaging events to people. Most natural hazards are insufficiently well understood, or manageable, to be physically suppressed at source through

environmental control engineering. On the other hand, hazard-resistant design and emergency measures can be applied to safeguard lives and property from selected phenomena in certain high-risk settings.

- *Modifying human vulnerability*: This is the largest group of responses. It includes all measures designed to reduce hazard losses by adjusting people to damaging events. This is achieved mainly through preparedness programmes aimed at changing human behaviour, through the installation of forecasting and warning schemes and through long-term planning to encourage more appropriate land use allocations.

Hazard Resistance

This requires hazard-resistant design and retrofitting. In most countries, public facilities such as dams, bridges and pipelines are likely to be hazard-resistant because they would have been designed by professional engineers. The same is true for structures such as nuclear power plants and chemical factories. However, design protection can never be total. Building failure will be greatest in data-poor, hazard-prone environments where many structures – notably private residences, older and smaller buildings – would have been constructed without any thought to hazard impact. The two most common natural hazards considered in building codes are earthquakes and windstorms, although the adjustment is also applicable to other hazards such as floods.

In the case of earthquakes, hazard resistance begins with geotechnical engineers who apply the principles of rock and soil mechanics to the safe designing of earth-supported structures. Other things being equal, buildings on solid rock are less likely to suffer damage than those built on clays or softer foundations. Microzonation maps of the substrate can be prepared which suggest appropriate local variations in building strength and ensure that major buildings are not located over faults or areas of unconsolidated material. Local building codes tend to undergo continuous improvement and the adoption and enforcement of such improvements is the primary means of increasing the hazard resistance of structures. In fact, proper hazard-resistant design could make structural property safe and its insurance will also provide damage loss, if any.

Retrofitting is the act of modifying an existing building to protect it, or its contents, from a damaging event. Many individual measures can be taken. In the case of earthquakes, for example, brick chimneys could be reinforced and braced to structural elements to prevent collapse into a living area. Unreinforced masonry walls could be strengthened and tied to adequate footings while closets and heavy furniture could be strapped to the walls whenever they constitute a danger or contain valuable items. To protect against floods, measures include making the walls watertight and fitting flood-resistant doors and windows.

Community Preparedness

Disaster preparedness my be defined as 'the pre-arranged emergency measures which are to be taken to minimize the loss of life and property damage following the onset of disaster'. It involves the detailed planning and testing of prompt and efficient responses, by both individuals and groups, once a hazard is either forecast or has actually occurred. Measures include the promotion of public education and awareness programmes, the development of evacuation plans (often in response to a warning message), the local supply of medical aid and the preparation of emergency food and shelter for evacuees.

Long-term community preparedness programmes have proved very successful in reducing deaths from hazards in many MDCs. They are a particularly important adjustments for hazards, like earthquakes and volcanic eruptions, for which reliable short-term forecasts and warnings are not routinely available. Within individual countries, the degree of preparedness usually depends on the arrangements which have been made for civil defence.

For a successful operation of any disaster plan, public understanding and cooperation are vital elements. Most of the people are not prepared to handle the consequences of a damaging event. Simple preparedness at domestic level such as a battery-operated radio, a working torch and a first aid box will help a family during any emergency.

Good educational programmes are a key element in ensuring that widespread public support is available for hazard mitigation, as well as facilitating the transfer of disaster experience from one community to another. Ill-conceived awareness exercises may create panic and other counter-productive attitudes, so it is important that such programmes are well based in terms of social feasibility and an understanding of the probable economic consequences of the mitigation measures recommended to the public. Recent experience suggests that carefully prepared advice need to be distributed, both widely and often, to the public through the media and other channels. All preparedness advice should come from an authoritative government agency and should be locally endorsed by well-known local officials and public figures. Each measure should be accompanied with a brief explanation which justifies the recommended actions and details on how it can best be implemented.

Forecasting and Warning

Forecasting and warning systems have become important in recent decades owing to scientific advances, for example in weather forecasting, and to associated improvements in information and communication technology such as satellites. Most warnings of a future environmental hazard are based on forecasts. But some threats, notably earthquakes and droughts, are insufficiently understood for forecasts to be routinely issued and reliance for warning has to be placed on predictions instead. It is important to understand the difference between predictions, forecasts and warnings.

A prediction is largely based on statistical theory and uses the historical record of past events to estimate the future probability or recurrence of similar event. Because the results are expressed in terms of average probability, there is no precise indication of when any particular event may occur. By definition, predictions tend to be relatively long-term. For some hazards, such as earthquakes, it is not possible to specify the location or the magnitude of the event with much precision. A forecast, on the other hand, depends on the detection and evaluation of an individual event as it evolves through a sequence of environmental processes which are relatively well understood. This means that, depending on the ease with which such individual events can be monitored, it is often possible to specify very precisely the timing, location and magnitude of an impending event. Forecasts are based on scientifically observable facts and offer no prescription as to how people should respond. They tend to be short-term. Indeed, contrary to predictions, the limited lead time for issuing forecasts often restricts the effectiveness of warnings.

A warning is a message which advises the public at risk about an impending hazard and advises what steps should be taken to minimize losses. All warning are based on either predictions or forecasts but for many agencies, such as those involved with national weather services, very few of the routine forecasts are followed by warnings. All warning systems consist of three inter-related stages of evaluation, dissemination/warning and response.

- Evaluation: As shown in Figure 10.6 the evaluation stage includes several sub-steps from trained observers, first detecting an environmental change that could cause a threat, through to estimating the nature and scale of the hazard and finally deciding to issue a warning to an endangered community. The priority at this stage is to improve the accuracy of the

Figure 10.6

Hazard Forecasting and Warning System

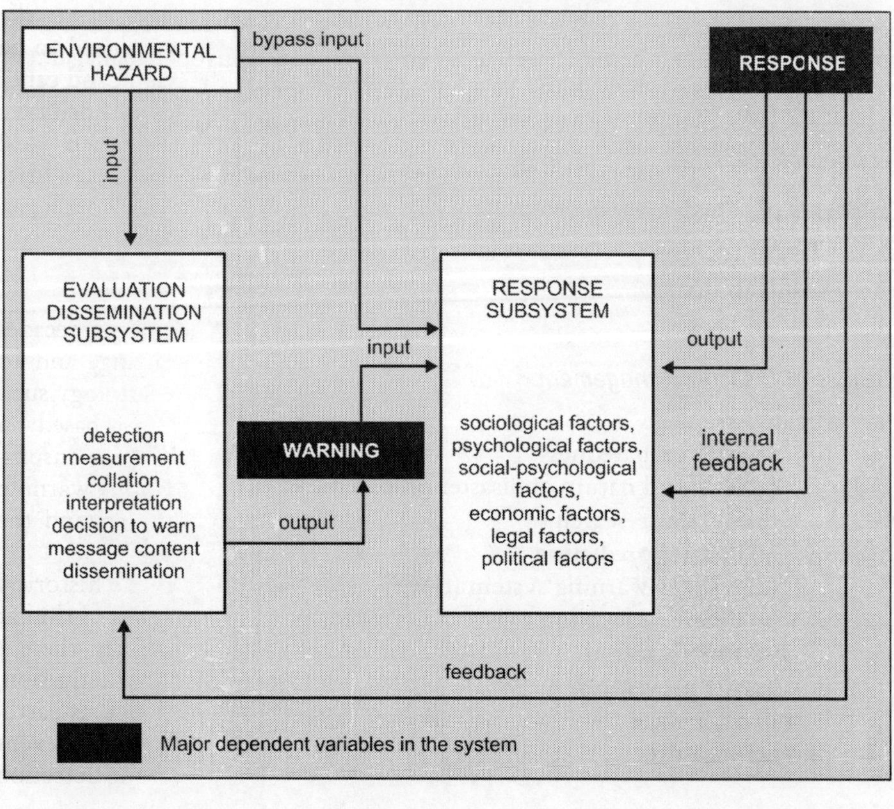

forecast and to increase the lead time between issue of the warning and the onset of the hazardous event.

- Dissemination: The dissemination stage occurs when the warning message is transmitted from the specialist forecasters to the hazard-zone occupants. The message is likely to be formulated and conveyed by a third party through an intermediate medium which may involve different communication methods, such as radio or television, and different personnel, such as the police or neighbours. Again, the stage contains several components, like the content of the message or the way in which it is conveyed, which are known to affect the eventual outcome.
- Response: The whole purpose of the two preceding stages is to activate the response stage, which is where loss-reducing action are expected to be taken by the recipients of the warning message.

One important purpose of warning systems is to enable people to evacuate an area in advance of an approaching hazard. Often there is a reluctance to evacuate. This is not because people are paralysed in the face of danger but more likely because the message fails to specify this action, or people feel that they can cope or because they fear looting of their empty house. There is a considerable natural attachment to the home environment which has to be overcome before people will evacuate.

Natural Disaster Hazard Management

Since natural disasters and hazards involved heavy destruction and loss of property and human being as well as to environment, its proper management is necessary not only for the reduction of effects but also for the protection measures. A systematic process of disaster management is based on following elements/stages:

Elements of Disaster Management

1. Hazard analysis
2. Vulnerability
3. Risk analysis

Stages of Disaster Management

1. Pre-disaster stage
   (i) disaster preparedness
      (a) level and nature of disaster preparedness
      (b) disaster research
      (c) disaster prediction
      (d) disaster warning system (DWS)
      (e) disaster education
   (ii) disaster mitigation
   (iii) disaster prevention
2. Post-disaster stage
   (i) relief measures

(a) social response
(b) participation in relief work
(c) relief materials
(d) execution of relief work
(ii) rehabilitation
(iii) disaster recovery

Hazard Analysis

Hazard analysis involves the histogenisis of different hazards, nature and mode of occurrence of a particular hazard in a specific region with the following objectives

1. to ascertain the number of hazards likely to occur,
2. to ascertain the degree of intensity and severity,
3. to understand the likely adverse effects of a disaster on human society
4. to understand the mode of occurrences of different hazards.

Vulnerability

The WCDR (World Conference on Disaster Reduction, 2005) has adopted the following definition of vulnerability:

Vulnerability denotes the 'conditions determined by physical, social, economic and environmental factors or processes, which increase susceptibility of a community to the impacts of hazards. Vulnerability is thus the probability of being damaged, destroyed or lost because of a natural hazard.' Vulnerability is far from being a static process; it is a dynamic process that keeps on changing the probability of process of loss and damage of all the elements exposed to disasters.' Vulnerability can be grouped as physical and social. The social or community vulnerability includes the consideration of susceptibility of a particular community or social group to a disaster. Geographical Information System (GIS) and satellite images help in identifying and mapping of vulnerable areas, for which planning and management strategy can be developed.

Figure 10.7
Risk Management

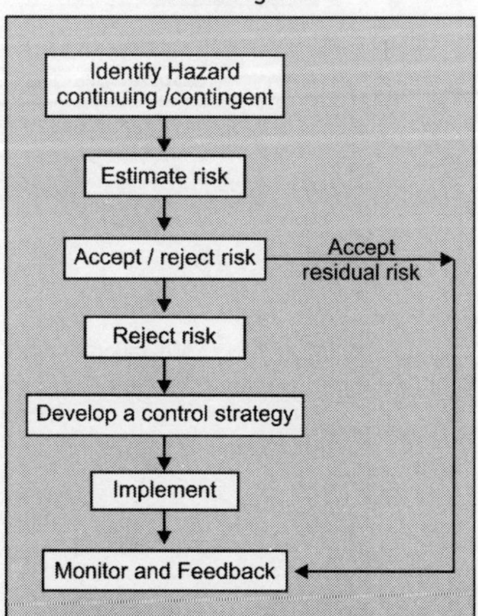

Risk Analysis

Hazard/disaster risk is a probability or likelihood that a hazard will happen or, more precisely, the likelihood that an event will coincide with elements which can be affected. The risk management is a very significant aspect of natural disaster reduction and management, which includes the aspects shown in Figure 10.7.

Disaster Management

Disaster Management is a systematic process in which proper measures have to be taken in every stage. The three stages of disaster management are:

1. *Pre-disaster* or *Anticipatory* or *Preparedness* stage
2. *On-disaster* or *Participatory* or *Mitigation* stage
3. *Post-disaster* or *Recovery* stage

Figure 10.8

Disaster Management Cycle

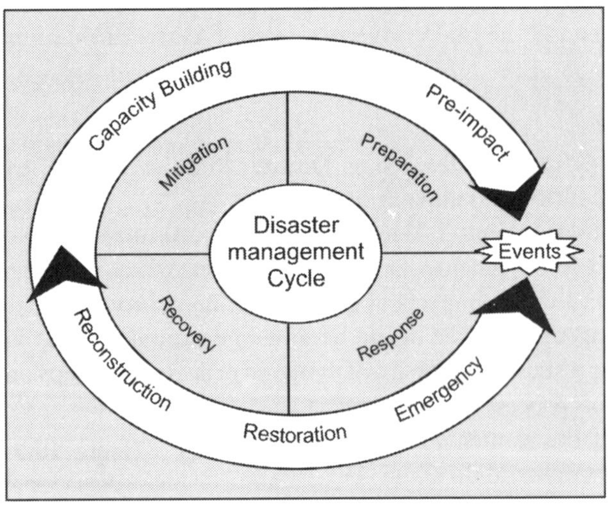

Figure 10.9

Disaster Event and Management Cycle

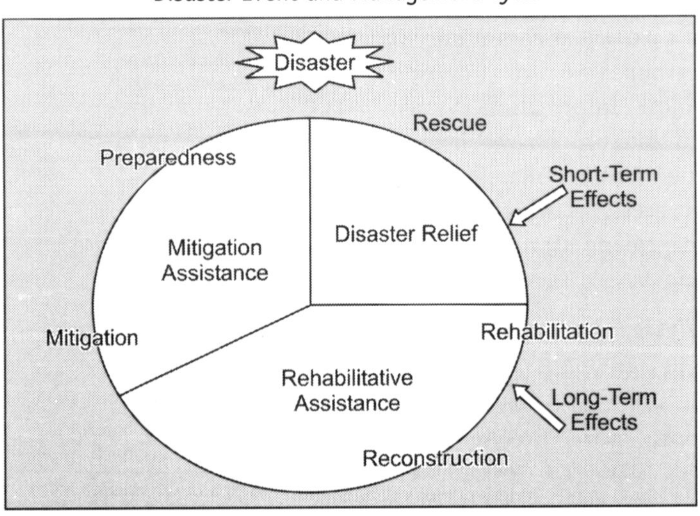

These three stages are often grouped into two stages, i.e. (i) Pre-disaster and (ii) Post-disaster stage. Disaster management is closely related with disaster event cycle and also management cycle as shown in Figure 10.8 and 10.9.

Pre Disaster Stage

Pre-disaster stage primarily aims at reducing the adverse impacts of hazards and disasters by adopting certain measures. There are certain hazards, the time of occurrence of which cannot be precisely known before hand such as seismic events (earthquakes), volcanic eruption, tsunami, tornadoes, but there are certain hazards and disasters of which occurrence time can be known well in advance such as tropical cyclones, floods, droughts, etc. If a long-range planning is formulated and executed, the occurrences of such hazards though cannot be prevented, but, their adverse impacts can be reduced to a certain degree. Pre-disaster stage involves three steps, viz. preparedness, mitigation and prevention.

Preparedness means necessary arrangements to meet any eventuality if a hazard strikes a locality and community. This involves the study of the vulnerability of a locality or a region to certain hazards and disasters; to identify hazards and to study the nature, severity, and mode of occurrence of particular hazard; to assess the risks of individual hazard; to determine the magnitude of a hazard; and quantum of damage to be done by such hazard; to prepare vulnerability maps of the concerned area; to educate the people about the nature and likely adverse impacts of hazard through various methods; to educate the people about avoidance of hazards and taking safety measures in the event of occurrence of a disaster, etc. The following measures should be adopted to make the disaster preparedness more effective.

1. Research, which includes the following:
   (i) mapping and preparation of disaster prone (vulnerable) areas and risk magnitude zone maps,
   (ii) study of nature (types, characteristics, origin), magnitude and severity of hazards and disasters,
   (iii) risk assessment of disasters,
   (iv) study of past history and recurrence intervals of hazards and disasters,
   (v) study of psychological, economic and social implications,
   (vi) prediction techniques, etc.
2. Preparation of contingency plans of infra structural facilities for relief work, recovery and rehabilitation activities to be followed in the post-disaster stage of disaster management. The needed infrastructural facilities include the following :
   (i) rescue operation,
   (ii) shelter and food,
   (iii) safe drinking water,
   (iv) electricity supply,
   (v) transport system,

(vi) language efficiency,
(vii) medical facilities, etc.

3. Adequate support systems are required to meet out the requirements of above mentioned infrastructural facilities. The following support systems are required:
   (i) financial support,
   (ii) material support,
   (iii) equipment support,
   (iv) administrative support,
   (v) medical support,
   (vi) social support, etc.
4. Disaster education and awareness
5. Disaster management training programmes
6. Disaster warning system
7. Dissemination of disaster information and knowledge
8. Formation of disaster reduction plans and projects for reducing poverty of the affected people due to loss of their life support systems, and disabilities caused by the disasters.

Mitigation includes all those measures, which enable the people to cope with the adverse situation created by disasters. The objective of disaster mitigation is to minimize the disruptive forces of disasters and to reduce the magnitude of disasters. It's primary goal is to save human lives from disasters as well as to reduce economic losses. The two basic approaches to meet out the disaster mitigation programmes are:

The *top-down approach* of disaster mitigation entirely depends on different measures of disaster management initiated by the government establishments at different hierarchical orders such as at national, state, tahsil, block levels etc. There are many bureaucratic hurdles and complications at government levels in implementing disaster mitigation programmes and hence actual victims of disaster-hit areas do not get the benefit of disaster mitigation measures. Sometimes, there is absence of such schemes in many countries.

The alternative disaster mitigation approach is community-based *bottom-up approach* wherein there is direct involvement of local people who participate in disaster mitigation mission. The community participation involves from local community to world community as follows:

- local community level participation
- non-governmental organization (NGOs)
- gram panchayat and block-level participation
- state-level participation
- central government level participation
- world community level participation

There is a need to associate both the government based and community-based mitigation programmes, so that the benefits of disaster mitigation measures reach the victims. Petak and Atkinson (1982) proposed the following seven categories of mitigation measures:

1. Approaches that involves measures to minimize the probability of hazard occurrence and/or to protect areas and building sites from hazards.
2. To focus on strengthening buildings exposed to the hazard or on the design of site level systems for protecting buildings from hazards.
3. To give attention to site development schemes for protecting structures from hazards.
4. Approaches that involve the identification of hazard prone sites and the application of measures to prevent or restrict their development and use.
5. Loss recovery, relief and community rehabilitation.
6. Hazard warning and population evacuation systems.
7. Approaches that provide policy makers with the information and decision assistance tools that facilitate rational and effective hazard decision-making.

Post-Disaster Stage

Post-disaster stage of management is a very important phase because it involves reconstruction, restoration, and rehabilitation measures both short-term and long-term. These measures can be grouped in following three categories:

1. relief measures,
2. recovery from shocks of disasters, and
3. rehabilitation of displaced communities.

Relief Measures

The moment a disaster occurs in an area, the immediate steps to be taken are to rescue people buried under debris, to provide food, drinking water, shelter, medicines, to restore power and water supplies and communication and transport systems. The following are the significant components of relief measures at the time of occurrence of a disaster:

- social response and action,
- participants in relief work,
- organizational involvement,
- correct picture of nature, magnitude and severity of disasters and necessary requirements,
- priority fixation of relief measures,
- zonation of priority on the basis of magnitude of damages and requirements, and
- active participation in relief work.

The important aspect of relief measures is the execution of the relief work with the help of individuals, local communities, voluntary social organizations and government officials. Similarly, priority must be decided before undertaking the remedial and relief measures.

Disaster Recovery

Recovery process includes all those actions and activities which help the individuals and communities to adjust with the effects of all sorts left behind by a particular disaster. Thus recovery process involves multiple strategies to get rid off the adverse effects of disasters. The adversaries of a severe disaster may include physical and mental injuries, loss of property, separation of family members, mental agony of social evils committed by antisocial elements, loss of territory, loss of native culture, etc.

Since disaster recovery is a community-based process, community participation is a prerequisite condition to recover the individuals and the community from the agony of the trauma of disasters. The first and foremost step in disaster recovery process is to rebuild the confidence level of disaster victims by bringing them in the mainstream of community programme, by injecting a sense of belonging and brotherhood, by making them active in the recovery process so that they may forget the fear psychosis created by the disaster, by diverting their mind from the sad memory of the destruction caused by the disaster.

Rehabilitation

It may be mentioned that recovery and rehabilitation are time consuming, lengthy and costly processes as these require long time and huge amount of money for successful execution of recovery and rehabilitation programmes. Rehabilitation process involves renewal of disrupted systems, repair of damaged systems and reconstruction of houses, buildings, roads, rails, bridges and other items of utilities, relief employment, etc. The success of rehabilitation of distressed and displaced people depends upon the economic condition of the concerned country. Similarly, the relief, recovery and rehabilitation cannot be successful on local resources alone. Thus, monetary, administrative and technological assistance from government departments and even international communities becomes necessary for natural disaster reduction.

In brief, environmental hazards are prevalent throughout the world and are the result of natural phenomena as well as of human activities. There is a need to prepare both short-term and long-term plans so that the loss can be minimized. A proper management system, broad community support and a strong political-will is very necessary for the adjustment and control, whenever a natural hazard occurs.

11

Population and Ecological Crisis

According to ecologists, population has been defined as a 'collective group of organisms of the same species or other groups within which individuals may exchange genetic information occupying a particular space.'[1] Thus, each and every organism is an integral part of the ecosystem and has its impact over it. In the present context, population means 'human population', which not only uses the environment as habitat but through its multifarious activities also, changes certain natural processes in the ecosystem, thus becoming a cause of the ecological crisis. The situation becomes more alarming due to the population pressure and overexploitation of environment.

The interaction between population and environment is very complex and dynamic. This complicated interrelationship is not yet fully understood or spelt out at micro-level, nor is the spatio-temporal variations over the nation or the globe understood in its entirety. But, there is no two opinion about the fact that the environment is constantly changing due to human activities, which leads to

Figure 11.1

Population and Eco-crisis

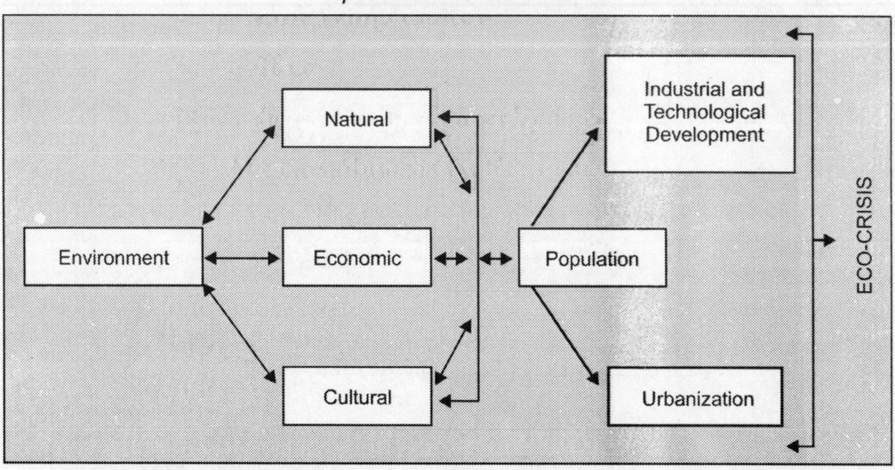

problems like soil erosion, devastating floods, acute droughts, unforeseen climatic changes, desertification and general degradation of the environment. Figure 11.1 indicates the relationship between population and eco-crisis.

The environment – natural, economic and cultural – acts and reacts in a composite manner and determines the growth and spatial pattern of the population. On the other hand, human activities create chain reaction in the ecosystem. These reactions, pollution and disturbances are now evoking great concern. All these aspects of the population and its impact over the environment need analysis so that this problem can be understood and steps be suggested for future planning.

Population Growth and Distribution

The growth in human population is a continuous process. From the dawn of human history until the beginning of the 1800s, population increased slowly and variably, with periodic setbacks. It was roughly around 1830 that the world population reached the one billion mark. By 1930, however, just 100 years later, the population had doubled to 2 billion. Barely 30 years later, in 1960, it reached 3 billion, and in only 15 more years, by 1975, it had climbed to 4 billion. Thus, the population doubled in just 45 years, from 1930 to 1975. Then, 12 years later, in 1987, it crossed the 5 billion mark! In 2001, world population passed 6 billion, and in 2011 according to World Bank, was 6,973,738,433 an now it is at 7 billion mark. (Table 11.1). On the basis of current trends, the U.N. population division projects that world population will pass the 8 billion mark in 2028.

According to United Nations estimates, the population growth rate at present hangs around 2.5 per cent annually. The rate of increase is staggering. Every 30 seconds, about 117 live babies are born and some 46 people die, for a net increase of 71 new inhabitants. This means, over 2,00,000 new mouths to

Figure 11.2

Trends in Growth of World Population

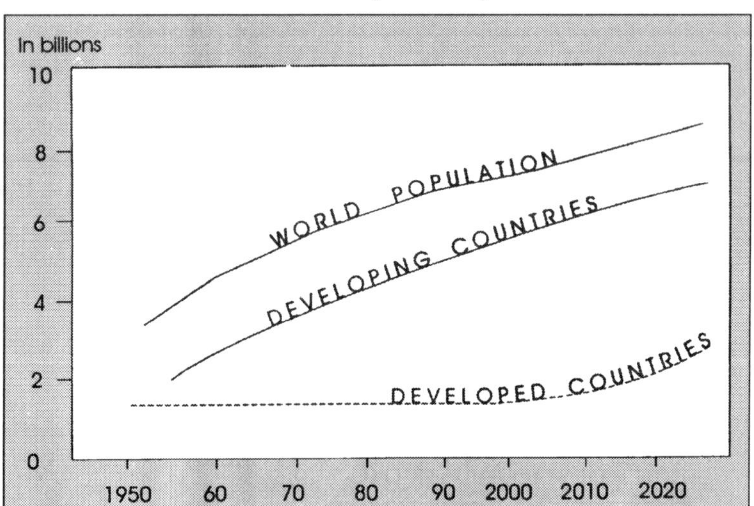

be fed each day or nearly 75 million a year. In spite of the implementation of population control programmes, in the next few decades there will be high growth rate, because of the effects of declining fertility. After that, if current programmes of family limitations succeed, there are reasons to believe that growth rate will slacken and the population will stabilize in the distant future. The growth rate is much higher in developing countries than in developed nations as shown in Figure 11.2.

Table 11.1

Growth of World's Population (1650–1990)

Year	*Population (in millions)*	*Average Annual Growth Rate*	
		(in millions)	*Percentage*
1650	545	1.8	0.29
1750	728	3.5	0.44
1800	906	5.3	0.51
1900	1,171	8.7	0.64
1920	1,810	21.8	1.11
1940	2,246	24.7	1.05
1950	2,493	49.1	1.81
1960	2,984	64.8	2.15
1969	3,788	145.8	2.23
1981	4,492	70.4	1.7
1990	5,293	80.1	1.76
2001	6,137	84.4	1.6
2011	6,973		

Another aspect of the population, which is considered to be effective in terms of its impact over the ecosystem is its spatial patterns both at macro and micro level. The characteristic feature of the distribution pattern of world population is the concentration of population in limited areas. The continent wise population is given in Table 11.2.

Table 11.2

Population of the Continents

Continents	*Population (in million)*
Asia	3913.84
Africa	891.43
Europe	729.34
North America	512.42
South America	371.72
Australia	327.44
Antarctica	Nil

This inequal distribution of population is the main cause of several environmental problems. On the one hand, there are countries like China and India, where population pressure is maximum and resources are limited, and on the other hand, are countries like USA and Canada, where population is limited and more resources are present. In fact, the positive correlation between population and resources is a key component for sustainable development.

The Demographic Transition

Generally, nations go through a four-stage change process over time, known as the 'demographic transition' as depicted in Figure 11.3.

Figure 11.3

Demographic Transition Model

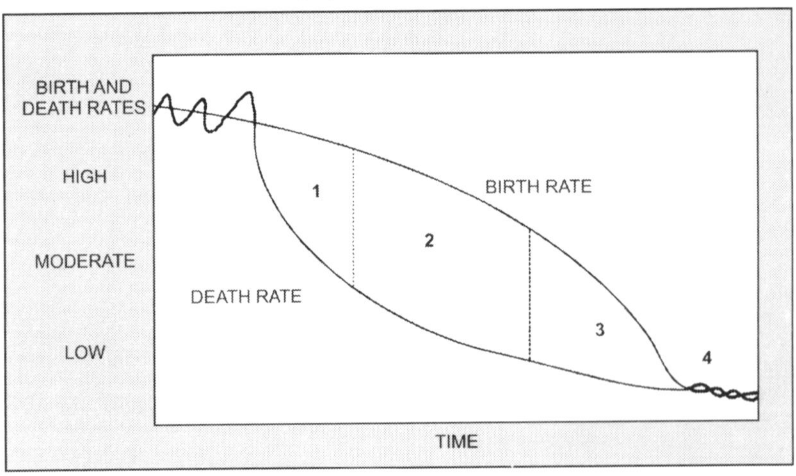

Stage I (no growth): Both birth and death rates are high, resulting in little or no population increase. There are no clear examples of such nations in present times.

Stage II (rapid growth): The birth rate is still high, but the death rate is falling due to advancement in public health. Many developing countries are in this category, the prominent ones are China, India, Bolivia, and Saudi Arabia.

Stage III (slow growth): The death rate is relatively low and the birth rate is declining such as in China and Indonesia.

Stage IV (stagnation): Birth and death rates are both low, giving a very slow rate of increase. Most of the European countries are in this category.

In fact, the rapid growth of population leads to chronic problems like malnutrition, unemployment, disease and of course, degradation of environment.

Ecological Checks on Growth of Population and Carrying Capacity of Environment

Although growth of population is a continuous process, there are certain natural checks which have limiting effects over unlimited growth. The famous demographer, Thomas Robert Malthus, in his writing *Principles of Population* (1798), had pointed out the dire ecological consequences in the continuing growth of population. He claimed that population has a tendency to increase geometrically (in series 1,2,4,8,16 ...), while the food resources for that population, even with improving agricultural methods, increase only arithmetically (in series 10,20,30,40,50, ...). When growth reaches that point, it can be kept in check, according to Malthus, only by 'war, vice and misery'. But, in 1817, Malthus paid more attention to curtailment of population increase through birth control than to the gloomy devices of war, vice and human misery.

Figure 11.4
Environmental Constraints on Growth of Population

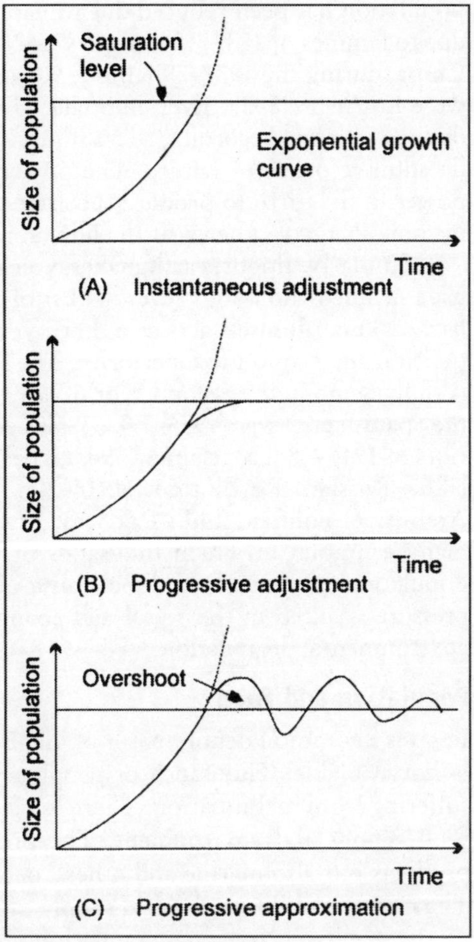

The carrying capacity of environment means the maximum number that the local environment can afford; or the biological capacity to provide food to them. This is represented in Figure 11.4 by three graphs illustrating hypothetical relations.

When population growth approaches a point of ceiling, the three situations are: first, the rate of increase may be unchanged until the ceiling is reached, and then abruptly drop to zero; second, the rate of increase may decline and third, the population may overshoot the ceiling.

The three possible adjustments are instantaneous adjustment, progressive adjustment and progressive approximation. In fact, the carrying capacity of the environment also changes with time due to technological and scientific developments.

The gist of Malthus' thoughts is, that the growth of population, if goes beyond the level of saturation, then nature makes adjustments in its own way. We can say that growth of population will be steady and smooth only when there is an adjustment between environment and growth. Increase in population leads to overexploitation of environment and becomes a cause of eco-crisis, ultimately leading to large scale deaths due to famines, floods, earthquakes, landslides, etc. and also sometimes due to epidemics. The war between nations and other conflicts are also sometimes due to overpopulation.

In the history of mankind, there are several instances when the population has been reduced due to natural calamities. The death toll in India due to famines of 1837, 1863 and 1876–78 was 8, 10 and 50 lakh respectively. In China, during the 1877–79 famine, 90 lakh lives were lost and another 30 lakh were lost in 1928–29. Even in Soviet Union, in 1932–34, 40 lakh people died due to famines. According to Malthus, 'famine seems to be the first, the most dreadful resource of nature. The power of population is so superior to the power in the earth to produce subsistence for man that premature death must in some shape or other visit the human race'.

Similarly, floods, earthquakes, volcanic eruptions, landslides, storms, etc. have claimed thousands (even lakhs) of lives every year. Apart from natural hazards, man himself acts as a destroyer either for his selfish motives or for political and economic superiority. The First World War (1914–18) claimed 72 lakh lives and in the Second World War (1939–45), the number was 73 lakh. In the Spanish civil war (1936–39), 63 lakh people died. While in India, communal riots of 1946–48 have claimed 59 lakh deaths, American civil war (1861–65), 58 lakh, Russian Revolution (1918–20), 57 lakh, etc. The Vietnam war, Arab-Israel conflict, India-Pakistan, India-China war, Gulf war, etc. are so many examples in which thousands of lives have been lost. The growth of population may not be a direct cause of it, but it is certain that population pressure resulted in the social and economic tensions and became a cause of environmental degradation.

Population and Food

Food is a common denominator of life. But food for many people on the earth is not available. Thousands of people are dying of hunger and millions are suffering from malnutrition. There is a close relationship between population, food supply and environment. The production of food, not only of farm products but also marine and others, depends on suitable climatic conditions. The non-availability or limited supply of food is a cause of tension and conflict on the one hand, and overexploitation of resources on the other.

The world food supply, like population, is unevenly distributed. Most of the developed countries either have less population and surplus food supply or are endowed with enough resources to obtain food from outside. On the other hand, in developing countries, food supply is limited and population is more, resulting in hunger and insufficiency of essential nutrients causing infant mortality, impaired development and disease.

The most important question related with population and food supply is the nutritional value of the food measured in terms of caloric intake. The UN World Food Survey has pointed out that 2,550–2,650 calories is the minimum requirement for each individual. A balanced meal means:

1. About 1,200 to 1,800 calories from cereals,
2. Roughly 100 to 200 calories from starch,
3. Around 100 to 150 or 200 calories from fats,

4. Nearly 250 to 300 calories from meat and pulses,
5. About 100 calories from fruits, and
6. Around 300 to 400 calories from milk and milk products.

Figure 11.5

World Pattern of Caloric Status

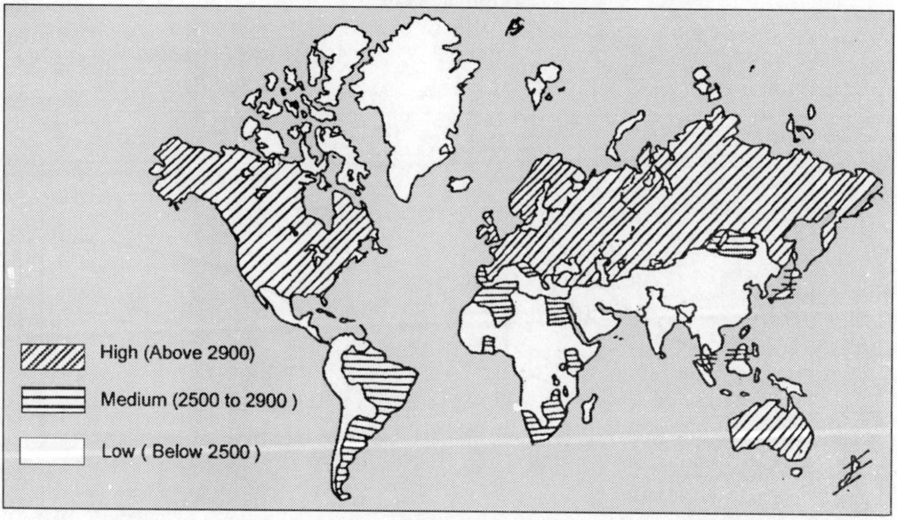

If we look at the pattern of the world calorie intake status (as shown in Figure 11.5), it becomes clear that most of the countries of Asia, Africa and Latin America have very low, i.e. less than 2,500 calories intake per day per person. Some of the countries have a medium level intake (2,500 to 2,900 calories) and only USA, Canada, countries of Europe and Russian Federation, Australia and New Zealand have a high calorie intake food. The pressure to produce food caused by a growing population or by crises such as droughts or floods can result in long-term damage to environment.

John Bongaarts, in March 1994 issue of *Scientific American*, states:

> Feeding a growing world population a diet that improves over time in quality and quantity is technologically feasible. But the economic and environmental costs incurred through bolstering food production may well prove too great for many poor countries. The course of events will depend crucially on their governments' ability to design and enforce effective policies that address the challenges posed by mounting human numbers, rising poverty and environmental degradation. Whatever the outcome, the task ahead will be made more difficult if population growth rates cannot be reduced.

Urban Ecology

Rapid urbanization is a worldwide phenomenon. According to an estimate in 1935, the number of 'million cities' was 56, which has now crossed the number 125. In year 1800, only 1.7 per cent population was living in cities, which became 5.5 per cent in 1900 and 25.8 per cent in 1975 and by the end of this century, it will be about 39.5 per cent or even more. Trends in the rate of urbanization vary from country to country. For example, the United States has more than 70 per cent urban population, while in India, its percentage in 1991 was 25.7 and in 2001 was 27.8. The pattern of urban population growth in developed and developing countries has been shown in Figure 11.6.

Figure 11.6

Actual and Projected Urban Population Growth (1950–2025) | *Actual and Projected Percentage Urban Population*

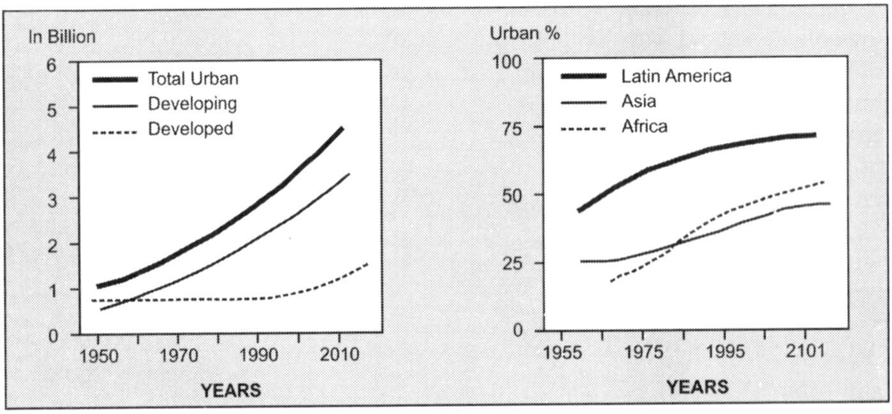

Source: Prospects of World Urbanization (U.N. 1989).

Urbanization continues to increase because the benefits gained from high density crowding are the so-called 'agglomeration economies'. While having a positive impact on income levels, employment, production and market economics, urbanization has brought with it many problems, including a shortage of housing, inadequate water supply, sanitation and waste disposal facilities, congestion, traffic problems, air, water and noise pollution, as well as unsafe social environment. The overall quality of urban environment has deteriorated over the years with the largest cities reaching saturation points and being unable to cope with the increasing pressure on their infrastructures. While cities are considered 'engines of economic growth', given the fact that they contribute nearly 60 per cent of the national income, they are also inherently unsustainable in environmental terms. The situation has become worse in developing countries where cities are growing at a very fast rate and very little efforts have been done to safeguard the urban ecology. In brief, it can be stated that nearly four-fifths of the total investment is cornered for the development of the urban habitat yet it has not been able to meet even the basic needs of its dwellers. About 30 to 60 per cent people are bound to live in clumsy houses and slums. Now urbanization is posing a threat to environment. The process of urbanization, coupled with industrialization, has caused urban atrophy, urban congestion, growth of stinking slums, and impurification of air and water. The urban slum people are denied minimum shelter space and even minimum amount of water, sanitation and electricity. There is a need to develop new kinds of technology to reconstruct industries in order to protect urban ecology. The process of urbanization should be channelized in such a way that the entire urban area will remain pollution free and become an environmental friendly zone.

Effects of Population Growth

According to an estimate, over 8 million people are added to world's population every year. That means more food must be produced, more water for drinking and agricultural purposes, more materials to provide shelter, and so on. Bigger population has put bigger pressure on environment and increased the demand for earth's resources. The collective impact of over 6 billion people living on earth is tremendous. Besides, the geographic distribution of people, the concentration of people in urban areas and the demographic characteristics of regional populations also have an important influence in the effects of human activity on the environment. Population growth has significant environmental consequences that lead to resource shortage for the world population. These include less land for farming and grain production, deforestation and water shortage.

As described earlier, there is great regional disparity in population size, which, in turn, has important implications for the environmental impact of human activity. More than 90 per cent of population growth is among less developed regions. As the residents of less developed countries embrace the living standards of the industrialized world, per capita consumption of energy and other goods will also increase at a faster rate resulting in a much greater impact on environment.

The pressures of population growth have led to force the loss of prime lands either through conversion to urban uses or through degradation by imprudent agricultural methods. In many cases, human activity has degraded land beyond recovery. A number of studies have revealed that increasing population and population density have negative impact on soil quality. A study of 38 sub-Saharan African countries found that the population growth tended to increase cropping frequency, land degradation and breakdown in community land management controls.

Increasing population exerts pressure on forest lands too. Recent studies by the World Resources Institute and the World Conservation Monitoring Centre have revealed that nearly one-half of the forests that covered the earth have now gone. Tropical forests constituting one-half of the world's forest cover, declined by about 450 million hectares between 1960 and 2000.

Forest loss leads to habitat loss. The earth has around 14 million different species, of which only about 1.7 million have been identified and described. Many of these species are threatened with extinction primarily because of habitat loss. An estimated 1 to 11 per cent of the world's species are facing the threat of extinction. Population growth and density contributed to deforestation in many countries.

Population growth also results in consuming metals and minerals in larger quantities. Global consumption of iron and other metals rose more than three times during the last four decades because of growing demand of goods. Population growth has also led to massive energy consumption. Energy consumption increased to 144 per cent in Africa and 185 per cent in Asia during the period 1975–2000. Oil reserves are depleting and new oil reserves have hardly been discovered in recent years.

Water available for human use is also shrinking because of pollution from agriculture and other human activities. Underground deposits of water are being pumped out faster than it is being replenished by surface water. Safe drinking water is still not within the reach of most of the people in many geographical regions. About 1.2 billion people in developing countries lacked safe drinking water supplies and about three billion people lacked access to sanitation services by the end of the last century.

Population density plays a vital role in environmental degradation. Due to migration from rural to urban areas, 90 per cent of future population growth and a major share of economic growth is expected to occur in cities. Urban growth leads to encroachments of farmlands, destroys habitats and threatens sensitive ecosystems and inshore fisheries. It also leads to improper atmosphere that promotes environmental degradation. In brief, population pressure leads to several environmental problems as shown in Figure 11.7.

Figure 11.7

Population Pressure and Environmental Problems

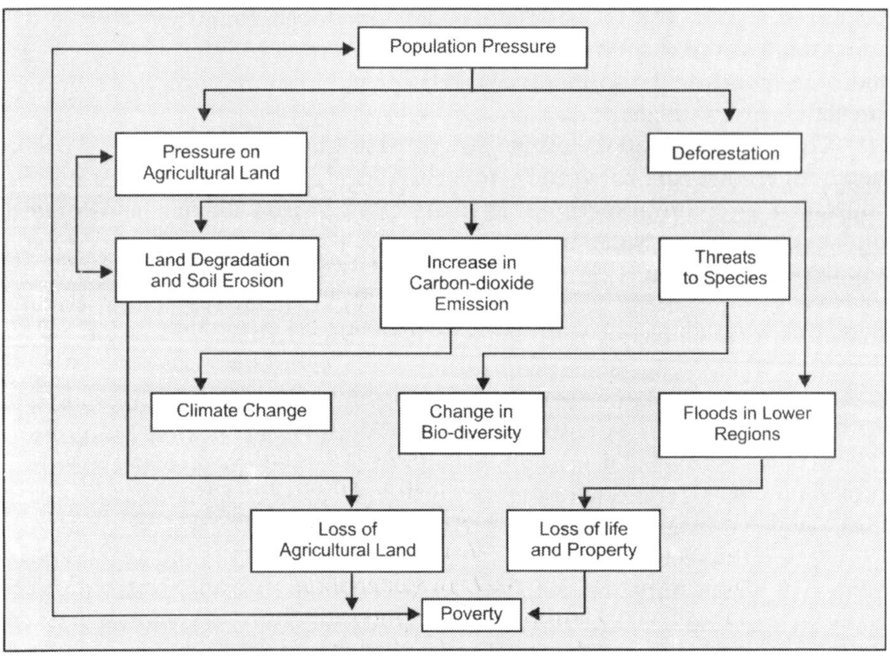

Population and Sustainable Development

Both population and environment cannot be viewed in isolation and their impact on development is also apparent. Actually, population, environment and development should be treated as interacting aspects of the global phenomena. Several aspects of population growth and its impact on environment have already been discussed. Similarly, development is done at the cost of natural resources and is responsible for the degradation of environment. It is also true that development is a continuous process. But, at the same time, it is also necessary to create a balance between growth of population, development and quality of environment.

The United Nations, realising that the environment was under great stress, and air and water pollutions were on the increase and the human environment was in a bad shape, particularly in developing countries, where two-thirds of the population lived in poverty, organized a World Summit on Human Environment in 1972 in Stockholm. This was the beginning of the worldwide awareness towards the various environmental problems, including problems related to population growth and its impact on environment. The environmental problems are often related with poverty and the problems could be overcome by development. The United Nations established a World Commission on Environment and Development in 1983 which published its report under the title 'Our Common Future' in 1987. The report considered the problems of sustainable development, food security, the urban challenge, development and the environment. In this direction U.N. Conference on Environment and Development at Rio de Janeiro was organized in 1992. In this conference, *Agenda 21* was adopted, along with several issues, emphasis was also placed on sustainable development.

Similarly, the World Population Conference, held in 1974 in Bucharest, recommended that the developing countries should reduce the rate of population growth and modernize methods of family planning. After a decade of Bucharest conference, the second World Population Conference was held in Mexico in 1984 with a different outlook. The conference emphasized that the development should be linked with population, physical resources, environment and the standard of living. Family planning may not be fully effective unless empowerment of women is undertaken and achieved with the sustained development of rational utilization of resources. In 1994, another World Population Conference was held in Cairo, in which emphasis was given on women's empowerment, which would help in reducing the rate of growth of population and also in protecting environment.

The limits of growth is a subject of debate. A group of intellectuals of the Club of Rome undertook a project called 'The Predicament of Mankind' and its report prepared in 1972 under the title, 'the Limits to Growth'. The report concluded that the continued growth would lead to exhaustion of natural resources, collapse of industrial and agricultural production owing to shortage of resources and fertilisers and a decline in population caused by famines and epidemics. It was also stressed that reduction of production and pollution would not solve the problem of exponential growth of population. The only alternative according to report is a 'no growth' policy for both industry and population. The population growth should be restricted to 'zero' level and development be done with minimum damage to environment.

Critical Issues

Some of the critical issues related with the problem of population and eco-crisis are as follows:

1. Both the population and environment are complex systems and their interactions are also dynamic, complex and inconstant with time and space. Therefore, planning for each country/region should be based on realistic research.
2. The production systems should be taken into consideration as interaction between population and environment varies with production system, i.e. capitalistic, socialistic or mixed.
3. The high population growth and high fertility among the poor masses, leads to more pressure upon the world's resources and deterioration of the environment.
4. The migration of population from one natural environment to another or from rural to urban areas is responsible for rapid environmental degradation.
5. Along with environmental issues, proper attention should be given to relieving hunger, poverty and malnutrition. Only then, the quality of environment can be improved in developing countries.
6. For growing population a three-fold development policy should be adopted:
   (i) Social welfare (food, water, shelter, health and literacy);
   (ii) Self-reliance (technological, economic and strategic); and
   (iii) Advanced science and technology (frontier level areas).

   This can be achieved through:
   (i) strategy for resource allocation,
   (ii) manpower development,
   (iii) extension and transfer facilities, and
   (iv) technological management.

Indian Scenario

India is the best example of the relationship between population and environment. Right from ancient times, too much importance has been given to nature or environment. The environment not only directs but also determines the growth of population, its distribution and functional structure. There was complete coordination and co-existence between man and nature but with the rapid growth of population, this balance has not only been disturbed but has become a threat to the ecosystem and is now responsible for the degradation of the environment.

India's population has very rapidly grown since the demographic watershed of 1921, when medical treatment, sanitary improvement, etc. started reducing mortality drastically and led to concomitant population explosion. In 1941, there were about 319 million Indians, which rose to 683 million in 1981 and 846 million in 1991. In 2001, India's population was 1028 million which reached 1210 million in 2011 (Table 11.3).

Table 11.3
Growth of Population in India

Census Year	*Population (in millions)*	*Percentage of Decadal Growth Rate*
1901	238.39	—
1911	252.09	+5.75
1921	251.32	-0.31
1931	278.97	+11.00
1941	318.66	+14.22
1951	361.08	+13.31
1961	439.24	+21.51
1971	548.15	+24.80
1981	683.32	+24.66
1991	846.30	+23.85
2001	1,027.015	+21.34
2011	1,210,195,000	+17.64

Over the past few decades, the average rate of population growth has roughly been about 2.5 per cent annually. If current high levels of fertility are drastically reduced to half (which is not possible), even then India's total population was 10,27,015,247 in 2001. It is estimated that in mid–1995, India's population had already crossed 1210 million. To provide food, clothing, shelter, education, employment and health care to this colossal mass of humanity would be a herculean task. The impact of such a growing population on the environment is also a cause of great concern to all of us and all efforts should be taken for the harmonious relationship between population and environment.

The growth of population in India has created several environmental problems, some of which are:

1. overall reduction in agricultural land,
2. lower productivity,
3. deforestation,
4. soil erosion, land slides, and expansion of wasteland,
5. problem of drinking water,
6. malnutrition,
7. gradual change in climatic conditions,
8. increasing frequency of floods and droughts,
9. increasing level of pollution, and
10. adverse effects on human health and quality of life.

In 1901, only 5.81 per cent population was living in cities, which increased to 18.92 per cent in 1951 and 25.63 per cent in 1971. In 1981, the percentage of urban population was 26.91, in 1991 census 25.7, while in 2001 the percentage of urban population was 27.8. The number of million cities in India in 1991 was 23. The process of migration from rural to urban areas has created several socio-economic problems. The overall quality of urban environment has deteriorated over the years and now has reached a point where immediate attention is needed not only by the government but people also. The situation in metropolises like Mumbai, Kolkata, Delhi, Madras, Bangalore, Hyderabad, Ahmedabad, Kanpur, etc. is becoming worse year by year, where even minimum shelter for a large section of population is not available.

Due to uncontrolled urbanization in India, environmental degradation has been occurring very rapidly and causing shortages of housing, worsening water quality, excessive air pollution, noise, dust and heat, and the problems of disposal of solid wastes and hazardous wastes. Urbanization, associated with industrialization, has created several environmental problems such as:

1. growth of slums;
2. industrial pollution (air, water, soil);
3. air and noise pollution by vehicular traffic;
4. problem of solid waste disposal;
5. sub-standard and unhygienic residential conditions;
6. problem of safe drinking water;
7. destruction of natural sites, vegetation, etc.; and
8. problem of waste water, etc.

Environment policies for cities in 1990s outline some basic principles for innovative policies, that are applicable to India. These are:

1. There is a need to intensify cooperation between different levels of government – local, regional and national – and each be made responsible to protect the urban environment.
2. Cooperation between the public and private sector must grow in designing solutions to environmental problems.
3. Minimum environmental standards must be established and enforced.
4. The local community must take part in devising and implementing environmental policies. Local initiatives stimulate greater awareness of local problems and potential.
5. Prices must incorporate environmental and social costs. Local taxes and user fees are most appropriate for demand management and lead to resource conservation.
6. Synergies that can contribute to environmental protection must be identified. In a synergetic relationship, mutually enhancing activities are more productive when they are coordinated.
7. Short-term decisions must be taken within the framework of long-term strategies. Environmental strategy based urban planning can provide a long-term strategy within which short run decisions should be taken.

There is a need to develop the environment friendly city. With systematic efforts by the government and communities including NGOs, each city can be developed as a model environmental city.

Tasks Ahead

We have discussed some measures for the improvement of urban environment. Apart from this, there is a need to plan strategy to limit population as well for a balanced ecosystem. Such measures are urgently required for countries having larger populations and also for all the developing countries. Some measures or action programmes are as follows:

1. Fertility reduction programmes should be intensified to limit population growth.
2. Population welfare programmes like child care, immunization, malnutrition control, safe drinking water, etc. and also programmes to relieve poverty, hunger and to raise socio-economic status of women should be implemented properly.
3. Rural development programmes related to massive employment in rural non-agricultural sectors, to conserve rivers, ponds, forests and grazing land in villages – To enhance the quality of life in rural areas by innovative technology for making cheap and durable housing, sanitation, drainage, water supply systems, etc.
4. Agriculture improvement programmes should be started along with horticulture, fishing, dairy and poultry.
5. Knowledge of land-soil water management be given to villagers so that they can adopt soil conservation measures and can use available water resources.
6. The use of alternative sources of power such as bio-gas as well as alternative technology for bio-fertilizers be popularized so that quality of environment can be improved.
7. Widespread social forestry and agro-forestry programmes should be implemented.
8. Conservation of forests and massive afforestation programmes should be part and parcel of a national programme.
9. For urban areas, apart from suggested measures, new urban land use policy, new mode of urban transport and rapid transit systems and urban slum development programmes be formulated for the improvement of quality of life in cities.
10. Strict pollution control measures should be adopted without any political interference.

The above suggested steps are preliminary and achievable, if there is political will, economic discipline, distributive justice and true commitment to ensure planning for balanced ecosystem.

Note

1. Odum, Eugene P. (1971), *Fundamentals of Ecology*, Saunders, Philadelphia, p. 162.

12

Conservation of Natural Resources

The modern economy is a resource-based economy and all the development of a country or a region is based on resources, especially on natural resources. Nature has given us abundant resources in the form of water, air, heat, natural vegetation, soil, wild animals, metals, fossil fuels, etc. and man by his technical skill and knowledge is using all these resources in some way or the other right from the dawn of civilization. 'Resource' is a wide term and its dictionary definitions are:

— that upon which one relies for aid, support, or supply;
— means to attain given ends; and
— the capacity to take advantage of opportunities or to extricate oneself from difficulties.

According to Zimmermann, 'the word "resource" does not refer to a thing or a substance but to a function which a thing or a substance may perform or to an operation in which it may take part'.[1] In fact, resource is an abstraction reflecting human appraisal and relating to a function or operation. In the process of the utilization of natural resources, man often misuses its availability, thus affecting the natural ecosystem. Earlier, this aspect was not considered properly by man, which resulted in the overexploitation of natural resources, but soon he realized that these resources are not unlimited and unless properly used, the day is not far off when he will not only be deprived of some of the natural resources but will have to face ecological consequences. This apprehension has given the concept of 'conservation'. Prior to a detailed discussion on conservation, it is necessary to know the types of resources.

Natural Resources: Classification

Any element of our natural environment such as soil, water, forest, wildlife, minerals, etc. that man can utilize to promote his welfare may be identified as a natural resource. Natural resources vary greatly in quantity, mutability and reusability, therefore, can be classified into the following categories:

Inexhaustible

1. *Immutable*: Seemingly incapable of much adverse change through man's activities, such as:
   (i) wind power
   (ii) precipitation
   (iii) tidal power
   (iv) atomic energy
2. *Misusable*: Little danger of complete exhaustion, but when improperly used, their resource quality may be impaired, such as:
   (i) solar power
   (ii) atmosphere
   (iii) ocean
   (iv) water power of flowing streams

Exhaustible

1. *Maintainable:* Those resources in which permanency is dependent upon the method of use by man.
   (i) *Renewable*: The living or dynamic resources whose perpetual harvest is dependent upon proper planning and management of man. These are:
      (a) water in place
      (b) soil fertility
      (c) products of the land – agricultural products, forests, forage land, wild animals
      (d) products of lakes, streams and impoundments – freshwater fish, black bass, lake trout, catfish
      (e) products of oceans – marine fish, marine mammals
      (f) human resources
   (ii) *Non-renewable*: Once gone there is no hope of replacement:
      (b) mineral resources
      (b) fossil fuels
      (c) species of wildlife

In general, the concept of conservation was started for non-renewable resources but for renewable resources also, proper management is necessary, which is a part of conservation.

Concept of Resource Conservation

The word 'conservation' is derived from two Latin words, '*con*' meaning 'together' and '*servare*', meaning to 'keep' or guard. Literally, therefore, conservation means 'to keep together'. The word was coined by Gifford Pinchot shortly after the White House Conference of 1908. Some of the definitions of resource conservation are:

> Resource conservation is the scheduling of resource use so as to provide the greatest yield for the greatest number over the longest time period (P. Haggett).

> Conservation means a sacrifice of the present generation to future generations, (Ely).
>
> The optimum allocation of natural, human and cultural resources in the scheme of national development, whereby maximum economic and social security will be assured (Harold M. Rose).

John F. Kennedy, in a special message to the Congress in 1962, interpreted conservation as 'the wise use of our natural environment, the prevention of waste and despoilment while preserving, improving and renewing the quality and usefulness of all our resources'.

The conservation of natural resources is a concept that deals with the rational use of resources so that a harmony between man's resource requirements and its availability can be established. While renewable resources need a proper resource management system (a part of conservation), the non-renewable resources require a long-term strategy for their proper use. This is not only necessary to assure long-term resource use but also to protect the ecosystem. Nowadays, every country is very particular regarding its natural resources and tries to genuinely adopt conservation policies. During the colonial period of 18th century to mid–20th century, most of the countries of Asia, Africa and Latin America have suffered a lot due to their resource exploitation by colonial powers. Gradually, all these countries got their freedom but either for economic needs or for political adjustment, their natural resources have been exploited unwisely, resulting in not only the loss of several resources but also was responsible for many problems of environmental degradation.

The conservation movement, in its present form, was first started in USA under the guidance of Gifford Pinchot, Head of the US Forest Service from 1898 to 1910, with active support from Franklin D. Roosevelt. The first wave of conservation was actually started when Roosevelt called the White House Conference on Natural Resources in 1908, followed by a number of important events in several subsequent decades, the enactment of several laws which regulated the manner in which the natural resources of the country can be protected. As a result of the conference, a fifty-men National Conservation Commission was formed composed of scientists, statesmen and businessmen. After 1930, in USA, several agencies have been established and several acts have been passed to protect forests, soils, wildlife and other natural resources. This tradition was followed by other US Presidents like John. F. Kennedy, Lyndon B. Johnson, Richard M. Nixon and is still in force to this day. In 1970, Nixon set up the 'National Environmental Protection Agency'.

The conservation wave was not limited to the US only but started in European countries as well as in Russia and soon all the countries of the world realized the need to conserve natural resources. Therefore, apart from national efforts by each country, efforts on global level have also been started under United Nations. Different agencies of the UN are now playing a key role in setting the world on the long road towards conservation of natural resources and protection of the environment. The increase in public support for conservation measures both at international and local levels has given this movement a new life, but still there is a long way to go in this direction.

Conservation of Important Natural Resources

Where geographical conditions permit vegetation to take the form of trees, the forest is one of the major forms of the natural landscape. The forest resources are valuable as an integral part of the ecosystem, from the commercial point of view, and as providers of shelter to wildlife. Today, forests provide the raw materials for over 5,000 products worth about 23 million dollars. They support industry which employs 1.3 million people. In fact, forests are still the natural habitat of several species of plants and animals, as well as of several tribal groups of the world. But the most unfortunate setback came in the form of commercial exploitation, which resulted in mass destruction of forest cover year after year. Originally, over two-fifths of the land area of the earth, exclusive of the polar regions, or about 1,200 million hectares was covered with natural forests. But now more than one-third of this area has been robbed by man of its natural protective cover and has been turned into barren land.

The history of the exploitation of forests is as old as man himself, but during earlier times it was balanced through a natural growth process because at that time forest cutting was done for personal or community use only. But with the expansion of agriculture, forest lands have been cleared. More destruction has been done after industrial revolution and urbanization. During the colonial period, commercial exploitation began and this was the main cause of the depletion of forests. The commercial use of forests nowadays has reached such an extent that it has become a threat to the environment in the form of:

1. increase in temperature,
2. lesser precipitation,
3. increased rate of soil erosion,
4. increase in frequency and volume of floods,
5. loss of soil productivity,
6. extinction of several species,
7. non-availability of several essential forest products, and
8. imbalance in ecosystem.

The harmful effects of deforestation are so much that all over the world, people and authorities have realized that forest resources must be conserved properly in order to protect the ecosystem.

The forest is a national resource and a social asset. It yields a great social profit which lies wholly outside the realm of business. But, at present, most of the forests of the world are so used that experts predict dire calamities in the not-too-distant future and irreparable damage on a catastrophic scale. If properly used and put on a sustained-yield basis, it will be one of the man's greatest resources and for this, conservation of forest is the only alternative.

Methods of Forest Conservation

The following steps should be taken for the conservation of forests:

Regulated and Planned Cutting of Trees

One of the main reasons of deforestation is the commercial felling of trees. According to an estimate, about 1,600 million cubic metres of wood have been used for various purposes in the world. Although trees are considered as perennial resources, when exploited on a very large scale, their revival cannot be possible. Therefore, cutting should be regulated by adopting methods like:

1. clear cutting,
2. selective cutting, and
3. shelter wood cutting.

The clear cutting method is useful for those areas where the same type of trees are available over a large area. In that case, trees of same age group can be cut down in a selected area and then marked for replantation. In selective cutting, only mature trees are selected for cutting. This process is to be followed in rotation. Shelter wood cutting is where first of all useless trees have been cut down followed by medium and best quality timber trees. The time gap between these cuttings is helpful in regrowth of trees. In regulated cutting, only one-tenth of the forest area is selected for use and rotational system is always followed for their protection. The forest can be managed in such away that a timber crop may be harvested indefinitely year after year without being depleted. This technique is called the 'sustained yield' method adopted by many countries of the world.

Control over Forest Fire

Destruction or loss of forest by fire is fairly common, because trees are highly exposed to fire and once started it becomes difficult to control. Sometimes, the fire starts by natural process, i.e. by lightning or by friction between trees during speedy winds, while in most cases it is also by man either intentionally or unintentionally. According to an estimate, during the period from 1940 to 1950, in the US alone, fires consumed an average of 21.5 million acres of timber yearly and as many as 1,175,664 cases of forest fires occurred during 1955 to 1964 period. Throughout the world, forest fire is common and in most cases, they were begun by man. As John D. Guthrie, former fire inspector of US Forest Service, has written: 'To stage a Forest Fire you need only few things – a forest, the right atmospheric conditions, and a spark either from a lightning bolt or a match in the hands of a fool or a knave. The formula is simple – the larger the forest, the drier the air, the bigger the fool, the bigger the fire you will have'.

In order to save forests from fire it is necessary to adopt latest techniques of firefighting. Some of the fire suppression techniques are to develop three metre wide fire lanes around the periphery of the fire, back fires, arrangement of water spray, fire retardant chemicals should be sprayed from back tank and if possible by helicopters. There must be a trained staff of fire fighters to control the fire.

Reforestation and Afforestation

The sustained yield concept dictates that whenever timber is removed, either by block cutting or by selective cutting, the denuded area must be reforested. This may be done by natural or artificial methods. Similarly, any forested land which has been destroyed by fire or mining activities should be reforested. In rugged terrain, aerial seeding is the method of choice.

Besides all this, fresh afforestation programmes should be started. New plantations will not only increase the forest cover but also help in making up the eco-balance. For afforestation, selection of trees should be done according to local geographical conditions and care must be taken during initial growth of the trees.

Check over Forest Clearance for Agricultural and Habitation Purposes

Most of the present-day agricultural land was once forested and then cleared for the use of agriculture. But now it has reached the stage where further clearance will be dangerous for the entire ecosystem. There are tribals in some parts of Asia, Africa and South America, where shifting cultivation is still a part of their system of land procurement. According to an estimate, about 40 million sq km of land is used for this purpose by 200 million tribals of the world. For the conservation of forest, this should be checked and an alternative system should be suggested to them. Similarly, for the development of villages, towns and cities, forest lands have been cleared and this process continues to this day causing loss of forest cover. This also should be checked and green belts around cities be developed.

Protection of Forest

The existing forests should be protected. Apart from commercial cutting, unorganized grazing is also one of the reasons. There are several forest diseases resulting from parasitic fungi, rusts, mistletoes, viruses and nematodes which cause the destruction of trees. The forests should be protected either by use of chemical spray, antibiotics or by development of disease resistant strains of trees.

Proper Utilization of Forest Products and Forests

Generally, forests have been cut for logs and rest of the tree – stump, limbs, branches and foliage, etc. left out in the forest as worthless debris. Further waste occurs at the sawmill. There is a need to use all this waste material. Now several uses have been developed and products like waterproof glues, boards, etc. can be obtained.

Similarly, forests can easily be used or developed as tourist centres. By using them as tourist centres, the country can earn substantial foreign exchange. This practice has been adopted by many countries, both developed and developing. The concepts of 'national park' and 'game sanctuary' have now become popular and every country has developed its unique forest area as a 'national park'. In India alone, there are as many as 20 national parks. This scheme is a good method of forest conservation.

Role of Government in Forest Conservation

Although the government of every country is very particular about conservation of its forest resources and have several rules and laws for the protection of forest, but their effective implementation is necessary. Both national and provincial governments can take some steps in this direction, such as:

1. pass acts for the conservation of forests,
2. survey of the forest resources,
3. categorization of forest areas and proper delimitation of reserved forest areas,
4. find out the areas where reforestation can be done,
5. regulate the commercial use of forest products,
6. protect forest from fire, mining and other natural calamities,
7. develop national parks,
8. encourage forests' developmental activities like social forestry, agro-forestry, etc. and
9. prepare master plans, both for long-term and short-term periods, etc.

Forest Management

Management of forest resources is the key to all conservation efforts. In forest management, the following aspects should be taken into consideration:

1. survey of forest,
2. categorization of forest,
3. economic use of forest,
4. administrative setting for forest management,
5. training programmes for persons engaged in forest conservation activities,
6. use of forest land as tourist centres,
7. social and agro-forestry,
8. development of new techniques for the conservation of forests,
9. research for efficient use and conservation of forest, and
10. policy-decisions and their proper implementation.

In brief, conservation of forest resources can be done by cooperative efforts of the government, non-government organizations and the public through a proper management system.

Soil Conservation

Soil is the uppermost layer of earth's surface which supports all forms of life. It is in the soil that all plants have their roots and from which they absorb life-sustaining moisture and nutrients. Man, in turn, feeds directly upon these plants or upon plant eating animals. Thus, virtually all terrestrial life ultimately derives from the soil. As a nation's soil resources are fertile and abundant, in like measure will that state have vigour and stability. When this resource is exhausted because of mounting demands of a swelling population or long mismanagement, the nation's survival is in jeopardy. Some authorities believe that one of the causes of the decline of the Roman Empire was deterioration of soils. The loss of soil, either by natural process or by mismanagement, is harmful not only for agricultural development but to the whole ecosystem, therefore, its conservation has now become a matter of grave concern.

The main reason for the loss of soil is soil erosion which is of four types as under:

1. normal soil erosion,
2. soil erosion by water,
3. wind soil erosion, and
4. accelerated soil erosion.

Normal soil erosion is a natural and a very slow process, therefore, not harmful, rather it helps in soil formation. Soil erosion by water is quite widespread and a main reason of soil loss and loss of fertility. Water soil erosion is of four types, viz. sheet erosion, rill erosion, gully erosion and riparian erosion. The third agent of soil erosion is the wind. Winds are more effective in arid and semi-arid regions. While the accelerated soil erosion is done by human activities, which accelerate the process of soil erosion by deforestation, grazing, mining, defective agricultural practices, etc. landslide and leaching is also a reason of soil loss. The problem of soil loss now has become a worldwide problem and its conservation is necessary.

Techniques of Soil Conservation

Soil erosion can be controlled by adopting land management practices and also by changing the pattern of some human activities which accelerate soil erosion. The following techniques can help in controlling soil erosion:

(i) *Contour Farming*: Contour farming may be defined as ploughing, seeding, cultivating and harvesting across the slope, rather than with it. Thomas Jefferson, who wrote in 1813 'we now plough horizontally, following the curvature of the hills ... . Every furrow thus acts as a reservoir to ... retain waters ... scarcely an ounce of soil is now carried off.'

1. *Strip Cropping*: Planting crops on contour strips will be an effective erosion deterrent. For effective control the width of the contour strip should vary inversely with the length of the slope. Strip cropping should be combined with crop rotation, so that a strip planted to a soil depleting, erosion-facilitating corn crop one year will be sown to a soil-enriching and protecting strip of legumes the next.
2. *Terracing*: The practice of terracing has been common in ancient China. The flat, step-like bench terraces are now not useful. The modern terrace is an embankment of earth constructed across a slope in such a way as to control water run-off and minimize erosion. To be effective, terrace must check water flow before it attains sufficient velocity to loosen and transport soil.
3. *Gully Reclamation*: Gullies are danger signals that indicate land is eroding rapidly and may become a wasteland as in the case of vast areas along the

rivers Chambal and Yamuna. If a gully is small, it may be ploughed in and then seeded to quick growing crop like barley, maize, jowar, wheat in order to check erosion. In case of severe gullying, small check dams of manure and straw constructed at 5 metre intervals may be effective, because silt will collect behind the dams and gradually fill in the channel. Earthen, stone and even concrete dams may be built at intervals along the gully. Once dams have been constructed and water run-off has been restrained, soil may be stabilized.

4. *Shelter Belts*: These are the 'green belts' of trees which help to break the force of strong winds and thus prevent or cut to a minimum, the blowing away of the loose top-soil. In areas where wind erosion is more, rows of trees may be helpful to check the flow of winds. Apart from this, these trees will also add colour to the landscape and help to control the desert spread. Soil blowing away can also be controlled if local shrubs and small trees are planted in a systematic way. Even useful trees can be planted and harvested after a regular interval of two to three years.

Other measures of soil conservation are:

1. expansion of vegetative cover and protective afforestation,
2. controlled grazing,
3. flood control,
4. prohibition of shifting cultivation,
5. proper land utilization,
6. maintenance of soil fertility,
7. land reforms, reclamation of wasteland,
8. establishment of soil research institute and training of soil scientists, and
9. effective agencies for soil management.

Conservation of Water

Water is an indispensable component of our ecosystem and the essence of our life. On an average, each individual uses 50 litres of water per day in rural areas and 150 litres in urban areas. Apart from drinking purposes, water is used for irrigation as well as in industries. Wherever available, about 12,000 m^3 to 14,000 m^3 water is used annually to irrigate one hectare of land. The main source of water is rainfall occurring through 'hydrological cycle', while other sources include underground water and water from the melting of snow. The oceans are full of water but because of salinity, this water is not usable, so also with several inland seas. In fact, usable water is very limited and there are areas in the world where water is a scarce commodity. The unfortunate situation is that wherever water is available, users never think about its proper use and they not only misuse it but waste a large quantity of it too. Therefore, conservation of water is a necessity for all the countries of the world so that minimum water requirement can be fulfilled. A few steps in this direction are:

Proper Distribution System of Water

The water distribution system should be such that loss of water should be minimum and it should remain free from pollution. According to local/regional conditions, the distribution system should be developed. A few general steps for distribution of water are:

1. As far as possible, water distribution be done through pipes so that loss will be minimum and water will remain free from pollutants.
2. If water storage place is small, it should be cemented and if its size is big, the selection of site be done where non-porous rocks exist.
3. The channels and distributors of canals should be cemented, and in fields also, either pipes be used or cemented channels (*nala*) be constructed.
4. For irrigation, sprinkle and drip system be used.
5. If possible, water tanks should be covered to avoid evaporation and also for cleanliness.

Rational Use of Underground Water

Underground water is also a main source of water specially in arid and semi-arid regions. The overexploitation of underground water often results in the lowering of water table, intrusion of saline water, subsidence of land, etc. The prime need is the proper and limited use of water and also geological and hydrological survey of the region for assessment of water availability.

Purification of Waste Water

The system of the recycling of water be used so that waste water can be used for irrigation purposes. There are several mechanical devices for the waste water treatment; some methods are expensive also. But the government agencies and industrial establishments should establish treatment plants. This will also help in control of water pollution.

Control over Deforestation

The loss of vegetation is also a cause of drought and reduction of rainfall and indirectly a cause of lowering of the water table. In order to maintain humidity in the atmosphere which helps in rainfall and to minimize evaporation rate, forest cover should be maintained.

Protection of Water from Pollution

Apart from availability of water, quality of water is also important, in other words, water should be free from pollution. The nature, form and types of water pollution and its effects have been discussed in detail in Chapter 7. Nowadays, several techniques such as physico-chemical purification methods – hydrolysis, electrolysis, ion-exchange, absorption, chlorination, ozonization, etc. are in use. Similarly, biological purification method is also in use.

The conservation of water can be done only by water management system. Apart from above mentioned measures, water can be obtained from desalination of sea water and even by artificial rain-making. Apart from conservation of water, mutual cooperation in use of water resources is necessary.

Wildlife Conservation

Like forests, wildlife is also a national resource which not only helps in maintaining the ecological balance but is also beneficial from economic, recreational and aesthetic points of view. There was a time when human interference was minimum, the number of wild animals were quite high and there was no problem of their protection or conservation. But, with the expansion of agriculture, settlement, industrial and other developmental activities and mainly due to the greed of man, the number of wild animals gradually became lesser and lesser. With the result that several species of animals have been pronounced extinct, and several others are at the verge of being done so. Deforestation is also one of the main reasons of the loss of wildlife. Mass scale killings of wild animals for their meat, bones, fur, teeth, hair, skin, etc. are going on throughout the world. Therefore, the need for wildlife conservation has now become a necessity.

Population growth, the expansion of agriculture and livestock raising, the building of cities and roads, and pollution are among the many pressures on the natural habitat of wildlife. Along with illegal hunting, habitat reduction and its degradation has threatened the biodiversity of the regions where these are rampant. Preservation of wildlife does not mean a blanket protection to all faunal and floral species; rather it implies a proper, judicious control over the multiplication of plants and animals which interact together to provide a proper environment to man whose very existence is in peril today. Due to the irrational use of natural and biotic resources of the earth in the past, most of the wildlife has been destroyed beyond retrieval. It is our urgent duty to protect the natural splendour of ecosystems and to evolve a system of co-existence with every living creature upon the earth.

Although countries of the world are very particular regarding conservation of wildlife, the number of wild animals is reducing day by day. World Wild Life Fund is the international agency doing commendable work in promoting the protection of wildlife. There are national agencies also engaged in conservation of wildlife. Some steps in the direction of wildlife conservation are as follows:

1. To survey and collect all the information about wildlife, especially, their number and growth.
2. To protect habitat by protecting forests.
3. To delimit the areas of their natural habitat.
4. To protect wildlife from pollution and from natural hazards.
5. To impose complete restriction on hunting and capturing of wildlife.
6. To impose restrictions on export and import of wildlife products and severe punishment to be given to those who indulge in this activity.
7. To develop game sanctuaries for specific wild animals or for general wildlife.
8. To make special arrangements to protect those species whose number is very limited.

9. To develop general awareness at national and international level regarding protection of wildlife.
10. A system of wildlife management be adopted through trained personnel.

India is a good example where several steps have been taken for wildlife conservation. It is a country of varied wildlife, where more than 500 types of wild animals, 2,100 types of birds and about 20,000 types of reptiles and fishes have been found. According to an estimate, in India about 200 species of wild animals and birds have been made extinct and another 2,500 are on the verge of extinction. Some of them are black buck, chinkara, wolf, swamp deer, nilgai, Indian gazelle, antelope, tiger, rhinoceros, gir lion, crocodile, flamingo, pelican, bustard, white crane, grey heron, mountain quail, etc. In India, the government and NGOs are taking keen interest in the protection of wildlife. The Wild Life Protection Act, 1972, has several provisions for the conservation of wildlife. As many as 165 game sanctuaries and 21 national parks have been developed to protect the natural habitat and wild animals. Apart from this, a Wild Life Conservation Week is also celebrated from the 7 October every year. But still, there is a long way to go in this direction.

Conservation of Energy Resources

Almost 90 per cent of the world's current energy supply is based on fossil or mineral sources, oil, coal, gas and uranium. Only 10 per cent is from renewable sources, especially hydropower and biomass, while the share of solar and wind energy is below 01 per cent. The following table indicates the pattern of primary consumption in the world:

Table 12.1

Primary Energy Consumption in the World

Energy Source	*World-wide Consumption*	*Percentage*
Oil	21678.10^6 barrels	34.2
Coal	4765.10^6 tons	30.2
Gas	1923.10^9 cbm	19.1
Nuclear	1670 TWH	5.0
Hydro-power	2050 TWH	6.1
Wood/biomass	1219.10^6 tons	5.3
Wind	3 TWH	1.10^{-2}
Solar	0.1 TWH	1.10^{-3}
Geo-thermal energy	76 TWH	0.1

It is undisputed that this structure of supply involves three problems: if consumption rates do not change, reserves of non-renewable energy will largely run out in one or two generations; the waste products caused by consumption in the water, the earth and in the atmosphere threaten mankind and the environment, and finally the cost of this shortage and the effects on consumption will be too much. The use of energy is increasing day by day and according to an estimate, the oil and gas will be in shortage after 2000 AD and coal reserves will also fall short after 2200 AD. Even today, there are problems in coal mining. The world supply of natural gas is also limited.

Nuclear power, which amounted to over 5 per cent of global energy production accounts for very limited use and after the Chernobyl accident, more technological precautions are needed. At present, there is a need for adopting methods for conservation of energy resources and also to develop and use more and more renewable energy sources.

The conservation methods should be applied for coal, petroleum and natural gas in order to preserve these resources, so that their use may be possible for a longer period and their environmental effects can be minimized. The coal mines have become deeper in most of the countries, therefore, new techniques should be applied. The new methods like gasification may help in deeper mining operations. The coal should be classified and cleared at production centres itself. The use of secondary coal products like coaltar, ammonium sulphate, gas, light oil, benzol, sulphur, etc. should be done. A better transportation and marketing system as well as research in use and mining of coal should be done.

The conservation of petroleum or mineral oil is the need of the times because the availability of petro-reserves is limited and the use of oil is increasing day by day especially in the field of transportation. The following steps may help in conservation of petroleum:

1. reduction of the consumption of petroleum,
2. development of such engines which consume lesser petroleum,
3. intensive survey for new resources,
4. exploitation of oil resources from ocean floors,
5. correct estimation of available resources,
6. new techniques for oil drilling and refining should be developed,
7. to reduce petroleum consumption in certain areas such as for military purposes,
7. to develop alternative source of energy for automobiles, etc.
8. international cooperation in reduction of the consumption of petro-products, and
8. research institutes working in this field should intensify their work.

The energy crisis can be solved to some extent by proper use of renewable sources of energy. The renewable energy sources include solar energy, wind energy, tidal energy, hydropower, ocean energy and geo-thermal energy. Renewable energies are also called the energies of the future, because in principal, they are inexhaustible and neutral in terms of CO_2. The renewable energy sources utilized by man come from three primary sources:

— isotopic disintegration in the earth's interior,
— thermo-nuclear conversion (fusion) in the sun, and
— the movement of planets, combined with mass attraction.

The greatest of the renewable energy sources is solar energy or solar radiation reaching the earth's atmosphere year after year. Every 29 seconds, the solar energy that falls on our planet is equivalent to human energy needs for one day at the 1980 consumption level. In case of solar-thermal utilization, the solar energy is converted to heat through absorption. This can occur through passive systems, i.e. through constructional measures for collecting, storing and distributing solar energy which renounce thermal installations to a considerable degree or through active system, i.e. collectors or absorbers. The advantages of using solar energy are – limitless supply, no air pollution, no water pollution, no harmful wastes, no possibility of a large scale explosion or disaster. There is a need to evolve cheaper technology to harvest solar energy. At least for tropical regions, it is the greatest source of energy.

There is no need to stress the importance of water energy or hydro-power. Water energy is already in use in developed countries, but in developing countries, vast energy potentials exist.

There is enough wind energy available to supply the world's energy needs, and it is technically feasible to build windmills capable of producing electricity. These can be built either as small home-sized units or as large central generators. Besides the conversion of rotation energy into electrical energy, the conversion of rotation energy into mechanical energy also finds application in pumping of water, etc.

Ocean energy refers to a number of different renewable energy sources – tidal energy, wave energy, ocean currents and ocean heat. Today, the significant attempts at harnessing the energy of the ocean have been related to tidal power. In many coastal regions, the flow of the tide naturally funnels through narrow entrances into bays and estuaries, and strong currents can be used to generate power.

Energy derived from the heat of the earth's crust is called geo-thermal energy. In various places on the globe, hot springs, geysers and underground steam are available, which can be used for power generation.

Bio-energy can be produced by organic compounds through chemical and biological processes. The combustion, gasification and liquification helps in production of gas which can be used as a source of energy. The bio-gas plant has now become a common feature in rural India.

Viewed in the long-term perspective, utilization of renewable energy sources is of major significance, in terms of both the conservation of the reserves of fossil fuels and the reduction of pollutant emissions caused by the conversion of energy from coal, oil and gas. The renewable energy sources have good chances for the future energy supply. How large their contribution will be, depends upon the question of how safe, economical and environment friendly each particular 'new' energy is.

Notes

1. Zimmermann, E.W. (1951), *World Resources and Industries*, Harper and Row, London, p. 7.

13

Management of Environment: Approaches and Issues

In the last decade of the present century, environmental degradation has emerged as a major global concern for human survival. A large number of international and global organizations with the collective wisdom of scientists, economists and planners have come up to settle the environmental issues confronting the nations and the physical world. The environmental crisis has convinced the world to use technology and resources to repair the damage already done to our environment and also to patent substitutes for certain harmful chemicals in order to protect and preserve nature and natural resources. The last two decades have witnessed tremendous awareness in the field of environment all over the world. The dictionary of environment is renewed every time with new terminologies like clean technologies, environmental auditing, environmental-friendly products (EFP), etc. Similarly, the concept of environmental management has also emerged and accepted as a tool for sustainable development. In fact, proper use of technology, natural resources and human resources can be done through proper management. Many problems related with environment and its degradation can be solved with the help of its proper management. The developed countries of the world have already developed a system of environmental management at all levels, i.e. national, provincial and local as well as domestic. But, in developing countries, this has to be done both through legislation as well as through people's participation.

Why Management of Environment?

Environment involves everything, living or non-living, that surrounds us, for example air, land, water, people, plants, animals, minerals, automobiles, factories, etc. therefore, an overall system of management is required to integrate our efforts in this direction.

Everyone of us has a right to live in an environment of quality. Therefore, it is essential to understand the function and interaction of physical and biological elements of the environment and apply this knowledge in sound management programmes to conserve the natural resources and culture. Hence, management of renewable resources in industry, agriculture, recreation, health, forestry, fisheries, education or urban development is the primary need of the times. Although man has continuously tried to improve his understanding about his environment, his efforts to manage it have fallen short of preventing the deterioration and depletion of resources due to his continuous indulgence in the exploitation of nature for immediate gains. As such, opportunities to design strategies for the effective management of our environment are unlimited. The complex environmental problems not only require imaginative and innovative specialists and technology but also the people's participation. Environmental management is 'a process of planning, review, assessment, decision-making and the like which is essential in the real life situation of limited resources and changing priorities'. This is not only essential for our present generation but also for future generations. The conservation of resources and maintenance of the quality of environment can be possible only through proper management. This is more true and essential for developing countries, where degradation of environment is going on at a faster rate due to overexploitation of natural resources and adoption of new technology.

The important aspects of environmental management are:

1. to create a pollution free environment,
2. to protect man and animals from pollution,
3. to protect the biodiversity of the world,
4. to establish coordination between government and non-government organizations in protecting the environment,
5. to analyze the impact of developmental plans on environment,
6. to help in formulation of national and regional environmental policies,
7. to monitor the ongoing plans for the quality of environment,
8. to sponsor awareness programmes,
9. to educate people at all levels, i.e. environmental education at different levels,
10. to develop long-term and short-term plans for the conservation of the environment,
11. to examine the efforts made under environment management and its results and also fix responsibility for non-implementation, if any,
12. to encourage research in the various fields of environment, and
13. to suggest guidelines to the government for the improvement of the quality of environment.

Approaches to Environmental Management

Environmental management has now become a priority throughout the world because of the serious consequences that have either been experienced by mankind or are likely to happen in future. There is a growing urgency of the management of land, water, minerals, soil, grasslands, forests, animal life, atmosphere, etc. in other words, the whole environment. Any plan envisaging management of the environment must be based on the fact that each area has certain characteristics and can tolerate a range of physical conditions to a limited extent and every human activity affects the environment favourably or unfavourably. Therefore, two courses are open to us in using our surroundings:

1. We adopt our needs and demands to the capabilities of each area for example some recreations, urban development, waste disposal without overloading the water or the air.
2. We modify an area to change its capabilities. To achieve this, we may take up 'intensive management', controlling the physical features of the environment (e.g. agriculture plantation, social forestry, pond-fish culture, etc). In less intensive management, we modify the environment to favour selected types of living beings present in the area (e.g. wildlife management in sanctuaries, extensive forestry). Regulation of harvest from the available population (e.g. plucking or gathering fruits, firewood in the forest) is also possible.

The best approach to environmental management is an integrated approach in which all the components of the environment have been taken into consideration and its proper management as a whole, is done. While doing so, the following aspects should be taken into consideration:

1. Perception and awareness of environment
   - (i) Source of perception and awareness,
   - (ii) Level of perception, and
   - (iii) Role of perception in environmental planning.
2. Environmental education and training
   - (i) At school, college and university levels,
   - (ii) Environmental education through media, and
   - (iii) Research and training institutes.
3. Resource Management
   - (i) Classification of natural resources,
   - (ii) Survey and evaluation of eco-resources,
   - (iii) Conservation of energy, mineral, forest, soil and water resources, and
   - (iv) Proper utilization of human resources.
4. Environmental Impact Assessment
   - (i) Assessment of present conditions,
   - (ii) Assessment of the impact of industrial and technological developments,
   - (iii) Assessment of the efforts done for eco-balance, and
   - (iv) Development of eco-friendly techniques.
5. Control over environmental degradation and pollution
   - (i) Purification of degraded environment,
   - (ii) Pollution control,
   - (iii) Monitoring, and
   - (iv) Forecasting of natural hazards and to minimize the losses.

There are two approaches of managing the environment, viz. (i) preservative approach, and (ii) conservative approach. According to the first approach, man should not disturb the natural system and should adjust according to it. But this is not possible because for all types of developmental activities, he will have to use nature and its components, resulting in eco-imbalance of varied intensity. The conservative approach is that there should not be overexploitation of nature and conservation of natural resources is essential for sustainable development. In fact, proper utilization and conservation of resources is the prime objective of environmental management. In the World Conservation Strategy (1980), three main objectives stated were: the maintenance of essential ecological processes, the preservation of genetic diversity, and the sustainable utilization of natural resources.

In recent years, several approaches have been developed for the management of environment. Regional planners often adopt a human ecology approach, while other planners suggest a system analysis or an ecosystem approach. Doxiadis (1977) has developed a science of planning settlement in balance with nature and termed it as *ekistics*. The various approaches developed for the environmental management are as follows:

- *Ad hoc approach*: developed in reaction to a specific situation.
- *Problem-solving approach*: for identification of problems and needs and implement solutions.
- *Systems approach*: such as ecosystem, agro-ecosystem, etc.
- *Regional approach*: based on ecological zones such as watershed, river basin, coastal zone, command area development, island, etc.
- *Specialist discipline approach*: often adopted by professionals for air, water and land management, urban management, tourism management, and environmental health.

Figure 13.1

Environmental Management System Approach

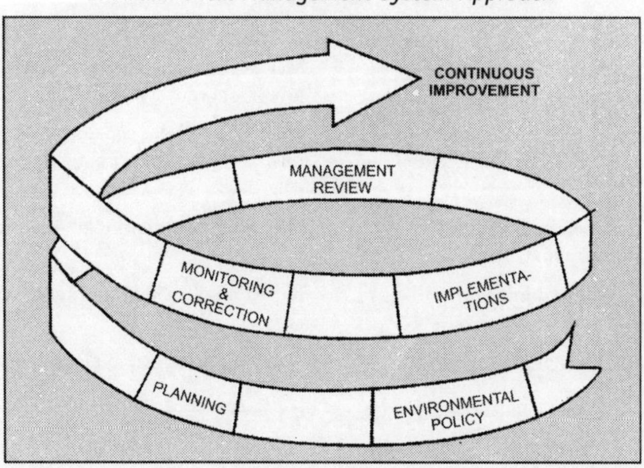

- *Voluntary sector approach*: encouraged and supported by NGOs.
- *Commercial approach*: environmental management for business.
- *Human ecology approach*: study of relations between humans or society and nature.
- *Political ecology approach*: to develop policies and laws.

Environmental management systems (EMS) have been developed as an integrated and proactive approach to environmental issues. They are means for helping industry or other bodies and are designed to ensure that an environmental policy and environmental objectives are adopted and followed. Figure 13.1 illustrates a basic EMS approach developed by Hunt and Johnson.

Thus, the system approach helps environmental management to:

— develop a proactive environment approach;
— ensure a balanced view across all functions;
— enable effective, directed environmental goal setting; and
— make the environmental auditing process effective.

The Ecosystem Approach

The ecosystem concept became a widely used conceptual tool for research after 1945. Now-a-days, ecologists often adopt an ecosystem approach when seeking to understand and monitor a given situation. The ecosystem concept allows the environmental manager to look at portions of complex nature as an integrated system. The concept may be applied to cities or agriculture (urban ecosystems and agro-ecosystems respectively). An ecosystem approach allows a holistic view of how components work together, in other words, it can incorporate human dimensions into biosphere functioning. The ecosystem approach helps in defining the temporal and spatial scale of management, thus, requires a multidisciplinary approach in order to deal with the complexities of ecosystem function and usage. The ecosystem approach has both advantages and disadvantages as listed below:

Advantages and Disadvantages of the Ecosystem Approach

Advantages	*Disadvantages*
Comprehensive, holistic approach for understanding whole system.	May neglect socio-cultural issues such as politics, power and equality.
Draws on theory and methods from different fields to generate models and hypotheses.	Nebulous: a vague, superorganismic theory of poor empirical foundation.
Contributes to understanding limits, complexity, stresses and dynamics.	Non-standard definition of 'ecosystem'.
Encourages preventive thinking by placing people within nature.	Reflection of analytical systems.
Facilities locally appropriate, self-reliant, sustainable action.	Narrow spatial focus on local ecosystem.

Cont'd...

...Cont'd

Requires recognition of mutual dependence on all parts of system, e.g. natural/ cultural, person/family.	Duplicates and/or overlaps other disciplines without special contribution of its own.
Results in criteria for management actions.	If ecosystem approaches can apply to everything they may be meaningless.
Facilitates studies that integrate a range of disciplines.	

Source: Based on D.S. Slocombe (1993).

Despite certain disadvantages, the ecosystem approach can best be used in (i) range management, (ii) forest management, and (iii) conservation management.

Human Ecology Approach

Human ecology is the study of relations between humans or society and nature, through a multidisciplinary approach. Similar to that of ecosystem approach, the main emphasis is on social relationship with environment which is a primary aspect of any management for planning and development. The scale of approach may be local to global, and it supports holistic study. The concept of Social Impact Assessment (SIA) seeks to assess whether a proposed development alters quality of life and sense of well-being and how individuals, groups and communities will adopt to change caused by development. The socio-economic and biophysical aspects of the environment are interconnected, therefore, for environmental management, human ecology approach should be adopted along with political ecology approach.

Political Ecology Approach

Political ecology also studies relationships between society and nature. It holds that radical changes in human habits are required in order to counter environmental degradation and achieve sustainable development. These are likely to be different perceptions of environmental needs and problems between planners, policy makers, ministers, various departments of government, etc. All this can be effectively done by adopting political ecological approach of environment management.

Commercial Approach

Now-a-days, there has been an increasing emphasis on environmental management for business as well as role of business houses in environmental protection. The emphasis is on:

— green corporate environmental management;
— green business ethics;
— eco-auditing;
— impact assessment, hazard and risk assessment;
— green marketing, labelling;
— recycling and waste disposal;

— environmentally sound investment and funding; and
— total quality management.

The commercial approach involves corporate priority, employee education, customer advice, transfer of technology, prior assessment, facilities and operations, research and compliance and reporting.

Integrated Environmental Management

The environmental problems often cross political boundaries between air, water and land; they also involve different disciplines; therefore, there is a need for integrated environmental management. The efforts should be properly coordinated between policy makers, planners, environmental scientists, businessman as well as public. The roots of integrated environmental management lie in integrated area development approaches and comprehensive regional planning. According to Born and Sonzogni (1995), the key elements of integrated environmental management are:

— coordinated control, direction, or influence of all human activities in a defined environmental system (such as a river basin or a watershed) to achieve and balance the broadest possible range of long-term and short-term objectives;
— a process of formulating and implementing a course of action involving human and natural resources in an ecosystem, taking into account the social, political, economic and institutional factors operating within the ecosystem in order to achieve social objectives; and
— an inclusive approach that takes into account the scope and scale of environmental and human issues and their interconnections.

The management of environment should be planned and implemented in a hierarchical order, i.e. international, national, regional and local; and priorities should be fixed according to the nature and state of environment and their associated problems. In fact, an integrated and balanced system is necessary. Similarly, implementation of policies be done by trained personnel, who have a proper perspective and dedication for the maintenance of quality of environment.

Management of Environment: Priority Sectors

Environment, being so vast and covering an exhaustive list of issues at the global and national levels, its analysis from the point of view of management needs several volumes. Several environment problems like air, water, land, noise and radioactive pollution; their management and conservation aspects have been discussed in earlier chapters. But, there are certain priorities for global and national levels, which need special mention. Mary H. Cooper, in a report on 'Setting Environmental Priorities' (1988), lists the following as the most important global environmental issues:

1. greenhouse effect and consequent climatic changes,
2. ozone depletion,
3. acid rain,

4. air pollution,
5. indoor air pollution,
6. municipal and industrial solid wastes,
7. likely escape of toxic/hazardous chemicals,
8. nuclear wastes,
9. arresting pace of deforestation and restoring balanced minimum tree cover areas,
10. habitat and ecological diversity protection, and
11. land, water and natural resource management at national level priority.

Environmental issues which need proper management are:

(i) balanced land use,
(ii) appropriate water use or water management,
(iii) maintenance of an effective and appropriate tree cover,
(iv) environmentally compatible siting and planning of mining and industrial units, human settlements and other developmental projects,
(v) effective abatement and control of pollution entering water bodies and the atmosphere, and
(vi) effective recycling or proper sustainable disposal of major solid wastes particularly toxic wastes, etc.

Developing countries like India are facing a number of environmental problems and all needs proper management. Some of the important aspects which can be dealt very briefly are as follows:

Agriculture

Now-a-days, agriculture has evolved beyond 'crop culture'. It has become an environmental technology with its prime focus on the management of air, water, land and biological resources to produce food and fibre and to preserve the natural resources. Through proper management and careful use we can successfully make permanent use of renewable resources (particularly land) and improve the quality of the resources to increase our production and income.

There is a need to emphasize on the conservation practices for soil, water, effective rotation and cropping practices. Misuse of renewable resources may lead to disasters, droughts and even desertification. Agricultural waste, including manure and crop residue, act as environmental pollutants – recycling of wastes in the farm minimizes the problem. Use of chemicals like pesticides, fertilizers, hormones are becoming popular. Farmers should be educated regarding the dangers of indiscriminate use of such chemicals.

Forestry

The importance of forests and their conservation measures have been discussed in earlier chapters. The forest resources of the world are now in danger due to overexploitation for commercial use. It is high time that management practices should be applied not only for the maintenance of minimum forest cover but also for the further growth of trees. It is concerned with the protection, manipulation and use of vegetation. This also involves fuel and timber production, wild life conservation, recreation, protection of scenic spots, watershed management, etc.

Apart from afforestation, the technique of 'coppicing' be applied, in which trees are cut in such a manner that other trees replace them, thus, a minimum tree cover can be maintained. Similarly, 'selective cutting' can also help in supply of timber wood. 'Social forestry' is a very successful method for the growth of forests. Similarly, more emphasis is given on 'agricultural forestry' in which trees or a belt of trees can be developed along field boundaries. The growth of 'national parks' is another aspect of forest management. Many countries of the world have developed national parks not only to protect biodiversity but also as tourist centres providing regular income. First of all, 'forest survey' be done properly and then proper management system be applied for their conservation and development.

Wildlife

Wildlife management includes both wild animals and their natural habitat or forests. The main objectives of wildlife management include:

1. the maintenance of non-game species;
2. the production and harvest of game species;
3. control of damage to plants, man and animals (domesticated);
4. control of damage by wildlife to crops, forests, livestock, human beings;
5. development and growth of game sanctuaries as tourist spots;
6. restrictions on hunting;
7. to conduct census of wildlife in order to provide them protection; and
8. the protection of wildlife from the adverse effects of pollution and forest fire.

Soils

In order to protect the environment, soil management is also necessary. Soil erosion and loss of fertility are the two main problems. These problems can be solved to a great extent through proper land utilization and by adopting proper farming techniques. Both organic and mechanical methods be applied to control the loss of soil fertility. Some of the organic methods which can be used according to regional conditions are:

1. strip cropping
2. crop rotation
3. use of manure
4. shelter belt, and
5. vegetation cover

Soil erosion can be checked by adopting methods of contouring, terracing and control of channels. Through proper soil management, problems of waste land, waterlogging, salinity, etc. can be solved. The excessive use of chemical fertilizers is also becoming a cause of the loss of soil fertility, therefore, farmers be given a proper guidance regarding the use of chemicals in agriculture.

Management for Pollution Control

Pollution has become the greatest threat to man and environment, throughout the world. Various types of pollution, i.e. air, water, land, noise, radioactive have already been discussed in detail. Pollution control requires a high degree of management, because it directly affects all of us. Throughout the world, there are several acts to control pollution. In India, the comprehensive Environmental Protection Act of 1986 provides ample power for the protection of all components of the environment. The environmental acts, not only in India but in other countries, also need proper enforcement. This can be possible only by adopting following measures:

1. There is a need to have sizeable technically and scientifically trained personnel capable of handling environmental legislation which demands a specialized integrated knowledge.
2. The polluters should also be enlightened and made conscious of their social obligations commensurate to profit motive.
3. Public awareness should have a scientific temper devoid of emotional surcharge.
4. An enlightened judiciary.
5. A political will.

There is a need to control each type of pollution through proper management. The measures taken to control water, air, noise, land and radioactive pollution need separate management system on one hand, and an integrated system on the other. This can be done from local to regional and national to international levels. The main questions related to pollution management are: What is to be controlled? How much is to be controlled? What will be the pace and phases of control? Where should the control be done, i.e. control sites? and How to control? The following steps may be helpful in integrated management:

1. Strengthening of monitoring network to be able to establish periodically the quality status of water bodies and ambient air along with regulating the compliance of emission norms from the polluting source.
2. Generation of base-line data along with collection, compilation and simulation of available data. This will help in the planning process.
3. Environmental information system be established where all types of data related to pollution will be available.
4. Both micro and macro level studies of pollution be conducted for each region.
5. Mass awareness programmes be conducted and full scientific knowledge be given to the people not only regarding adverse effects of pollution but also for its control.

6. Environmental research, including pollution, should be encouraged at all levels, for which priority areas should be identified.
7. Indigenous methods of pollution control be developed and popularized, wherever possible.
8. One of the objectives of the management of pollution is to maximize people's participation in pollution control.

Health

Environment plays a vital role in controlling human health. Society, therefore, should be equally concerned with the environmental health as well as health of the people. Man affects his environment, changes it and in the process transforms himself also. Types and frequency of various diseases are related to environmental conditions. The climate, food, soil, water, air, etc. affect the health, physical stature and behaviour of the people.

Rapid industrialization and urbanization have resulted in degradation of the environment which is the cause of several diseases, mental tensions and other miseries. Therefore, efforts should be made to ensure that the common man gets pure water, pure air and pure food. Re-introduction of some values and amenities should be made. There is a need to fight against adulteration, contamination, malnutrition, filth, pollution and disease. Maintenance of sanitary conditions, both in urban and rural areas, is necessary.

Population and Urbanization

The problems associated with population explosion and its impact on environment are varied and have been discussed in Chapter 9. Many environmental problems in developing countries are mainly due to rapid growth of population which has resulted in the overexploitation of existing natural resources. The cities are growing at a fast rate and the rapid growth of urban population is responsible for the lower quality of life, growth of slums and other associated problems. The need for houses, transportation, factories, schools, playgrounds, offices and other infrastructure is growing day by day. Human activities in these places produce lot of wastes and waste disposal, both municipal and other types, is the greatest problem of the modern cities. The waste products act as pollutants, thus affecting the environment and man adversely. Therefore, proper waste disposal system, including recycling of waste is an immediate need. Proper use of land, water and air has to be the main concern of the city planners.

Management of Power and Mineral Resources

The power and mineral resources are the building blocks of modern industrial and economic development. On one hand, these are the sources of development but on the other hand, generation of power through coal, oil and radioactive minerals are also the causes of environmental degradation. Similarly, mining has become a threat to the environment in many parts of the world. Therefore, both these resources require management. In management of power resources, the following steps may be helpful:

1. Limited use of coal and petroleum.
2. To plan the use of coal and petroleum in such a way that its impact on environment can be checked.
3. The sites of hydro-dams be selected with full consideration of environmental conditions.
4. The establishment of atomic reactor be done at proper sites and strict measures be taken to check leakage of radioactive elements.
5. Urgent need to develop non-traditional alternative sources of power such as solar energy, wind energy, tidal energy, etc.
6. Conservation measures be adopted to control wastage of power.

In mining, first of all scientific methods be used and should be controlled. The measures to control pollution created by mining activities be taken. Along with mining, waste treatment techniques be applied. Mining has had adverse effects on forest, land, soil and also is a cause of air pollution. Special attention should be given for the health of mine workers. The use of synthetic substances such as plastic, plywood, etc. be made popular. There must be a national policy and national plan for the management of power and mineral resources.

Environmental Education

One of the most important aspects of environmental management is environmental education. This mainly concerns itself with the various aspects of environment and ecology. Its primary objective is to teach the components of environment, ecology and ecosystems, man-environment relations, degradation of environment including various forms of pollution and their impact, environment versus development, population, urbanization and environment, environmental management and contemporary issues related to environment.

Although some aspects of environment were also a part of the several physical and social sciences, they do not provide an overall picture of this diversified subject. Therefore, during the last 50 years, it has been realized that environmental education be given not only to the students but to the public also, so that they can play a positive role in environmental protection. Environmental education is necessary because:

1. it will provide a complete knowledge of the components of environment;
2. the knowledge of ecosystems and interaction can be given;
3. it will give complete knowledge of various types of pollution, their causes, effects and measures of control;
4. it will explain the role of man and his activities in degradation and maintenance of environment;
5. it will provide knowledge of the latest research in the field of environment;
6. it will help in environmental awareness;
7. it will pursue the people's participation in environmental issues; and
8. it will prepare research workers and social workers to work for the conservation of the environment.

In brief, environmental education is the need of the hour and priority should be given to it so that in phases the entire society will become aware of environmental issues. This will be the real success of environmental management programme. In 1981, Indian Environmental Institute suggested an outline for environmental education. The main points included in it, are:

1. The pattern of environmental education be such that it should help in formulation of environmental policies and it reflects the relationship between man and environment.
2. Environmental education be included in secondary and university level syllabus, so that awareness among students can be developed.
3. This should include various principles and concepts of environment.
4. At university level, knowledge of regional environment be given and a report should be prepared on any aspect of environment of that region by a group of students.
5. Training to the students be provided so that they can educate the public about environment.
6. An integrated rural development concept of environment be applied.
7. Orientation and reorientation programmes be organized not only for teachers but for doctors, engineers, planners, social workers and politicians to give them complete and up-to-date knowledge of environment.

Throughout the world, environmental education has now become popular. There are several research institutions engaged in the field of research in environment. Similarly, several world organizations, national organizations and non-governmental bodies are engaged in environmental education. But still, there is a long way to go in countries like India, where 80 per cent of the population is living in rural areas and their literacy level is very low. The media can help in environmental education for the masses.

Environmental Legislation

The conservation and protection of environment in general as well as of its specific variables like air, water, forests, sea, wildlife, etc. is controlled through several international and national laws. Thus, implementation of these laws and to develop public awareness about them is an aspect of management. The laws to protect environment are essential, because with the growth of commercialization and also due to man's greed, overexploitation of environment has become a common feature. This can be checked only through legislation.

Initially, the subject of protection of the environment was dealt with in a fragmentary manner under International Law through various international conventions and treaties. In 1963, the Nuclear Weapons Tests Ban Treaty and in 1968 Treaty on the Non-Proliferation of Nuclear Weapons were signed. Treaties on Outer Space (1971), Oil Pollution in Seas (1969), etc. have also been made. A few important international conventions/laws are:

1. International Plant Protection Convention, 1951.

2. The International Convention for the Prevention of the Pollution of the Sea by Oil, 1954.
3. The African Convention on the Conservation of Natural Resources, 1968.
4. The Convention of Wetlands of International Importance, 1971.
5. The Convention of the Protection of the World Cultural and Natural Heritage, 1972.

The year 1972 was a landmark in the field of environment, when United Nations Conference on the Human Environment was held at Stockholm (Sweden) from 5 to 16 June, in which 'Declaration on the Human Environment' was adopted. This may be considered as the beginning of environment movement in the world. UNO has also adopted a programme known as 'UNEP' (United Nations Environmental Programme) and also 'Earthwatch' and 'Global Environment Monitoring System' and declared 5 June as 'World Environment Day'. A new international machinery known as 'Habitat' was established in Nairobi (Kenya) and in 1982, the tenth anniversary of Stockholm Conference was celebrated in Nairobi, followed by the 'Nairobi Declaration'. In fact, a number of conventions, treaties and instruments have been concluded or adopted dealing with the problem of the control of environmental pollution in one way or other, which shows that all over the world environment and its protection has become an issue of concern and all the international bodies have shown their concern over it.

Apart from international laws, every country has enacted laws regarding environment protection/pollution control, etc. In India, there are several acts for environment protection. Even in Indian Constitution, three clauses, viz. 47, 48A and 51A have been introduced for the protection of the environment. There is no need to elaborate all the acts. The four notable acts of great importance are:

1. The Forest Conservation Act, 1980.
2. The Prevention of Air and Water Pollution Act, 1974. 1981.
3. The Air Prevention and Control of Pollution Act, 1981.
4. The Environmental Protection Act, 1986.

It is the duty of the management personnel to have complete knowledge of all the laws and help to develop an awareness among the people about the provisions available under various laws.

Monitoring and Mapping of Environment

Monitoring of environment, pollution and its associated aspects are essential for environment management. Monitoring simply means checking of the environment, pollution, etc. with special reference to the techniques applied to check its degradation. In other words, quality of environment or quantity of pollutants is to be measured in order to find out the effectiveness of the measures taken for its improvement. Thus, monitoring is a continuous process and requires trained personnel. The five categories of monitoring are:

1. *Instrumental system* in which physical characteristics are measured by instruments, such as:
   (i) sound level metre,
   (ii) Geiger counter,
   (iii) thermometer,
   (iv) oxygen metre,
   (v) hydrometer, and
   (vi) clorimeter.
2. *Chemical analysis* through mechanical devices is done by several methods such as:
   (i) atomic absorption spectroscopy,
   (ii) infra-red spectra photometry, and
   (iii) gas-chromatography.
3. *Chemical analysis* in which volumetric, gravimetric and calorimetric analysis is done.
4. *Biological system of monitoring*, which is done by examining the nature of certain plants and other microorganisms.
5. *Hybrid system* is the combination of any two systems mentioned above.

Nowadays monitoring of environment has become popular even in developing countries because it helps in setting standards and limits of tolerance as well as for examining the effectiveness of the measures taken for improvement.

Environmental mapping is an emerging aspect of management. Since a lot of information regarding environmental variables and pollution is available through GIS and satellite, its mapping is very essential for proper planning and control. Through maps, one can get to know the spatial dimensions of the problem. The mapping of environmental aspects has gained impetus and now there is a separate branch of cartography known as 'Environmental Cartography'. The landsat imageries received through satellite provides lot of information regarding forest cover, water resources, soil degradation, land utilization, etc. which provide a base for computer analysis and interpretation, thus help not only in knowing the existing position but also for its planning.

Biosphere Reserves: Environmental Management in Action

A good example of environmental management is the 'biosphere reserve', a concept which is wider than that of a national park. According to Batisse (1985), biosphere reserves, 'combine nature conservation with scientific research, environmental monitoring, training, demonstration and environmental education'. Today, there are 243 biosphere reserves in a total of 65 countries. In USA alone, the number of biosphere reserves is 41 covering a total area of 1,08,111 sq km. The basic idea is that representative areas of ecological importance should be linked together through a network of information. Each reserve would contain at its core a minimally disturbed ecosystem. The preservation of genetic richness in this system would be able to withstand the processes of attrition which have affected protected areas in most parts of the world. The core area is surrounded by buffer zones in which only limited agricultural activities are allowed and research stations are located. Minimum disturbance of the buffer zones enables the protected areas to survive intact, but an accommodation has to be found with the local human population who habitually use the reserves.

There is a need for effective management 'on the ground', not simply on paper. Goodland (1988) has observed that mere 'declaration of intent to protect wildlands or wildlife does not ensure effective management unless specific supporting measures are implemented'.

In developing countries, the pressure on biosphere reserves from 'the poverty of the poor and the greed of the rich' (Goodland, 1988: 12) is much more than in developed countries.

References

Batisse, M. (1985), 'Action Plan for Biosphere Reserves', *Environmental Conservation*, 12(1), p. 17.

Born, S.M. and Sonzogni, W.C. (1995), Integrated Environmental Management: Strengthening the Conceptualization. *Environmental Management*, 19(2), pp. 167–181.

Doxiadis, C.A. (1977), *Ecology and Ekistics*, Elek Books, London.

Goodland, R. and Ledec, G. (1988), *Wildland: Their Protection and Management in Economic Development,* The World Bank, Washington, DC, p. 11.

Slocombe, D.S. (1993), Environmental Planning Ecosystem Science and Ecosystem Approaches for Integrated Environment and Development, *Environmental Management*, 17(3), pp. 289–303.

14

Environmental Impact Assessment and Eco-Auditing

The present world is facing multifarious problems of environmental degradation due to technological and industrial development as well as an explosive growth of population, which has caused enormous strain on environmental resources. Thus, there is an urgent need for maintaining a balance between the capacity of environment and the quantum of sustainable utilization. It is possible only by understanding environment in its entirety and the principles of its scientific management. In the preceding chapter, we have discussed various approaches to environmental management, and also a suitable approach for implementation. There is now a growing concern about 'Environmental Impact Assessment' (EIA) all over the world, specially after the US National Environmental Policy Act of 1969 in which it was stated that all federal agencies should conduct an EIA on all major developments before approval would be given to developers. The present chapter has been designed to discuss the EIA and also major issues and strategies for proper environmental management.

Environmental Impact Assessment

Nature and Characteristics

EIA is the prior assessment of the future impact of the consequences of any decision on the quality of the total human environment on which man largely depends for his well-being. There is no universal definition of what exactly EIA is, but a simple explanation might be 'an approach which seeks to improve development by a *priori* assessment'. It is difficult to provide a clear definition of an EIA, although its characteristics can be listed as follows:

- It is a systematic evaluation of all significant environmental consequences.
- It is a structured, systematic and comprehensive approach.
- It is a process which forces developers to reconsider proposals.
- It is a process leading to a statement to guide decision-makers.

- It is a process which has the potential to increase developer's accountability to the public.
- It should be subject to independent, objective review of results.
- It should include clear statement of identified impacts as well as possible alternative development options.
- It is the study of a proposed development project on the environment.
- It considers both environmental and economic, costs and benefits.
- It is a predictive mechanism and estimates the changes in the quality of the current environment as a result of the proposed action.

The growth of EIA is associated with the legislation passed in the United States on 1 January 1970, arising out of the National Environment Policy Act (NEPA) of 1969, requiring EIA for major projects. The complete report of the EIA is called an Environmental Impact Statement (EIS), but it is not necessary that all EIAs lead to an EIS. In general, an EIA is an objective evaluation of the environmental problems that may arise from the implementation of a project, and it should provide the public with this information, clearly and completely.

Structure and Stages

The structure of an EIA varies according to the project, location and the authorship. EIA, which deals only with project impacts, includes an accurrent environmental conditions (known as the baseline study) and accounts of project impact as well as available alternatives to the proposed project. EIA differs in both structure and their emphasis. A general design can be constructed by dividing EIA into several sections. These are as follows:

1. Preliminary activities
2. Scoping
3. Baseline study
4. Environmental impact evaluation
5. Mitigation measures
6. Assessment of alternative measures
7. Preparation of final document
8. Decision-making
9. Monitoring of project implementation and impacts

A project can impact a given environment under three circumstances. In the first instance, it may arise from project location. This happens because the project would change the existing attributes of environment in the locality. Every such change would release specific process-bound changes. Second, some impacts may take place during construction of the project. These can produce temporary or permanent changes in the environment. Then, after completion of the given project, the manner of envisaged uses of it may cause impacts upon environment. All such expected impacts are to be stated in the EIA report. The essential steps in an EIA are:

1. Prediction of the anticipated changes in the environment.

2. Determination of the magnitude or scale of the particular change.
3. Application of an important significant factor for the change.

Figure 14.1 represents the stages involved in an EIA.

Figure 14.1
Environmental Impact Assessment

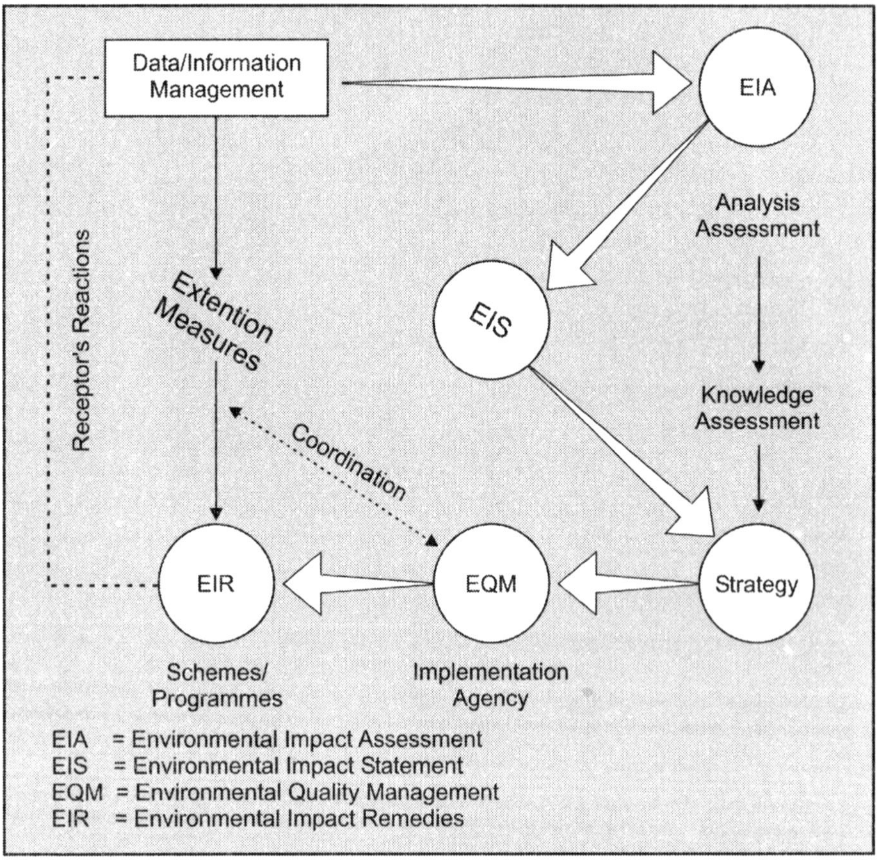

Techniques

Various techniques are used for identifying environmental impacts. These include:

(i) checklists, (ii) matrix, (iii) networks, and (iv) overlay techniques.

Checklists used for identifying project impacts are expected to encompass all possible impacts of this kind. Table 14.1 illustrates a section of a checklist designed for a small water power project.

Earlier, simple checklists which only included a list of the environmental indices that could be affected, were used. But now, detailed descriptions of the impact of each environmental factor are used in the form of questionnaire. In more sophisticated checklists, the relative importance of individual factors can be shown by varying weights being given to different factors along with the total ranking calculated for all the versions of the project.

Table 14.1

Parts of a Checklist for Evaluating a Water Power Project

A. **Physical Impacts**
   - change in sediment transport downstream
   - change in water quality
   - change in river hydrology
   - river erosion
   - slope instability
   - deforestation
   - increased erosion on slopes
   - increased seismic disturbance in the area
   - inundation of significant amounts of natural vegetation
   - disturbance of the local fauna
   - alteration of aquatic life
   - alteration of soil fertility
   - loss of rare or unique features or life forms

B. **Impact on Local Population**
   - inundation of settlements
   - resettlements of people
   - problems of sanitation
   - availability of river water downstream
   - spread of endemic diseases
   - introduction of new diseases
   - psychological stresses due to changes in lifestyle and economy

C. **Impact on Local Economy**
   - disturbances in local economy
   - inundation of existing transport network disruptions in river transport
   - submergence of objects of cultural heritage
   - inundation of historicity sites
   - availability of electricity
   - improvement of agriculture
   - industrial expansion

Matrixes are very suitable for EIA as they link a particular environmental factor to a specific action of the development project. Leopold and his associates, in the late 1960s, designed techniques for evaluating landscape. The Leopold matrix is a large arrangement with 88 environmental factors along the vertical axis and 100 development characteristics along the horizontal one, giving rise to 8,800 cells. Each cell is divided by a diagonal line, and magnitude and importance of the impact are written on a scale 1 to 10. This type of matrix is called an interaction matrix.

Figure 14.2
Use of the Overlay Technique in an EIA

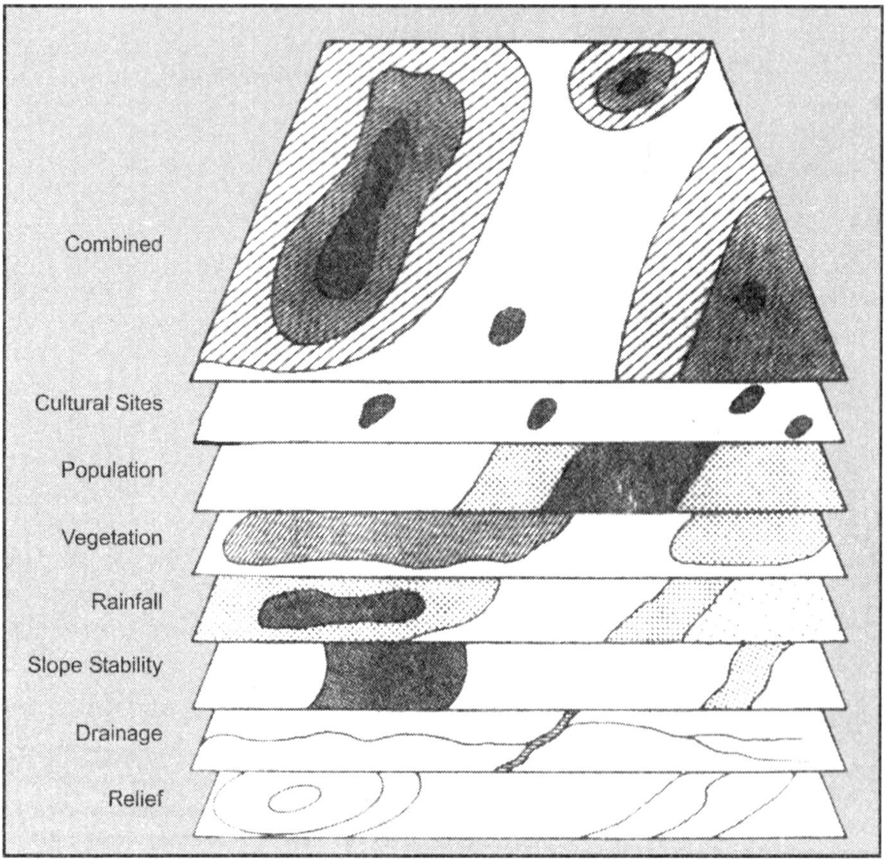

The Leopold matrix has following characteristics:

- It is excellent as a basic scanning tool.
- It allows the application of only a section of the matrix, as required.
- It can be used to indicate both beneficial as well as adverse impacts by writing a plus or minus sign in front of the entries in the cells.
- It can be used repeatedly throughout a project, to estimate the impact at different stages and also for different parts of the project areas.

There are also other types of matrix. At times, a particular development step may affect an environmental factor which in turn alters other factors. Such changes are examined by using a stepped interaction matrix.

Networks are used to show interrelationships among the different components of the environment of the area and also to indicate the flow of energy or impact throughout the environment such as an upland ecosystem or a drainage basin. Different types of network are known as sequence diagrams, directed diagrams or impact trees. Networks can be used to show both temporal and spatial flows of impacts.

Overlay techniques are used in planning before formal EIA process starts. It provide a base for the study of EIA. Individual impacts, such as affecting soil, water settlements and noise conditions, are separately summarized and mapped over the area using choropleths to indicate the intensity of the impact. These maps are transferred on to transparencies which are then laid over one another to produce a composite effect. Summing up the total impact as shown in Figure 14.2 with the advent of Geographic Information System (GIS) and computer technology, this system has now become more useful.

EIA and Planning

EIA is a systematic approach for planning. It can be a policy instrument and, as Graham Smith (1993: 12) argues, it should be treated as 'a bridge that integrates the science of environmental analysis with the policies of resource management'. Figure 14.3 below depicts how EIA fits into planning.

Figure 14.3
EIA and Planning

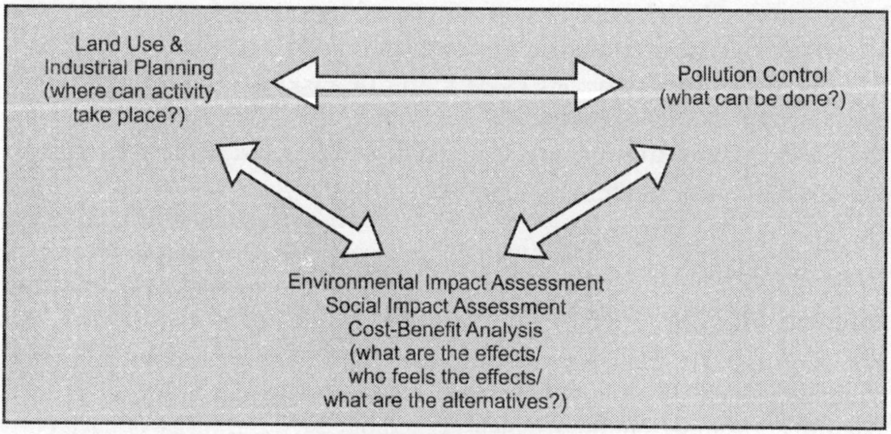

By improving understanding of the relationship between development and environment, EIA can lead to a better environmental management. If EIA is to become an integral part of planning, it must be applied before development decisions are made. While preparing an EIA report, it is useful to classify comprehensively the environmental resources upon which impacts are anticipated. These resources can be land (including soil), water (including surface and ground water), ambient air quality, ambient noise level, floral assemblages, faunal assemblages, human society and element of cultural heritage. For each category of resources to be impacted upon, the EIA report should specify the sources of the anticipated impacts and to suggest the required mitigation measures.

EIA cannot be taken in isolation, rather, it is related with other measures, i.e. state of environment reports, environmental audit, new system of national accounts, etc. This relationship has been shown in Figure 14.4.

Figure 14.4

EIA and its Relationship with other Aspects

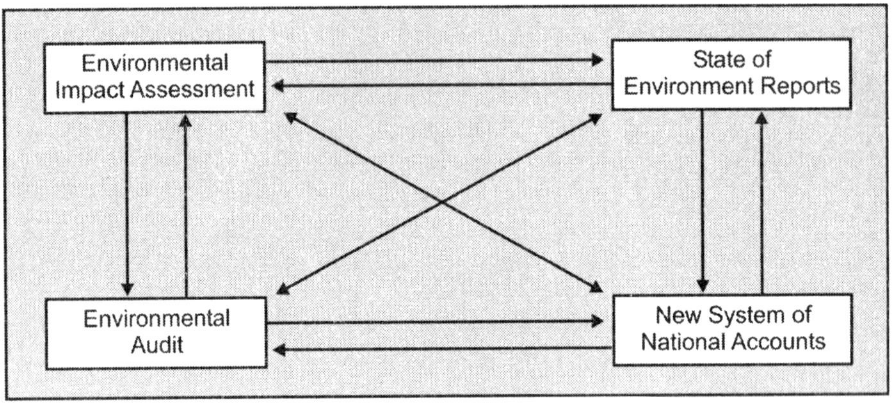

The studies related to EIA are very useful in policy making, planning and administration. It should be related with Social Impact Assessment (SIA). The integrated regional environmental assessment is useful to:

1. provide a broad, integrated perspective of a region about to undergo or undergoing developments;
2. identify cumulative impacts from multiple developments in the region;
3. establish priorities for environmental protection;
4. assess policy options; and
5. identify information gaps and research needs.

Problems with EIA

EIA is an essential component of a project but there are certain misunderstandings, which are often elaborated by those who are not in favour of EIA. The three misunderstandings which exist about EIA are: (i) EIAs have been interpreted as holding up development or at least delaying it. (ii) These are expensive exercises. (iii) This is just a bureaucratic process which does not really influence decision-making.

But none of these are correct. EIAs are designed not to delay development but to make sure that a large-scale development project does not degrade the environment severely and unnecessarily. EIAs are supposed to start very early at the pre-feasibility stage of the project, so that project design and EIA preparation can go simultaneously. Similarly, cost of EIA would range from 0.5 to 1 per cent of the project cost and this percentage comes down for bigger projects due to economies of scale. The cost of EIA is therefore reasonable, considering what it achieves in return. The third objection that it is simply a bureaucratic process is also not true, because EIA itself is not a formality but its recommendations are used in project design as well as by decision-making authorities.

The major problem with EIAs is size. Over time they have evolved into large reports which require the expenditure of considerable financial, technical and human resources. Such resources are not always available in a number of developing countries. In fact, all projects are not required intensive EIA. There are some donor agencies, such as the World Bank or UNDP, which provide financial assistance and some agencies have also designed concise environmental assessments, which are used to determine whether the project requires EIA or not.

Other practical problems that may come up include too many possible alternatives, a wide range of impacts, lack of necessary data, lack of technical expertise, difficulties with quantification of impacts, etc. But, in spite of these problems, EIA now has become a necessity for major projects, which have had their impact on environment.

Eco-Auditing or Environmental Audit

Eco-auditing or environmental audit is a process of extracting information about a company that provides a realistic assessment of how the company affects the environment and also a set of environmental objectives and targets to reduce the effects.

Eco-auditing is a systematic multidisciplinary method used periodically to assess the environmental performance of a project. Eco-auditing evolved as a management tool in the USA in 1980s. It has been promoted in Europe by the International Chamber of Commerce and by some multinational corporations as a means of getting effective environmental management. But, in developing countries, the eco-auditing concept is still a theoretical concept. However, India has modified its Companies Act to include a requirement for eco-audits.

The benefits are as under:

1. it is a means for ensuring the continual improvement in environmental management;
2. it is a good method of monitoring;
3. it can assist efforts for sustainable development;
4. it can help involve the people in environmental management;
5. it may reduce the need for government inspections;
6. it may help identify cost recovery through recycling and sale of by-products, etc.; and
7. it may generate valuable data for regional or national environment reports.

During the course of company's quest for an environmental system, two audits are done. The first is the environmental audit, which is done to know what a company actually does and how it affects the environment. The second audit is conducted to see whether an environmental system works according to its procedures and that objectives and targets are being achieved.

The eco-audit includes the following:

1. internal audit;
2. external audit;
3. aspects and impacts analysis;

4. applicable legislations;
5. environmental objectives and targets;
6. environmental policy statement; and
7. corporate environmental plan.

An *internal audit* is the first step to list the key factors that the internal analysis might encompass and then conduct an internal appraisal. The results of this internal analysis are used to form the basis of aspects and impacts analysis. The factors and areas of the internal audit are:

1. Financial:
   —Liquidity and profit margins
2. Personnel:
   —Management experience
   —Training and education
   —Motivation and attitudes
   —Workforce skills
3. Research and development (R&D) and design
4. Engineering and production
5. Environmental management

An *external audit* encompasses all the forces and factors outside a business that have some impact on it but that cannot influence or control it. In this audit, following points are taken into consideration:

1. Environmental pressure groups
2. Consumers:
   —Economic growth
   —Income levels
   —Interest rates
   —Exchange rates
   —Employment
   —Credit policies
   —Income distribution
   —Savings and debt
   —Taxation
3. Competitors:
   —Age structure of population
   —Attitude towards health and lifestyle
   —Changing work pattern
   —Culture, etc.
4. Operational policy:
   —Automation
   —New methods of travel

—New materials

—Improved communication

Aspects and impacts analysis is another phase of eco-auditing. ISO 14001 guidelines describe an environmental aspect as '... an element of an organization's activities, products or services that can have a beneficial or adverse impact on the environment ...' similarly an environmental impact is '... the change that takes place in the environment as a result of the aspect ...'

This analysis is done with the help of well-prepared matrixes. The result of the analysis indicates:

— those areas of company's operations that may harm the environment;
— the different aspects of the company's operations have been classified into levels of environmental impact, i.e. low, medium or high;
— a scale 1–5 has been used to determine each impact's probability of occurrence; and
— a matrix has been created to calculate the impact scores of each aspect.

A *register of environmental legislations* is a fundamental part of the planning and implementation process. In this, all those regulations are mentioned that apply to the environmental aspects of the company activities.

The setting of *environmental objectives and targets* is the main way in which a company's environmental performance can be improved. The environmental management plan is designed to be dynamic and open-ended list of objectives and goals for a company to achieve in accordance with its stated objectives and targets. The environmental plan should be regularly revised and updated by the environmental manager in consultation with the environmental working group.

Environmental policy statement states the company's commitment to the protection of the environment, prevention of pollution and continuous improvement of company environmental performance. For example, these aims could be to reduce:

— consumption of energy and water,
— generation of waste, and
— emissions.

The *corporate environmental plan* is the culmination of the auditing process. Having identified, listed, analysed and prioritized the operational aspects of the company, the environmental impacts, emergent targets and objectives will form the basis of the corporate environmental plan. Its prime objective should be the achievement of an environmental management standard.

The environmental audit will provide an insight into how a company operates. If an auditing process is already in place, it should be applied in environmental context. An audit is a natural management tool and it will become more effective when environmental audit is added to it. The need of the time is that all the developed and developing countries should adopt the eco-auditing system. For eco-audit, the establishment of standard is a prime task and most of the developed countries and a few developing countries have already established the 'standards' for the environmental protection, pollution management as well as for quality of life. In early 1992, the world's first eco-audit standard was published by British Standards Institute. A number of countries have adopted it which was revised in 1993 and 1994 to make it more compatible with the more recently introduced Eco-Management and Audit Scheme (EMAS). The International Standards Organisation (ISO) has also developed standards broadly compatible with EMAS. The need is that the system of eco-audit should be implemented as a part of environmental management.

References

Batisse, M. (1985), Action Plan for Biosphere Reserves, *Environmental Conservation*, 12(1), p. 17.

Dimento, J.F. (1989), Can Social Science Explain Organizational Non-compliance with Environmental Law, *Social Issues*, 45(1), pp. 109–112.

Goodland, R. and Ledec, G. (1988), *Wildland: Their Protection and Management in Economic Development*, The World Bank, Washington DC, p. 11.

Graham Smith, L.G. (1993), *Impact Assessment and Sustainable Management*, Longman, London, p. 13.

15
Biodiversity

Biodiversity or biological diversity is the foundation of life on earth. It is crucial for the functioning of ecosystems which provide us with products and services without which we couldn't live. Biodiversity is extremely complex, dynamic and varied like other features of the earth. Its innumerable plants, animals and microbes physically and chemically unite the atmosphere, geosphere and hydrosphere into one environmental system which makes it possible for millions of species, including people, to exist. At the same time, no other feature of the earth has been so dramatically influenced by man's activities. By changing biodiversity, we strongly affect human well-being and the well-being of every other living creature. The increasing interest in biodiversity is the result of concerns regarding species extinction, depletion of genetic diversity and disruptions to the atmosphere, water supplies, fisheries, and forests.

Definition

Biodiversity is a combination of two words, 'biological' and 'diversity'. Biologists most often define biodiversity as the 'totality of genes, species and ecosystems of a region'. A few definitions of biodiversity are as follows:

1. 'Biodiversity is the full variety of the biosphere – all the species of living organisms, including plants, animals and microorganisms. Biodiversity also includes the genetic variability within species, and the biotic communities in which they interact'. (Joseph Kerski and Simon Ross, 2005)
2. 'A term describing the variety of species, both flora and/or fauna, contained within an ecosystem'. (John Small and Michael Witherick, 1995)
3. 'Biological diversity – an enormous cornucopia of wild and cultivated species, diverse in form and function, with beauty and usefulness beyond the wildest imagination' (Iltis, 1988 quoted by Andrew Houdie, 1994)
4. United Nations Earth Summit (1992) defined biological diversity as:

'the variability among living organisms from all sources, including, "inter alia", terrestrial, marine and other aquatic ecosystems, and the ecological complexes of which they are part: this includes diversity within species, between species and of ecosystems'.

In brief, biodiversity is that part of nature which includes the differences in genes among individuals of a species; the variety and richness of all the plant and animal species at different scales in space – locally, in a region, in the country and the world; and the types of ecosystems, both terrestrial and aquatic, within a defined area.

The term 'biological diversity' was first used by wildlife scientist and conservationalist Ramond F. Dasmann in the 1968 lay book, *A Different Kind of Country*. The term was widely adopted only after more than a decade, when in the 1980s it came into common usage in science and environmental policy. Thomas Lovejoy, in the foreword of the book *Conservation Biology*, introduced the term to scientific community. Until then the term 'natural diversity' was common, introduced by The Science Division of *The Nature Conservancy* in an important 1975 study, 'The Preservation of Natural Diversity'. The term's contracted from biodiversity may have been coined by W.G. Rosen in 1985, while planning National Forum on Biological Diversity. It first appeared in publication in 1988 when socio biologist E.O. Wilson used it as the title of the proceedings of that forum. Since this period, the term has achieved widespread use among biologists, environmentalists, political leaders, etc.

Types or Levels of Biodiversity

Biodiversity is usually described at three levels:

1. Genetic diversity
2. Species diversity
3. Ecosystem diversity

In 2003 Anthony Campbell at Cardiff University, UK and the Darwin Centre Pembrokeshire, defined a fourth level: *Molecular diversity*.

Genetic Biodiversity

Each member of any animal or plant species differs widely from other individuals in its genetic makeup, owing to the large number of combinations possible in the genes that give every individual specific its characteristics. Thus, for example, each human being is very different from all others. This genetic variability is essential for a healthy breeding population of a species. If the number of breeding individuals is reduced, the dissimilarity of genetic makeup is reduced and in-breeding occurs. This leads to genetic anomalies and, eventually, to the extinction of that particular species. The diversity in wild species forms the 'gene pool' from which our crops and domestic animals have been developed over thousands of years. Today, the variety of nature's bounty is being further exploited by using wild relatives to create new varieties of more productive, disease-resistant crops and to breed superior domestic animals. Modern biotechnology also manipulates genes to develop better types of medicines and variety of industrial products.

Species Biodiversity

The number of species of plants and animals that are present in a region constitutes its species diversity. This diversity is seen both in natural ecosystems and in agricultural ecosystems. Some areas are richer in species than the other. For example, natural undisturbed tropical forests have much greater species richness than monoculture plantations developed by the Forest Department for timber production. A natural forest ecosystem provides a large number of Non-Timber Forest Products (NTFPs) that local people depend on, such as fruit, fuelwood, fodder, fibre, gum, resin and medicines. Timber plantations do not provide the large variety of goods that are essential for local consumption. In the long-term, the economic sustainable returns from NTFPs is said to be greater than the returns from felling a forest for its timber. Thus, the commercial value of a natural forest, with all its species richness, is much greater than a plantation. Modern intensive agricultural ecosystems have a relatively lower diversity of crops than traditional agro-pastoral farming systems, where multiple crops were planted. At present, conservation scientists have been able to identify and categorize about 1.8 million species on earth. However, this is only a fraction of what really exists. Many new species are being identified, especially in the flowering plants and insects. Areas that are rich in species diversity are called 'hotspots' of diversity; India is among the world's 15 nations that are exceptionally rich in species diversity.

Ecosystem Biodiversity

An ecosystem is a set of life forms (plants, animals, microorganisms) interacting with one another and with non-living elements (soil, air, water, minerals, etc.).

There are a larger variety of different ecosystems on earth, each having their own complement of distinctive interlinked species based on the differences in the habitat. Ecosystem diversity can be described for a specific geographical region, or a political entity such as a country, a state or a taluka. Distinctive ecosystems include landscapes like forests, grasslands, deserts, mountains, etc. as well as aquatic ecosystems like rivers, lakes, and seas. Each region also has man-modified areas such as farmland or grazing pastures.

Two other forms of biodiversity (although included in above mentioned levels) that need special mention are domesticated biodiversity and microorganism diversity.

Domesticated Biodiversity

When we think of biodiversity, we tend to think only of wild plants and animals. But there is also considerable diversity among domesticated plants and animals. Domesticated biodiversity may be the result of manipulation by humans, or of natural adaptations to different conditions over a period of time.

Since the dawn of agriculture, people in different parts of the world have developed different plant and animal varieties to meet certain needs and conditions. These include higher productivity, better taste, resistance to pests or diseases, and the ability to withstand adverse conditions like floods, drought or frost. When humans first started cultivating cereals, they must have chosen those with characteristics that were suitable for easier cultivation and harvesting; for example, varieties of grains that had larger kernels, more rows of kernels on each stalk, or those whose seeds did not disperse as soon as they were ripe. This selection was done by storing and sowing seeds of a few plants with the desired characteristics. Over a period of time, these developed into the ancestors of the cereal crops, we know today.

Over time, this kind of adaptation to natural conditions or to human manipulation has led to a large variety in the species of domesticated plants. The number of varieties of rice grown in India is estimated to have been between 50,000 and 60,000. Other crops with high diversity include mango, 1,000 varieties; sorghum, 5,000 varieties and pepper, 500 varieties.

Microorganism Biodiversity

When we think of biodiversity, we rarely think of the most abundant organisms on earth – microorganisms or microbes. Microbes include bacteria, viruses, protozoa, yeast, fungus, etc. and form a vital part of life on earth. Bacteria are the oldest life forms on earth. Microbes play an important role in various biogeochemical cycles. They live in the digestive tracts of most animals (including humans and insects), where they break down the food and facilitate digestion. Microbes that live in the roots of leguminous plants transform atmospheric nitrogen and make it available to the plants (nitrogen fixation). The soil contains thousands of species of microbes which decompose dead organic matter and help maintain soil structure. Just a teaspoonful of soil contains billions of these microscopic living organisms!

The number of microbial cells on earth is estimated to be about 4 to 6×10^{30}, containing nearly half of the total carbon and 90 per cent of the nitrogen and phosphate on this planet. Microorganisms are the only living forms which are present in the most extreme environments, such as salt pans, deep down inside rock and even in extremely cold places.

Evolution and the Genesis of Biodiversity

The origin of life on earth some three-and-a-half billion years ago was obscure. Life was probably initiated as a product of organic reactions in the earth's primordial seas. Alternative possibilities such as life beginning in muddy ooze, or of life having been seeded from outer space have also been suggested. Once life took hold on the planet, it began gradually to diversify. Unicellular unspecialized forms gradually evolved into complex multicellular plants and animals. Evolution is related to the ability of living organisms to adapt to changes in their environment. Thus, the abiotic changes in nature – such as climatic and atmospheric upheavals, repeated glaciations, continental drift and the formation of geographical barriers – segregated different communities of plants and animals and gradually led to the formation of new species over millions of years.

The history of biodiversity during the Phanerozoic (the last 540 million years), starts with rapid growth during the Cambrian explosion – a period during which nearly every phylum of multicellular organisms first appeared. Over the next 400 million years or so, invertebrate diversity showed little overall trend, and vertebrate diversity shows an overall exponential trend. This dramatic rise in diversity was marked by periodic massive losses of diversity classified as mass extinction events. A significant loss occurred when rainforests collapsed. The worst was the Permo-Triassic extinction, 251 million years ago. Vertebrates took 30 million years to recover from this event.

The fossil record suggests that the last few million years featured the greatest biodiversity in history. However, not all scientists support this view, since there is uncertainty as how strongly the fossil record is biased by the greater availability and preservation of recent geologic sections. Some scientists believe that corrected for sampling artifacts, modern biodiversity may not be much different from biodiversity 300 million years ago, whereas others consider the fossil record reasonably reflective of the diversification of life. Estimates of the present global macroscopic species diversity vary from 2 million to 100 million, with a best estimate of somewhere near 9 million, the vast majority arthropods. Diversity appears to increase continually in the absence of natural selection.

Most species appear to have a life span extending over several million years. Their adaptability to gradual changes in their habitat, and interactions with newly-formed species produced groups of interlinked organisms that continue to evolve together. Food chains, prey-predator relationships, parasitism (complete dependence on another species), and commensalisms (a partnership beneficial to both species), are some important examples. The behavioral patterns of the different species comprising a community of species link them to each other through their breeding biology, feeding patterns, migrations, etc. As ancient species became extinct due to geological upheavals, they left behind empty 'niches' in the habitat that stimulated existing species to fill them through the formation of new species. The earth's ancient history has seen periods of mega-extinctions, which have been followed by periods of formation of new species. Though these repeatedly led to a drastic reduction in the number of species, the diversity of life recuperated each time by gradually increasing the number of species existing on earth. This, however, took million years, as evolution is a very slow process. Thus, when man appeared on scene about 2 million years ago, the Earth was richer in species than ever before. During the recent past, however, extinctions due to the activities of modern man have begun to take place so rapidly that nature has had no time to evolve new species, earth is losing species more rapidly than ever before.

Importance of Biodiversity

Biodiversity provides a variety of environmental services from its species and ecosystems that are essential at the global, regional and local levels. The production of oxygen, reduction of carbon dioxide, maintaining the water cycle, and protecting soil are some important services. The world now acknowledges that the loss of biodiversity contributes to global climatic changes. Forests are the main mechanism for the conversion of carbon dioxide into carbon and oxygen. The loss of forest cover, coupled with the increasing release of carbon dioxide and other gases through industrialization, contribute to the 'greenhouse effect'. Global warming is melting ice caps, resulting in a rise in the sea-level, which will eventually submerge the low-lying areas in the world. It is also causing major atmospheric changes, leading to increased temperatures, serious droughts in some areas and unexpected floods in other areas.

Biological diversity is also essential for preserving ecological processes, such as fixing and recycling of nutrients, soil formation, circulation and cleansing of air and water, global life support (plants absorb CO_2, give out O_2), maintaining the water balance within ecosystems, watershed protection, maintaining stream and river flows throughout the year, erosion control and local flood reduction.

Food, clothing, housing, energy, medicines, are all resources that are directly or indirectly linked to the biological variety present in the biosphere. This is most obvious in the case of tribal communities who directly gather resources from the forest, or fisherfolk who catch fish in marine or freshwater ecosystems. For others, such as agricultural communities, biodiversity is used to grow their crops to suit the environment. Urban communities generally use the greatest amount of goods and services, which are all indirectly drawn from natural ecosystems.

Broadly speaking, biodiversity is having immense importance in man's life and in socio-economic development. The following facts illustrate the importance of biodiversity in various fields:

Consumptive Importance

The biodiversity contained in the ecosystem provides forest dwellers with all their daily needs, food, building material, fodder, medicines and a variety of other products. They know the qualities and different uses of wood from different species of trees, and collect a large number of local fruits, roots and plant material that they use as food, construction material or medicines. Fisherfolk are completely dependant on fishes and know where and how to catch fish and other edible aquatic animals and plants. Perhaps the most important value of biodiversity, particularly in a country like India, is that it meets the basic survival needs of a vast number of people. Even today, a large number of traditional communities depend, wholly or partially, on the surrounding natural resources for their daily needs of food, shelter, clothing, household goods, medicines, etc.

Productive Use Value

This category comprises of marketable goods. The biotechnologist uses bio-rich areas to 'prospect' and search for potential genetic properties in plants or animals that can be used to develop better varieties of crops for use in farming and plantation programs or to develop better livestock. To the pharmacists, biological diversity is the raw material from which new drugs can be identified from plant or animal products. To the industrialists, biodiversity is a rich storehouse from which to develop new products. For the agricultural scientists, the biodiversity in the wild relatives of crop plants is the basis for developing better crops.

Genetic diversity enables scientists and farmers to selectively develop better crops and domestic animals through careful breeding programmes. Earlier, this was done by selecting or pollinating crops artificially to get a more productive or disease-resistant strain. Today, it is increasingly being done by genetic engineering–selecting genes from one plant and introducing them into another. New crop varieties (cultivars) are being developed using the genetic material found in wild relatives of crop plants through biotechnology.

The preservation of biodiversity has now become essential for industrial growth and economic development. A variety of industries, like pharmaceuticals, are highly dependant on identifying compounds of great economic value from the wide variety of wild species of plants located in undisturbed natural forests. This is called 'biological prospecting'. Up to 80 per cent of the people in developing countries, depend, for primary health care on traditional medicine, most of which is derived from plants and some from animal and mineral sources. Even several modern drugs have been derived from plant sources. Some such drug; their sources and uses is given in Table 15.1.

Table 15.1

Commonly Used Modern Drugs and their Plant Sources

Drug	*Plant Source*	*Use*
Atropine	Belladonna	Anticholinergic; reduces intestinal pain in diarrhoea
Bromelain	Pineapple	Controls tissue inflammation due to infection
Caffeine	Tea, Coffee	Stimulant of the central nervous system
Camphor	Camphor tree	Rubefacient; increases local blood supply
Cocaine	Coco	Analgesic and local anaesthetic; reduces pain and prevents pain during surgery
Codeine	Opium poppy	Analgesic; reduces pain
Morphine	Opium poppy	Analgesic; controls pain
Colchicine	Autumn crocus	Anticancer agent
Digitoxin	Common foxglove	Cardiac stimulant used in heart diseases
Diosgenin	Wild yams	Source of female contraceptive; prevents pregnancy
L-Dopa	Velvet bean	Controls parkinson's disease, which leads to jerky movements of the hands
Ergotamine	Smut-of-rye or ergot	Control of haemorrhage and migraine headaches
Glaziovine	Ocotea glaziovii	Anti-depressant

Cont'd...

...Cont'd

Gossypol	Cotton	Male contraceptive
Indicine N-oxide	Heliotropium Indicum	Anticancer agent
Menthol	Mint	Rubefacient; increases local blood supply and reduces pain on local application
Monocrotaline	Cotolaria sessiliflora	Anticancer agent
Papain	Papaya	Dissolves excess protein and mucus, during digestion
Penicillin	Penicillium fungi	General antibiotic, kills bacteria and controls infection by various microorganisms
Quinine	Yellow cinchona	Antimalarial
Reserpine	Indian snakeroot	Reduces high blood pressure
Scopolamine	Thorn apple	Sedative
Taxol	Pacific yew	Anticancer (ovarian)
Vinblastine, Vincristine	Rosy periwinkle, (Vinca rosea) (Sadaphali)	Anticancer agent; controls cancer in children

Ethical, Religious and Cultural Importance

Ethical values related to biodiversity conservation are based on the importance of protecting all forms of life. Most religious and secular creeds believe that all forms of life have the right to exist on earth. Man is only a small part of the earth's great family of species; plants and animals have an equal right to live and exist on our planet. We do not know if life, as we know it, exists anywhere else in the universe. Do we have the right to destroy life-forms or do we have a duty to protect them?

Apart from the economic importance of conserving biodiversity, there are several cultural, moral and ethical values, which are associated with the sanctity of all forms of life. Indian civilization has, over several generations, preserved nature through local traditions. This has been an important part of the ancient philosophy of many of our cultures. We have in our country, a large number of sacred groves or 'deorais' preserved by tribal people in several states. These sacred groves around ancient sacred sites and temples act as gene banks for wild plants.

In India, many plants and animals have ritual significance and are associated with religious, spiritual and other cultural uses. Among the auspicious flowers offered in temples are hibiscus, offered to the goddess Kali, and *datura* flowers offered to Shiva. Various plant and animal species are considered sacred on account of their association with different deities. Some animal species are believed to be the *vahans* or vehicles of the deities and are hence venerated. Important among these are the bull of Shiva, the rat of Ganesha and the lion of Durga.

Aesthetic Value

The appreciation of the presence of biodiversity for its inherent value and beauty, as well as for the contribution it makes to our knowledge – our aesthetics, imagination and creativity – are other reasons to preserve it. Quite apart from killing wildlife for food, it is important as a tourist attraction. Biodiversity is a beautiful and wonderful aspect of nature. Sit in a forest and listen to the birds! Watch a spider weave it's complex web! Observe a fish feeding! It is magnificent and fascinating. Each species and ecosystem adds to the richness and beauty of life on earth. Perhaps no artificial medium can match the sheer joy of watching a sunset over an ocean, the sight of a leaping deer, the sound of a singing bird, or the smell of wet earth after the first rains. A natural ecosystem, once destroyed, is impossible to recreate.

Ecological Services

Biodiversity supports many ecosystem services that are often not readily visible. It plays a part in regulating the chemistry of our atmosphere and water supply. Biodiversity is directly involved in water purification, recycling nutrients and providing fertile soils.

Species evolve to fill particular niches (roles) in an ecosystem or habitat. Many species also depend on each other in intricate ways for survival. Destroying one species can lead to further extinctions or changes. Specific life forms present in a particular habitat help to create conditions for other life forms to live in that environment. For example, a single tree provides not only its products, which may have economic value, but it is also a habitat for innumerable living things. In addition, it also plays a vital role in conserving soil and water and helping to keep the air clean. Mangroves and coral reefs, apart from their normal ecosystem service role in preventing erosion, play a critical role in protecting offshore life during hurricanes and storms.

Apart from the above mentioned values of biodiversity, it also has relevance to human health. This issue is closely linked with the issue of climate change, as many of the anticipated health risks of climate change are associated with changes in biodiversity. Some of the health issues influenced by biodiversity include dietary health and nutrition security, infectious disease, medical science and medicinal resources, social and psychological health. Biodiversity is also known to have an important role in reducing disaster risk, and in post-disaster relief and recovery efforts.

Number of Species

How much biodiversity of species is there? Estimates range widely, and the fact is, no one knows. Today, the only two certainties are that 1.75 million species have been described and many more than that, exist. Most people are completely unaware of the great diversity of species within any given taxonomic category. Groups especially rich in species are the flowering plants (2,70,000 species) and the insects (9,50,000 species), but even less diverse groups, such as birds or ferns, are rich with species that are unknown to the most of the people.

Table 15.2 shows the know and estimated species on earth.

Table 15.2
Known and Estimated Species on Earth

Taxonomic Group	*Number of Known Species*	*Estimated Numbers*			*Accuracy*
		High	*Low*	*Working*	
Viruses	4,000	1,00,000	50,000	400,000	very poor
Bacteria	4,000	1,00,000	50,000	1,000,000	Very poor
Fungi	72,000	2,700,000	200,000	1,500,000	Moderate
Protozoa	40,000	200,000	60,000	200,000	Very poor
Algae	40,000	1,000,000	150,000	400,000	Very poor
Plants	2,70,000	500,000	300,000	320,000	Good
Nematodes	25,000	1,000,000	100,000	400,000	Poor
Arthropods					
Crustaceans	40,000	200,000	75,000	150,000	Moderate
Arachnids	75,000	1,000,000	300,000	750,000	Moderate
Insects	950,000	100,000,000	2,000,000	8,000,000	Moderate
Mollusks	70,000	200,000	100,000	200,000	Moderate
Chordates	45,000	55,000	50,000	50,000	Good
Others	115,000	800,000	200,000	250,000	Moderate
Total	1,750,000	111,655,000	3,635,000	13,620,000	Very poor

Source: United Nations Environment Programme, Global Biodiversity Assessment (Cambridge, UK: Cambridge University Press, 1995).

According to the Global Taxonomy Initiative and the European Distributed Institute of Taxonomy, the total number of species for some phyla may be much higher than what was known in 2010:

- 10–30 million insects; (of some 0.9 million we know today)
- 5–10 million bacteria;
- 1.5 million fungi; (of some 0.075 million we know today)
- 1 million mites;
- The number of microbial species is not reliably known, but the *Global Ocean Sampling Expedition* dramatically increased the estimates of genetic diversity by identifying and enormous number of new genes from near-surface *plankton* samples at various marine locations, initially over the 2004–06 period. The findings may eventually cause a significant change in the way science defines *species* and other taxonomic categories.

Taxonomists are aware that their work in finding and describing new species is incomplete. Whatever the number, the planet's biodiversity represents an amazing and diverse storehouse of biological wealth.

Distribution of Biodiversity

Biodiversity is not evenly distributed, rather it varies greatly across the globe as well as within regions. Among other factors, the diversity of all living things (*biota*) depends on temperature, precipitation, altitude, soils, geography and the presence of other species. The study of the spatial distribution of organisms, species, and ecosystems is the science of biogeography.

Diversity consistently measures higher in the tropics and in other localized regions such as the Cape Floristic Region and lower in polar regions Generally, rain forests that have had wet climates for a long time, such as Yasuni National Park in Ecuador, have particularly high biodiversity.

Terrestrial biodiversity is up to 25 times greater than ocean biodiversity. Although, a recent discovered method put the total number of species on earth at 8.7 million of which 2.1 million were estimated to live in the ocean.

Generally, there is an increase in biodiversity from the poles to the tropics. Thus localities at lower latitudes have more species than localities at higher latitudes. This is often referred to as the latitudinal gradient in species diversity. Several ecological mechanisms may contribute to the gradient, but the ultimate factor behind many of them is the greater mean temperature at the equator compared to that of the poles.

Even though terrestrial biodiversity declines from the equator to the poles, some studies claim that this characteristic is unverified in aquatic ecosystems especially in marine ecosystems. The latitudinal distribution of parasites does not follow this rule.

Tropical regions of the world are having the highest biodiversity. Even in tropics, three specific regions are having rich biodiversity. These regions are (i) tropical rainforests (ii) coral reefs and (iii) wetlands. Tropical rainforests are unique and the richest in terms of number of species. Covering just 7 per cent of the earth's surface, they are home to nearly half of all known species. According to the US National Academy of Sciences, a typical patch of rainforest about 10 sq km contains as many as 1,500 species of flowering plants, 750 tree species, 400 bird species, 150 butterfly species, 100 reptile species, 60 amphibious, and an unknown number of insect species. Biologist E.O. Wilson identified 43 species of ants on a single tree in a Peruvian rainforest. Unfortunately, the same tropical forests that hold such high biodiversity are also experiencing the highest rate of deforestation.

Coral reefs are also rich in biodiversity because of the abundance and diversity of life found there. Australian Great Barrier Reef is the world's largest coral reef ecosystem, containing more than 3,000 animal species. The richest, most diverse coral reefs in the world are found in the Indo-West Pacific. These reefs contain more than 2,000 species of fish, 5,000 species of molluscs, 700 species of corals and countless species of crabs, sea urchins, brittle stars, sea cucumbers and worms. Coral reefs provide habitats for many organisms that cannot live elsewhere. The coral reefs of Australia, the Philippines, Indonesia, Malaysia, Japan, Palau, Maldives, Tanzania, Seychelles, Belize and Florida are very rich in biodiversity, but at the same time, worst hit areas by variety of human activities.

Wetlands are transitional zones between land and water and are among the most biologically interesting and productive habitats. There are two types of wetlands: *inland* (fresh water ecosystems include marshes, swamps, riverine wetland and bogs) and *coastal* (either fresh or salt water and mangroves). Fresh water marshes are important habitats for wildlife, and migratory birds, while swamp are dominated by woody vegetation, reptiles, amphibians and fishes. Tidal salt marshes occur along the coastlines of mid-latitudes and dominated by mangroves in subtropics and tropics. Mangrove swamp is a dominant type of coastal wetland. These are forested wetland and cover approximately 16 million hectares world wide, with largest concentration in tropical Asia. Both coastal and inland wetlands are the reservoirs of biodiversity.

The fact is that most of the world's bio-rich nations are in the South, in other words, they are the 'developing' nations. In contrast, the majority of the countries capable of exploiting biodiversity are northern nations, in the economically 'developed' world.

India as a Megadiversity Nation

India is one of the world's 12 megadiversity countries. India contains 8.1 per cent of the world's biodiversity on 2.4 per cent of the earth's surface. It is estimated that 47,000 wild species of plants and over 89,450 wild species of animals occur in our country. India has a tremendous range of ecosystem, species and genetic diversity. The location of the Indian subcontinent at the confluence of three biogeographic realms has led to the presence of elements of African, European, and Chinese and Indo-Malayan characteristics in the flora and fauna. The diversity is also due to the fact that India has almost every major type of habitat and climatic condition – from alpine heights to coasts and plains; from areas of the heaviest rainfall to dry deserts.

India is a bio-rich country having great variety of plants and animals, many of which are not found elsewhere (Table 15.3).

Table 15.3

Bodiversity of India and its World Ranking

	India's World Ranking	*Number of Species in India*
Mammals	8th	350
Birds	8th	1,200
Reptiles	5th	453
Amphibians	15th	182
Angiosperms	15th–20th	14,500

India has 350 different mammals (rated the eighth highest in the world), 1,200 species of birds (eighth in the world), 453 species of reptiles (fifth in the world) and 45,000 plant species, of which most are angiosperms, (fifteenth in the world). These include especially high species diversity of ferns (1,022 species) and orchids (1,082 species). India has 50,000 known species of insects, including 13,000 butterflies and moths. It is estimated that the number of unknown species could be several times higher.

It is estimated that 18 per cent of Indian plants are endemic to the country and found nowhere else in the world. Among the plant species, the flowering plants have a much higher degree of endemism, a third of these are not found elsewhere in the world. Among amphibians found in India, 62 per cent are unique to this country. Among lizards, of the 153 species recorded, 50 per cent are endemic. High endemism has also been recorded for various groups of insects, marine worms, centipedes, mayflies and freshwater sponges.

Apart from high biodiversity in Indian wild plants and animals, there is also a great diversity of cultivated crops and breeds of domestic live stocks.

India is also considered one of the world's eight centres of origin of cultivated plants. It has 51 species of cereals and millets, 104 species of fruits, 27 species of spices and condiments, 55 species of vegetables and pulses, 24 species of fibre crops, 12 species of oilseeds, and various wild strains of tea, coffee, tobacco and sugar cane. India also has significant indigenous livestock diversity, with 27 breeds of cattle, 40 breeds of sheeps and 22 breeds of goats. For example, India's eight breeds of buffaloes represent the entire range of the genetic diversity of buffaloes in the world. Many of these today have died out or are dying out due to our misguided adoption of all 'foreign' things – Jerseys and Holsteins have largely replaced Indian cattle like Brahma bull, high yielding cultivars have eaten away into centuries old landraces of crops.

Decline of Biodiversity and Hot Spots

The extinction of several species or loss of biodiversity is a major problem of the international world. Species are disappearing rapidly and there is growing consensus within the international community that a system should be put in place to slow or halt the process of extinction. All the species are the integral part of ecosystem and extinction of some species diminishes the well-being of the remaining species, including human beings. The concern about biological diversity is the shrinking genetic pool. It is estimated that tropical forests contain at least 50 per cent and perhaps 90 per cent of the world's species. According to an estimate, 20 to 75 species are becoming extinct each day because of deforestation in the tropics.

Species do not persist indefinitely. Indeed, it was the realization that fossils are the remains or traces of organisms no longer to be found alive that led to the development of our present understanding of evolutionary processes. Estimating the number of species existing at present as a proportion of all the species that have ever lived is difficult, but the plants and animals alive today probably constitute around 2–4 per cent of all the species that have lived in the last 600 million years.

Extinctions have not been spread evenly. There is a general background rate at which species go extinct, but there have also been at least five mass extinction in which 65–85 per cent (and at the end of the Permian Period, around 225 million years ago, 95 per cent or more) of all species disappeared. Background extinctions have little or no effect on biodiversity, because their rate is matched or exceeded by that of speciation, in which one species divides into two and thus increases the number of species. Nor do mass extinction produce more than a temporary reduction in the number of species, because historically they have been followed by the rapid adaptive radiation of surviving species into the vacated niches. The extinction of the dinosaurs, for example, was followed by rapid speciation among mammals. Today, the number of land-dwelling species alive is about twice the average of the last 450 million years.

Man has begun to overuse or misuse most of these natural ecosystems. Due to this unsustainable resource use, once productive forests and grasslands have been turned into deserts and, wastelands have increased all over the world. Mangroves have been cleared for fuelwood and prawn farming, which has led to a decrease in the habitat essential for the breeding of marine fish. Wetlands have been drained to increase agricultural land. These change have grave economic implications in the longer term.

As mentioned earlier, there are about 1.8 million species of plants and animals known to science in the world at present. Unfortunately, at the present rate of extinction, about 25 per cent of the world's species will undergo extinction fairly rapidly. This may occur at the rate of 10–20,000 species per year, at a 1,000–1,00,000 faster rate than the expected natural rate. Human actions could well exterminate 25 per cent of the world's species within the next twenty or thirty years. Much of this mega extinction spasm is related to human population growth, industrialization and changes in landuse patterns. A major part of these extinctions will, of course, occur in bio-rich areas such as tropical forests, wetlands, and coral reefs. The loss of wild habitats, due to rapid human population growth and short-term economic development, is the contributor to the rapid global destruction of biodiversity.

India's rich biological diversity is rapidly eroding. At least 10 per cent of its recorded flora, and possibly, a large fraction of its wild fauna, is threatened. Many may be on the verge of extinction. In the last few decades, India has lost at least 50 per cent of its forests, polluted over 70 per cent of its water bodies; built, cultivated or otherwise encroached upon its grasslands; and degraded many coastal areas.

The cheetah and the pink-headed duck are amongst the conspicuous species that have become extinct. More than 150 of the known species of medicinal plants in India have already become extinct due to unsustainable methods of harvesting. Of the species of flora and fauna that remain, over 10 per cent of flowering plants, 21 per cent of mammals and 5 per cent of birds are believed to be under threat. India's domesticated biodiversity is also under threat. Hundreds of crop varieties have disappeared, and even their genes have not been preserved.

The World Conservation Monitoring Centre (WCMC) has evaluated and describe threats to about 60,000 plant and 2,000 animal species in its series of Red Data Books. Threatened species include those which are endangered, vulnerable, and rare in IUCN categories. The great majority of species on these lists of Red Data Books are plants. However, there are also species of fish (343), amphibians (50), reptiles (170), invertebrates (1,355), birds (1,037), and mammals (497). India contains 172 species of animals considered globally threatened by IUCN.

Biodiversity Hot Spots

A biodiversity hotspot is a region with a high level of *endemic* species that is under threat from humans. The term hotspot was introduced in 1998 by Sabina Virk. While hotspots are spread all over the world, the majority are forest areas and most are located in the tropics.

Brazil's Atlantic Forest is considered one such hotspot, containing roughly 20,000 plant species, 1,350 vertebrates, and million of insects, about half of which occur nowhere else. The island of Madagascar, particularly the unique *Madagascar dry deciduous forests* and lowland rainforests, possess a high ratio of *endemism.* Since the island separated from mainland *Africa* 65 million years ago, many species and ecosystems have evolved independently. *Indonesia's* 17,000 islands cover 7,35,355 square miles (19,04,560 km^2) contain 10 per cent of the world's flowering plants, 12 per cent of mammals and 17 per cent of *reptiles, amphibians* and *birds* – along with nearly 240 million people. Many regions of high biodiversity and/or endemism arise from specialized habitats which require unusual adaptations, for example alpine environments in high mountains, or Northern European peat bogs.

Figure 15.1

Biodiversity 'Hot Spots'

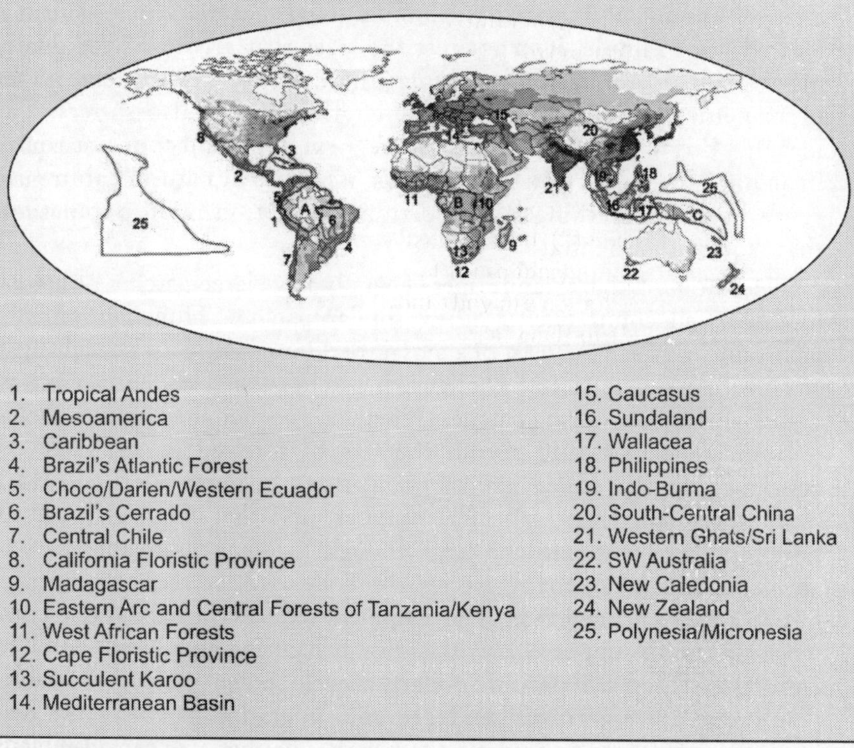

1. Tropical Andes
2. Mesoamerica
3. Caribbean
4. Brazil's Atlantic Forest
5. Choco/Darien/Western Ecuador
6. Brazil's Cerrado
7. Central Chile
8. California Floristic Province
9. Madagascar
10. Eastern Arc and Central Forests of Tanzania/Kenya
11. West African Forests
12. Cape Floristic Province
13. Succulent Karoo
14. Mediterranean Basin
15. Caucasus
16. Sundaland
17. Wallacea
18. Philippines
19. Indo-Burma
20. South-Central China
21. Western Ghats/Sri Lanka
22. SW Australia
23. New Caledonia
24. New Zealand
25. Polynesia/Micronesia

Note: Some 25 regions in which 60 per cent of the Earth's biodiversity has been located on just 1.4 per cent of the globe's land surface.

Source: Conservation International, 1999, NCGIA/CIESIN, 1995.

In the world, 25 regions have been identified as biodiversity 'Hot Spots' as shown in Figure 15.1.

Causes of the Decline or Loss of Biodiversity

The loss of biodiversity throughout the world has become a common phenomena. The known causes of extinctions are the destruction of habitats, introduction of foreign species and hunting, etc. Jared Diamond describes an 'Evil Quartet' of habitat destruction, overkill, introduced species, and secondary extinctions as the reasons for biodiversity loss. While Edward O. Wilson prefers the acronym HIPPO, standing for habitat destruction, invasive species, pollution, human overpopulation, and over-harvesting as main causes. Figure 15.2 depicts the causes of the loss of biodiversity as identified in Global Biodiversity Strategy WRI, IUCN, UNEP.

Habitat Destruction

One of the main causes of the loss of biodiversity is the physical alternation or destruction of habitats through the processes of conversion, fragmentation, and simplification. Habitat destruction has already been responsible for 36 per cent of the known extinctions.

1. *Conversion:* Natural areas are converted to farms, housing subdivisions, shopping malls, marinas, and industrial centers. When a forest is cleared, for example, it is not just the trees that are destroyed, every other plant or animal that occupies the destroyed ecosystem, either permanently or temporarily (e.g. migrating birds), also suffers. The idea that this wildlife simply will move 'next door' and continue to live in an undisturbed section is erroneous. Any loss of natural habitat can result in only one thing: a proportional reduction in all populations that require that habitat.
2. *Fragmentation:* Natural landscapes generally have large patches of habitat that are well connected to other, similar patches. Human-dominated landscapes, however, consist of a mosaic of different land uses, resulting in small, often geometrically configured, patches that frequently contrast highly with neighbouring patches. Whether it be ploughed fields or house lots, the tendency is to shrink the size of forested parcels. For the continued survival of any natural population, the number of individuals must never fall below a critical number, and that requires a certain minimum area that must be large enough to compensate for years of adverse weather. For many species, more area will be required during a dry year than a normal year. If development reduces the habitat to a point where it cannot support the critical number during an adverse year, the entire population will perish. Similarly, development (such as a highway) that fragments a territory and prevents migration between the two fragments will cause a population to perish if neither area can adequately support the critical number. Also, reducing the size of a habitat creates a greater proportion of edges, a situation that favors some species, but may be detrimental to others.

Figure 15.2
Causes and Mechanics of the Loss of Biodiversity

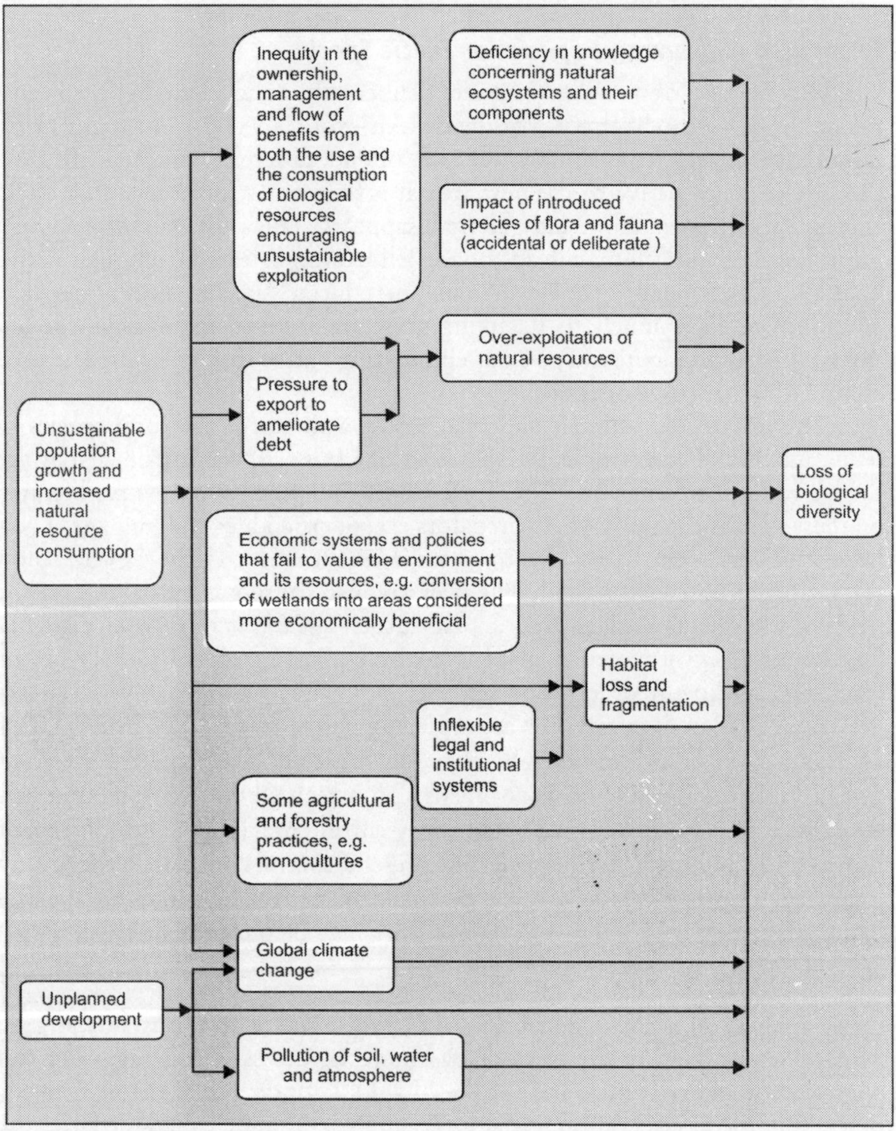

3. *Simplification:* Human use of habitats often simplifies them. Removing fallen logs and dead trees from woodlands for firewood, for example, diminishes an important microhabitat on which several species depend. When a forest is managed for the production of a few or one species of tree, tree diversity declines, and with it so does the diversity of the cluster of plant and animal species that depend on the less favoured trees. Streams are sometimes 'channelized' – their beds are cleared of fallen trees and riffles, and sometimes the stream is straightened out by dredging. Such

alternations inevitably reduce the diversity of fish and invertebrates that live in the stream.

Introduced and Invasive Species or Exotic Specie

The introduction of non-native species deliberately or accidentally, has been a major threat to biodiversity worldwide. An exotic species is a species introduced into an area from somewhere else, often a different continent. Because the species is not native to the new area, it is frequently unsuccessful in establishing a viable population and quietly disappears. This is the fate of many pet birds, reptiles, and fish that escape or are deliberately released from their native habitats. Occasionally, however, an introduced species finds the new environment very much to its liking and can become an invasive species, thriving, spreading out, and perhaps eliminating native species by predation or competition for space or food.

The introduced animals and plants pose a threat to the local species of fauna and flora. For example, the spotted deer (*Axis axis*) was introduced by the British in the Andaman and Nicobar islands. The deer have now proliferated on these islands as they have no predators except crocodiles and humans. Their large numbers seem to be affecting forest regeneration as they over-exploit certain forest species for food. The deer are also causing crop damage in the fields of the islands settlers. Exotic plants, such as *Lantana camara,* originally introduced as an ornamental plant from Brazil, are spreading rapidly in our forests at the expense of local species.

Exotics such as eucalyptus are replacing native tree species because they are fast growing and commercially valuable. It is estimated that 18 per cent of Indian flora comprises invasive aliens, of which about 55 per cent are American, 10 per cent Asian, 20 per cent Asian and Malaysian, and 15 per cent European and Central Asian species. At present, several countries have already imported so many exotic species, particularly agricultural and ornamental plants, that their own indigenous fauna/flora may be outnumbered.

Overexploitation

Overexploitation occurs when a resource is consumed at an unsustainable rate. This occurs on land in the form of overhunting, excessive logging, poor soil conservation in agriculture and the illegal wildlife trade. Joe Walston, director of the Wildlife Conservation Society's Asian programmes, called the latter the 'single largest threat' to biodiversity in Asia. The international trade of endangered species is second in size only to drug trafficking. About 25 per cent of world *fisheries* are now overfished to the point where their current biomass is less than the level that maximized their sustainable yield.

Many plant and animal species have been over exploited by humans, sometimes to the point of extinction. Many species such as tigers and elephants are killed or poached for their skin, tusks, claws, etc. which have high commercial value. Others, such as several snakes and bird species are caught and smuggled out for their curiosity value for collectors and as pets. Marine fauna are under great threat from over-exploitation, largely as a result of mechanized fishing and increasing international fishing operations in Indian waters.

Trade in exotics is another form of overuse. Trafficking in wildlife and in products derived from wild species, although illegal, but a main reason of the loss of several species.

It flourishes because some consumers are willing to pay exorbitant prices for such things as furniture made from tropical hardwoods (like teak), exotic pets, furs from wild animals, traditional medicines from animal parts, and innumerable other 'luxuries,' including polar-bear rugs, baskets made from elephant and rhinoceros feet, ivory-handled knives, and reptile-skin shoes and handbags. Tigers and rhinoceros populations in the wild have declined drastically in the last two decades because of the widespread, but unfounded, belief in far Eastern countries that parts of these animals have medicinal or aphrodisiac properties. Poor management represents another form of overuse leading to a loss of biodiversity. Forests and woodlands are overcut for firewood, grasslands are over grazed, game species are overhunted, fisheries are overexploited, and croplands are overcultivated.

Pollution: Another major factor that decreases biodiversity is pollution, which can directly kill many kinds of plants and animals, seriously reducing their populations. Soil, water and air pollution affects the functioning of ecosystems and may reduce or eliminate sensitive species. Several studies carried out in India have clearly traced the effects of pesticide pollution on populations. Every oil spill kills seabirds and, often sea mammals, sometimes by thousands. Human wastes can spread pathogenic microorganisms to wild species, a threat called 'pathogen pollution'.

Climate Change: Global climate change brought on by anthropogenic greenhouse gas emissions, is already having an impact on many species. In the polar north, polar bears are declining because the sea ice they depend on for hunting seals is melting at a rapid rate (up to 9% per decade). In the polar South, many penguin species are in decline as waters undergo unusual warming that disrupts the fish and invertebrate species the penguins depend on. Rising land temperatures are already encouraging the spread of vector-borne diseases (diseases carried by ticks, flies, mosquitoes, etc.) such as lyme disease, malaria, dengue, and west nile disease. Some of these are pathogens for both humans and wildlife.

In the coming years, climate change could also affect global biodiversity. There are several hypotheses about this. Different species of plants and animals may respond in different ways to an increase in global temperature. Many species which cannot adjust to warmer temperatures could be extinct. Species with specialized niches or rare species will be at greatest risk being most sensitive to any atmospheric change. A change in the climate may also result in a change in the characteristics of habitats, thereby affecting the species within those habitats. Some habitats such as islands and coastal systems, which are at risk of flooding and submergence due to rising sea levels, could suffer particularly high losses of biodiversity.

Population Growth: Past losses of biodiversity can be attributed to the expansion of the human population over the globe. Continuing human population growth will further alter natural ecosystems, resulting in the inevitable loss of more wild species and additional declines in populations. The losses will be greatest in the developing world, where biodiversity is greatest and human population growth is highest. Africa and Asia have lost almost two-thirds of their original natural habitat. People's desire for a better way of life, the desperate poverty of rural populations, and the global market for timber and other natural resources are powerful forces that will continue to draw down biological wealth on those continents.

From 1950 to 2011, world population increased from 2.5 billion to 7 billion and is forecast to reach a plateau of more than 9 billion during the 21st century. Sir David King, former chief scientific adviser to the UK Government, told in a parliamentary inquiry: 'It is self-evident that the massive growth in the *human population* through the 20th century has had more impact on biodiversity than any other single factor.

Other causes of the loss of biodiversity includes-hybridization, genetic pollution, loss of traditional knowledge, ineffective legal systems, changing agricultural and forestry practices, growing demands, international trade and management system based on exploitation of natural resources, etc.

Conservation of Biodiversity

The overall discussion on biodiversity reveals that this invaluable heritage is being destroyed at an alarming rate. All the nations of the world have shown their concern regarding conservation of biodiversity and several measures have also been taken in this direction. Prior to describe the measures of conservation, as special mention of International Union for Conservation of Nature and Natural Resources (IUCN), which is now known as World Conservation Union (WCU) established in 1988 is necessary, which has prepared and revised from time to time a list of threatened plants and animals. Such lists are called Red Data Books or Red Data Lists. The Red List of threatened species released by IUCN includes the following categories:

Extinct Species (EX), those species of plants and animals which have been completely eliminated.

Extinct in Wild (EW), those species which are eliminated but exist in wild areas.

Critically Endangered (CR) those species which have lost their 80 per cent population.

Endangered (EN) species are those which have lost their 70 per cent of population within last 10 years.

Vulnerable (VU) species, those which are vulnerable to extinction and have lost 50 per cent of population.

Near Threatened (NT) species, those species which are above the critically endangered and vulnerable species but the loss of its population is high.

Least Concern (LC), those species which are having abundant population and required least attention.

The conservation measures of biodiversity can be divided into two categories:

1. *In-situ* conservation
2. *Ex-situ* conservation

In-situ Conservation

In-situ conservation, as defined by the UNEP (United Nations Environment Programme), is 'the conservation of ecosystems and natural habitats and the maintenance and recovery of viable populations of species in their natural surroundings'. This is an effective conservation strategy since it ensures the maintenance of ecosystems and species in their natural conditions. Biodiversity at all its levels, as genetic species and as intact ecosystems, can be best preserved *in-situ* by setting aside an adequate representation of wilderness as Protected Areas. These should consist of a network of National Parks and Wildlife Sanctuaries with each distinctive ecosystem included in the network. Such a network would preserve the total diversity of life of a region.

India has taken several steps towards *in-situ* conservation of biodiversity. India's major attempt to conserve *in-situ* wild biodiversity has been through a network of protected areas are recognized by the legal system in India; national parks and wildlife sanctuaries. National parks are highly protected by law. No human habitation, private landholding or traditional human activity such as firewood collection or grazing is allowed within the park. Sanctuaries are also protected, but certain types of activities such as collection of firewood and grazing of cattle are permitted within these areas.

The first such initiative was the establishment of the Corbett National Park in 1936. As of 2003, India has 99 national parks and 513 wildlife sanctuaries occupying 1,56,000 sq km, or over 5 per cent of the country's area. These protected areas have helped in conserving habitats and their biodiversity. Several special projects have also been launched to save certain animal species which have been identified as needing concerted protection effort. These projects are designed to protect the species *in-situ*, by protecting and conserving their natural habitat. Project Tiger and Project Elephant are two such major initiatives. Other species' conservation projects include Project Crocodile Breeding and Management, which was stated in 1976 and now operated in 16 sanctuaries; and the Gir Lion Sanctuary Project in Gujarat, which aims at saving the Asiatic lion which once roamed over a wide stretch of the northern and central parts of the Indian subcontinent, but is today one of the most threatened species of animals in India.

To conserve medicinal plants, the departments of Andhra Pradesh, Karnataka, Kerala, Maharashtra and Tamil Nadu, with the help of the Foundation for the Revitalisation of Local Health Traditions (FRLHTs), have set aside 54 forest patches as Medicinal Plants Conservation Areas (MPCAs) measuring 200 to 500 ha each. These MPCAs represent all the forest types and climatic zones of the region and harbour many species of medicinal plants threatened with extinction.

Wildlife sanctuaries and national parks of India: There are 612 protected areas in India, of which 99 are national parks and 513 are wildlife sanctuaries. They include a variety of ecosystems and habitats. Some have been created in order to protect highly endangered species of wild plants and animals found nowhere else in the world.

The Great Himalayan National Park is the largest sanctuary in this ecosystem and is one of the last homes of the beautiful snow leopard. The Dachigam Sanctuary is the only place where the rare *Hangul* of Kashmir stag is found. There are several sanctuaries in the *Terai* region; the Kaziranga National Park is the most famous which has elephant, *gaur*, wild boar, swamp deer, and hog deer, in large numbers, as well as tigers and leopards. Its bird life is extremely rich and includes ducks, geese, pelicans and storks. The Manas Sanctuary, in addition to the above *Terai* species, also includes the rare golden *langur* and the very rare pygmy hog, the smallest wild boar in the world. The florican is found only in a few undisturbed grasslands in the *Terai* sanctuaries.

In the sal forests of Madhya Pradesh, there are several PAs. Kanha offers a wonderful opportunity to observe wild tigers from elephant's back. It is the only PA in which a sub-species of the *barasingha* is found.

Bharatpur is one of the most famous water-bird sanctuaries in the world. Thousands of ducks, geese, herons, and other wading birds can be seen here. This is the only home of the very rare Siberian crane, which migrates to India every winter. During the last 20 years, the 30 or 40 Siberian cranes have dwindled to only 2 or 3. During 2002–03, no cranes were seen and it is possible that this beautiful bird will never again come to India.

In the Thar Desert, the wildlife is protected in the Desert National Park. Here, large numbers of blackbuck, *nilgai* and *chinkara* can be seen. The Great Indian Bustard lives in these arid lands.

Ranthambore was the most well-known sanctuary for observing tigers in the wild till about 3 to 4 years ago. Since then, many tigers have been killed by poachers.

The Great and the Little Rann of Kutch have been made into sanctuaries to protect the very rare wild ass, the flamingo, the star tortoise and the desert fox.

In Gujarat, the Gir Sanctuary protects the last population of the majestic Asiatic lion. This thorn and deciduous forest is also the home of large herds of *chital*, *sambhar*, and *nilgai*.

The sanctuaries of the Western Ghats and associated hill ranges protect some of the most diverse forest types in the country. The few examples of highly-threatened species include the Malabar giant squirrel, the flying squirrel and a variety of hill birds, several species of amphibians, reptiles and insects. These regions are also rich in highly endemic plant life. Sanctuaries such as Bhimashankar, Koyana, Chandoli and Radhanagari preserve this rich flora in Maharashtra; Bandipur, Bhadra, Dandeli, Nagarhole, etc. in Karnataka, and Eravikulam, Parambikulam, Periyar, and the Silent Valley, in Kerala.

In the Nilgiri Hills, the rich forest sanctuaries protect some of the last pockets of the Indian elephant in South India. Examples include Bandipur, Mudumalai, Wynad and Bhadra. During the last 10 years, a large number of the great tusker elephants of this region have been ruthlessly killed for their ivory. Now, very few of these magnificent animals are left in these jungles.

Two important sanctuaries meant for preservation of coastal ecosystems are the Chilika Lake in Orissa and Point Calimere in Tamil Nadu. The Sunderbans protect the largest mangrove delta in India. The Marine National Park in Gujarat protects shallow areas in the sea, islands, coral reefs and extensive mudflats.

Over a hundred protected areas (PAs) have been created in the Andaman and Nicobar Islands to preserve their very special island ecosystems.

Integrated Protected Area System (IPAS)

For the conservation of biodiversity, Protected Areas (PAs) is the effective system in every bio-geographic region. A relatively larger representation must be included of highly fragile ecosystems, areas of high species diversity or high endemism. PAs must also be integrated with each other, by establishing corridors between adjacent areas wherever possible so that wildlife can move between them. The International Union for Conservation of Nature and Natural Resources (IUCN) states that it is essential to include at least 10 per cent of all ecosystems as Protected Areas if biodiversity is to be conserved in the long-term.

India has only 5 per cent of land in its 589 Protected Areas in 2004. However, much of this includes plantations of sal or teak, which were developed for timber in past and are thus relatively poor in diversity and have a low level of 'naturalness'. There are only a few good grasslands left in our country that are notified as PAs. Some are merely overgrazed wastelands in areas that were once flourishing grasslands. Most of these areas have a low biological value and need careful management to allow them to revert to a more 'natural' state, with their full complement of plants and animals. Only a few wetlands have been made into sanctuaries. These require better management.

A major strategy to reduce impacts on the biodiversity of the PAs should be to provide a sustainable source of resources for local people living around them. A PA curtails their traditional grazing practices and access to fuelwood sources. These resources must be provided by developing them in the buffer areas. Plantations of fuelwood and good grassland management in areas outside PAs can help in reducing reduce the pressure on the habitat of wildlife within the PA. Good management practices must ensure that local people derive a direct economic benefit from the presence of the PA. Involving local people in Protected Area Management and developing tourist facilities that support the income generation for local people help in enlisting their support for the PA.

Ex-Situ Conservation

Ex-situ conservation means, conserving the species outside its natural habitat in a carefully controlled situation such as a botanical garden for plants or a zoological park for animals, where there is expertise to multiply the species under artificially-managed conditions. These breeding programmes for rare plants and animals are however more expensive than managing a PA.

There is also another form of preserving a plant by preserving its germplasm in a gene bank so that it can be used if needed in future. This is even more expensive.

When an animal is on the brink of extinction, it must be carefully bred so that inbreeding does not lead to the genetic composition becoming weak. Breeding from the same stock can lead to poorly adapted progeny or even an inability to beget enough offspring.

Modern breeding programmes are done in zoos that provide for all the animal's needs, including enclosures that simulate their wild habitats. There may also be a need to assist breeding artificially.

A number of orchid sanctuaries and orchidariums have been established in orchid-rich habitats, like the foothills of the Himalayas, the Western Ghats, the western coastal region and the southern hill stations.

The Central Council for Research in Ayurveda and Siddha (CCRAS), established in 1978, has promoted the cultivation of medicinal plants at five herbal gardens across the country covering an area of 135 acre.

The pygmy hog is the smallest and rarest wild pig in the world. Very small populations survive in the wild but it is almost on the brink of extinction. A small population exists in the Manas National Park in Assam. The Pygmy Hog Conservation Programme (PHCP), a collaborative programme of the government and international specialist groups, focuses its work on trying to ensure that pygmy hogs are successfully bred in captivity and released in the wild in an effort to save them from extinction.

In India, successful *ex-situ* conservation programs have been done for all our three species of crocodiles and has been highly successful. The most successful example is the Madras Crocodile Trust Bank, which has grown from 10 to 8,035 crocodiles and the crocodiles here lay two clutches of eggs a year, compared to one in the wild. Another recent success has been the breeding of the very rare pygmy hog in the Guwahati Zoo. The Delhi Zoo has successfully bred the rare Manipur brow-antlered deer.

A lot of effort is underway to collect and preserve the genetic material of crop, animal, bird and fish species by the Government of India. This work is being done by institutions such as the National Bureau of Plant Genetic Resources (NBPGRs), New Delhi, and the National Bureau of Animal Genetic Resources (NBAGRs), Karnal.

Legislative Measures

Several legislative measures have been taken throughout the world to protect biodiversity. India has several acts in force, which have a bearing on conservation of biodiversity. These are:

The Environment Protection Act, 1986, relates to general measures to protect the environment, such as restrictions on industrial and other processes or activities in specified areas. It also deals with the prevention of and control over the manufacture, use, release and movement of hazardous substances.

The Fisheries Act, 1897, prohibits the use of explosives and poisons for fishing. It also regulates fishing in private waters.

The Forest Act, 1927, deals with the setting up and management of reserved, protected and village forests, and controls how selected products from the forest can be sold and at what price.

The Forest (Conservation) Act, 1980, primarily focuses on prohibiting or regulating non-forest use of forest lands.

The Wildlife (Protection) Act, 1972; The Wildlife (Protection) Amendment Act, 1991; and The Wildlife (Protection) Amendment Act, 2002, deal with the restriction and prohibition of the hunting of animals, and with the protection of specified plants. They also deal with the setting up and management of sanctuaries and national parks, setting up of the Central Zoos Authority, control of zoos and captive breeding. They also control trade and commerce in wild animals, animal articles and trophies. The Biodiversity Act, 2002, which was formulated after a long period of discussions and considerable public debate, is part of India's follow-up of the International Convention on Biological Diversity. The Act provides for a National Biodiversity Authority (NBA), which will screen proposals for the transfer of genetic resources abroad, and set up a system for genetic material from India to be sent to any other country. This is to ensure that there is a record of all genetic material going out of the country. The NBA will also advise the Central Government on measures for conservation, sustainable use, and benefit-sharing resulting from the use of the biodiversity with the communities which are its traditional custodians.

The National Biodiversity Strategy and Action Plan (NBSAP)

This plan was initiated in late 1999 by the Ministry of Environment and Forests (MoEF), Government of India, as part of India's commitment to the Convention on Biological Diversity. The broad purpose of this process was to produce an implementable action plan that would ensure the conservation of India's biodiversity, its sustainable use and equitable sharing of benefits arising from its use – the main tenets of the Convention on Biodiversity. The two prerequisites of this plan are the ecological security of the country or of any region within it, and the livelihood security of those most critically dependent on biodiversity and its components.

Other efforts for the conservation of biodiversity in India are –

1. Project 'Mangroves for Future' (MFF) for coastal ecosystem conservation.
2. Four major coral reefs areas identified for intensive conservation are Gulf of Mannar, Gulf of Kutch, Lakshadweep and Andaman and Nicobar Island.

3. UNDP-GOICCF II Programme on promoting conservation of Medicinal Plants and Traditional knowledge
4. Wildlife Crime Control Bureau and National Board for Wildlife, etc.
5. Biosphere Reserve programme is an effective method of conservation of abiotic and biodiversity of natural habitats. Biosphere Reserves are recognized within the framework of UNESCO's Man and Biosphere Programme. The world major ecosystem types and landscapes are represented in this network, which is devoted to conserving biological diversity. India has designated 17 biosphere reserves in different bio-geographical regions of the country, out of which 7 have been included in the world network of biosphere reserves of UNESCO.

Evolution of the Biodiversity Regime

The Stockholm Conference was not the first occasion at which issues related to biodiversity were placed on the international agenda. A variety of legal instruments already existed. Some of these were regional, such as the Convention on Nature Protection Wildlife Preservation in the Western Hemisphere (Washington, D.C., 1940) and the African Convention on the Conservation of Nature and Natural Resources (Algiers, 1968). Other measures dealt with specific species, such as birds and marine and polar region species. Prior to Stockholm, the only major international legal instrument for conserving biodiversity was the 1971 Convention on Wetlands of International Importance especially as Waterfowl Habitats. The 61 parties to this convention pledged to make an inventory of designated wetlands to promote their conservation, and to establish nature reserves in the important wetlands.

But, no comprehensive international treaty addressed the problem of biological diversity. Although Stockholm did not produce such a document, it was the first time the issue was addressed in such a holistic manner. The issue of biodiversity was debated, and the conference documents included provisions aimed at protecting biodiversity.

The area of genetic resources and biotechnology was already controversial in 1972, at the time of the Stockholm Conference. Developing countries were concerned about their lack of access to the biotechnology that was developed from the use of their resources, and they tried to use linkage politics to gain access to this technology. They pressed their demand for the transfer of technology on non-commercial terms. This demand was rejected by the developed countries, and the final statement of the conference diluted the language that addressed this linkage between economics and the environment.

Both the declaration of the United Nations Conference on the Human Environment and the Action Plan for the Human Environment included language that addressed biodiversity. Principles 2, 4, and 6 of the declaration included provisions for safeguarding flora and fauna and preserving ecosystems. The Action Plan generally addressed the importance of protecting forests, which among other things, were important habitats for wildlife; it recommended that wildlife be monitored to assess the impact of pollutants and that the economic value of wildlife be assessed. Governments were asked to take steps to protect ecosystems that have international significance. Recommendations 39 through 45 deal generally with the issue of preserving genetic resources.

In the period between the Stockholm and Rio conferences, the body of legislation concerning biodiversity continued to grow. Relevant major international legislations included the convention concerning the Protection of the World Culture and Natural Heritage (1972), the Convention on International Trade in Endangered Species of Wild Fauna and Flora (1973), and the Convention on Conservation of Migratory Species of Wild Animals (1979).

In the aftermath of Stockholm, developing countries continued their push for greater access to the biotechnology that resulted from their biological wealth. A non-binding International Undertaking on Plant Genetic Resources (Resolution 8/33) was passed by the FAO in 1983. It stated that access to genetic resources should not be restricted, and it declared that all seed resources were the common heritage of humanity; therefore, both primitive stocks and those developed by proprietary means should be free to all. Some developed world producers were opposed to this, and many industrialized countries were not parties to it because it was incompatible with their patent rights. Third World members, however, accepted its provisions.

A 1991 Costa Rica deal is considered one possible model for future bilateral arrangements. The parties to the deal were the National Biodiversity Institute, a non-profit Costa research centre, and Merck & Co., the world' larges pharmaceutical company. Merck agreed to pay the institute US$1.1 million, as well as royalties from any product developed, in return for plant, insect and microbe samples from all over Costa Rica. Costa Rica is estimated to have about 5 per cent of any royalties go directly into conservation. Although this agreement provides for compensation, some critics say it gives away Costa Rican resources for too little money. Nevertheless, many developing countries are studying this arrangement with interest.

The years between Stockholm and Rio saw incremental progress with regard to the items on the Third World's biodiversity agenda. Developing states wanted to make their own arrangements concerning resources access, and as the Merck deal illustrates, the slow process of norm changes had begun. But, the progress of these states in the campaign for transfer of technology and financial assistance is more difficult to assess. Although developing countries continued to push for change in these areas, developed countries maintained their determined resistance. These conflicting perspectives were reflected in the preparatory negotiations for the Rio Conference.

The declaration of UN Conference on the Human Environment and the Action Plan for the Human Environment addressed the importance of protecting forests, which among other things, are important habitats for wildlife. Other conventions of biodiversity include May 1992 at Nairobi; June 1992 at UNCED; while in December 1993 Convention on Biological Diversity entered into force. The highlights of the convention were:

1. National identification and monitoring of biological diversity.
2. The development of national strategies and programmes for conserving biological diversity.
3. National *in situ* and *ex situ* conservation measures.
4. Environmental assessment procedures to take into account the effects of projects on biological diversity.
5. National reports from parties on measures taken to implement the convention and the effectiveness of the measures.

The issue of biodiversity should be seen as a common problem; as such, action by any country has implications for any other country. The biodiversity of every country should be protected with mutual cooperation and exchange of biotechnology.

In fact, this is a common property resource concern, as decrease in biodiversity has consequences for all species. But, this common property resource has some distributional characteristics that makes it different from other environmental issues. The bulk of the world's biological diversity is found in tropical forests, which are located in the Third World. The loss of biological diversity has tremendous implications for agriculture. The range of food resources has declined over time. One estimate says that more than 3,000 plant species have been used for food throughout human history, however, there certainly has been a significant reduction in variety. Now, most of the world's food comes from 20 species. The issue of patents is at the core, nowadays, because several developing countries have a fear that the use of traditional plant species will be restricted by scientifically advanced countries. After Uruguay round of GATT seeks to standardize intellectual property rights, much debate has been done and still continues. The controversy over patent right of 'Neem' tree, which has been used by Indians for centuries, is one example.

To conclude, one can quote from a stanza in Isho-Upanishad which says: 'The Universe along with its creatures belongs to the land. No creature is superior to any other. Human beings should not be above nature. Let no one species encroach over the rights and privileges of the other species.'

References

Goudie, Andrew and Thomas, David S.G. (eds.) (1995), *The Encyclopedic Dictionary of Physical Geography*, Oxford, Blackwell, p. 56.

Joseph, Kerski and Simon Ross (2005), *The Essentials of Environment*, Hodder Arnold, London, p. 17.

Small, J. and Withael, M. (1995), A *Modern Dictionary of Geography*, Arnold, London, pp. 22–23.

Saxena, H.M. (2010), *Environmental Management*, Rawat Publication, Jaipur, pp. 303–312.

Government of India Publication (2013), *India 2013*, Publication Division, Government of India, New Delhi, pp. 291–302.

16

Climate Change

Climate change is the single biggest environmental and humanitarian crisis of our time. The earth's atmosphere is overloaded with heat-trapping carbon dioxide, which threatens large scale disruptions in climate with disastrous consequences. Global climate change has been a burning problem for environmentalists since the middle of the 20th century and has stirred up a serious debate among the researchers even in the contemporary scientific community. Small and Witherick (1995) have defined climate change as 'A change in the climate of a region or place usually reflected in the mean annual temperature or mean annual precipitation, over a time period that can vary from long term to short term'. Thus, climate change refers to a large scale change in one or more basic climate components, such as temperature or precipitation. Such a change would be significant enough to have profound impact on the natural environment and human society.

Climate change refers to the variation in the earth's global climate or in regional climate over time. It describes change in the variability or average state of the atmosphere over time scale ranging from decades to millions of years. These changes can be used by processes internal to the earth; external forces for example, variations in sunlight intensity or more recently, human activities. In recent usages, especially in the context of environmental policy, the term 'climate change' often refers to changes in modern climate which according to the Intergovernmental Panel on Climate Change (IPCC) are 90 to 95 per cent likely to have been in part caused by human action. Consequently, the term *anthropogenic climate change* is frequently adopted; this phenomenon is also referred as 'global warming'. Thus, now the term climate change has become synonymous with 'global warming'. Global warming is the increase of earth's average temperature due to effect of greenhouse gases, such as carbon dioxide emissions from burning fossils or deforestation, which trap heat that would otherwise escape from the earth. This is a type of *greenhouse effect.* This chapter presents the climate change in terms of global warming and greenhouse effect as main source of global warming/climate change. Both these aspects are inter-related with each other and are responsible for climate change.

Climate Change in the Past

The earth's temperature has not been constant in the past. Indeed our planet has gone through several periods of global cooling and global warming as it has moved in and out of what we call *ice ages*. Variability, in both time and space, is an inherent feature of climate, as the atmosphere is always in the state of turmoil and instability leading to variations in weather and climatic conditions. The climochronology (history of palaeoclimates) reveals the fact that climate have changed in geological past and hence it is opined that the world's climate have changed in the past, are changing now, and there is every reason to expect that it will change in future (J.E. Hobbs, 1980). The change is climate in a reality because climate of a region is not static rather it goes on changing some time in a cyclic manner, known as *climatic cycle*. This has happened in the past, evident from the occurrences of ice ages during (i) *pre-Cambrian period* (850–600 million years before present mbp), (ii) *Ordovician period (*450–430 mbp*), (iii) Carboniferous – Permian period* (300 mbp) (iv) *Pleistocene period* (2–3 mbp).

The final retreat and withdrawal of recent ice sheets started about 18,000 years before present and continued upto 10–12000 years before present, when most of the ice covers were withdrawn from the USA. The record shows that the climatic conditions of Europe in 1st century of Christian era was similar to that of present conditions. Europe and North America experienced dry phase resulting into semi-drought conditions in the 5th century AD. The climate in northern hemisphere became further harsh during 600–700 AD.

The period from 950 to 1250 AD is called as 'Phase of Little Climatic Optimum' in the climatic history of earth, when climate become warm and relatively dry as average temperature increased by 1° to 2° C from the present day global average temperature. The period from 1250 AD to 1450 AD was characterized by the reversal of mild climate of 10th to 13th centuries AD The climate continued to become further harsh for another 400 years (from 1450 AD to 1880 AD) as the temperature fell below freezing point, giving birth to another glacial period known as Little Ice Age. The coldest period culminated in the year 1876 when major parts of northern Europe and the USA did not experience summer season. The year 1816 is known as *the year without summer* in the climatic history. Climate began to improve due to rise in temperature and by the end of 19th century the ice age was terminated.

The records of global temperatures since Industrial Revolution (1860) denote slow but irregular trend of rise of temperature. R.Q. Barry and R.J. Chorley (2002) have identified the following trends in rise of temperatures after 1860:

— Highly irregular trend of rise in temperature ranging between 0.2°C to 0.4°C between 1860–1920 AD.
— Consistent trend of rise of temperature of mean value of 0.4°C between 1921 and 1945.
— Oscillating trend of rise of temperature between 1946 and 1975 AD around 0.4°C.

Figure 16.1
Increase in Global Temperatures

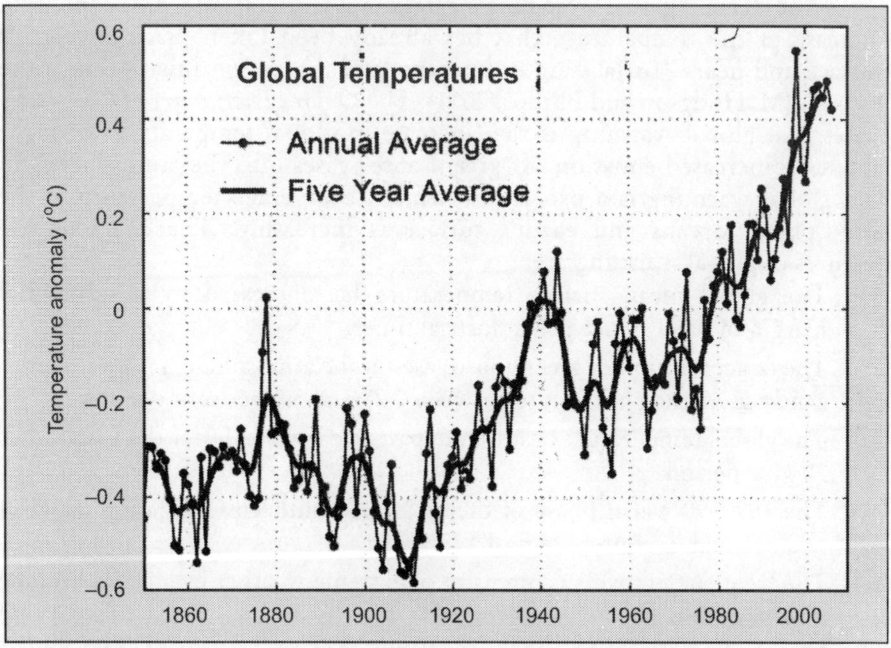

— Overall waring trend continued between 1976 and 1989 AD when temperature registered an upward rise by 0.2°C.
— The global rise of temperature reached its peak in the last decade of 20th century. Six out of seven warmest years on record since 1980.

A large number of temperature measurements indicate that the earth's average surface temperature rose about 0.75°C (about 1.3°F) compared with early twentieth century. The increase has not been smooth (Figure 16.1). It warmed noticeably from about 1910 to the mid 1940s, levelled off, and began increasing again in late 1970s. The 1990s were warmest of all, and 1998 was the warmest year. The year 2001 was the second warmest with an average temperature of 14.3°C (57.8°F) About two-thirds of the 20th century warming has occurred between 1975–2000.

Global Warming

As described above, temperature variations and climatic changes has been occurred during geological periods of the earth's history. Why, then, is so much attention being paid now to the phenomenon of global warming if it is something that has happened before? This is because some of the warming that is taking place now is believed to have different causes from previous warmings; it is what we call 'anthropogenic', which means it has been caused primarily by human activity, unlike previous global temperature shifts which were entirely due to natural causes. The scientific consensus is that human activities have been warming the world and it will continue to get hotter for the next hundred years or so, irrespective of whatever we choose to do now. In general , now the term global warming is used to explain the nature of climate change.

'The term "global warming" refers both to (a) the anthropogenic increase in the temperature that has already been taken place on earth's surface and near – surface air and oceans, and (b) its continuation into the future'. (M. Hodgson and Philips, 2011). The *Oxford Dictionary of Geography* states that global warming is the 'increase in global temperatures brought about by increased emission of "greenhouse" gases into the atmosphere'. In fact global warming is a process by which the average temperature of the atmosphere, oceans and earth's surface is increasing. The characteristic features of global warming are:

1. The global mean surface temperature has increased by a substantial margin of 0.6°C since pre-industrial times.
2. The concentration of greenhouse gases in the atmosphere has been significantly increasing at the rate between 0.25 and 5.0 per cent per annum.
3. The global temperature has risen in past two decades faster than any other 20 year period.
4. The 1990–99 warm phase of the El Nino/Southern Oscillation (ENSO) was unusual in context of past 120 years.
5. The frequency as well as intensity of extreme weather events has become more variable.
6. Magnitude and pattern of climate change and variability are not uniform throughout the world.
7. There has been marked to phosphoric depletion of ozone gases and increase in greenhouse gases near the earth's surface.
8. The frequency and intensity of extreme weather events have an increasing trend.
9. The diurnal range of temperature has increased.

The IPCC, the Intergovernmental Panel on Climate Change, was formed in 1988 by the World Meteorological Organization and the United Nations. Hundreds of scientists from 178 member nations do exceedingly careful reviews of climate data, and periodically report on their findings. A 1995 report said that there is a 'discernible human influence' on global warming. Then in 2001 after 3 years of further study, an IPCC report went further and attributed most warming of the past 50 years to human activities. Several of their conclusions are as follows:

- Atmospheric greenhouse gases continue to increase as a result of human activities.
- The fraction of warming due to carbon dioxide exceeds 50 per cent.
- Decrease of about 10 per cent in snow cover has occurred since the late 1960s.
- Reductions in the year's duration of lake and river ice cover in the northern hemisphere's mid and high latitudes occurred in the twentieth century.

- Increase in the heat contained in the world's oceans have occurred since the late 1950s.
- Between 0.1 and 0.2 m rise in global average sea levels occurred in the twentieth century.

IPCC scientists did not believe that variance in solar radiation, natural climate fluctuation, or poor climate models explained the warming. As IPCC Chairman Robert T. Watson said, 'We see changes in climate, we believe we humans are involved, and we're projecting future climate changes that will be much more significant in the next 100 years than in the past 100 years.' The panel projects that, if current trends continue, increase in atmospheric greenhouse gases in the twenty-first century will result in an average global temperature increase between 1.4 and 5.8°C (2.5 to 10.4°F). About 75 per cent of the warming is expected to be due specifically to carbon dioxide. They also project sea-level rises of another 0.1 to 0.9 m. A sober forecast is that climate change will persist for many centuries.

Causes of Global Warming

A number of causes are responsible for the global warming. Apart from natural phenomena, the main causes are 'anthropogenic', i.e. caused by human activities. Industrialization, urbanization, population explosion, shrinking agricultural land etc. have had their contribution in global warming but the main cause is the greenhouse effect or concentration of greenhouse gases. Therefore, it is necessary to explain the process of greenhouse and its effect on climate change/global warming.

Greenhouse Effect and Global Warming

The phenomenon commonly known as 'greenhouse effect' occurs due to emission of certain gaseous pollutants in the air which after the heating of the atmosphere, causes the average global temperature to rise. The effect produced by certain gases, such as carbon dioxide or water vapour, causes warming of the earth's atmosphere by absorption of infra-red radiation. The atmosphere protects us by serving as a light scattering and heat mediating blanket. A large portion of heat emitted from the earth is reabsorbed by the atmosphere and is, in effect, conserved, with the result that the surface of the earth is warmer than it otherwise would be. In simple words, this warming is called the greenhouse effect, by analogy with the ability of a greenhouse to keep its inside warmer than the outside during the day time (Figure 16.2).

The atmospheric concentration of these gases has relatively increased since the advent of the industrial era. There has emerged a growing consensus in the scientific community that the anthropogenic green house gas emission of carbon dioxide, nitrous oxide, CFCs, methane and low level ozone threatens to disrupt human societies and natural ecosystems.

Recently, the world has been getting hotter. Over the same period, the composition of the earth's atmosphere has been changed by increased human activity that produces CO_2 particularly by burning fossil fuels (in power

Figure 16.2
Greenhouse Effect

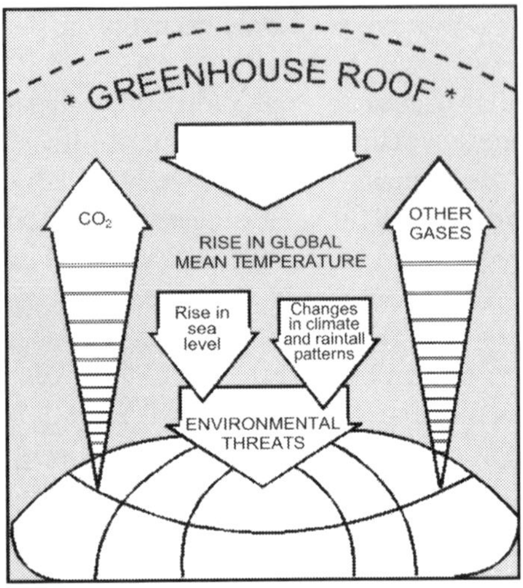

Figure 16.3
Annual Greenhouse Gas Emissions by Sector

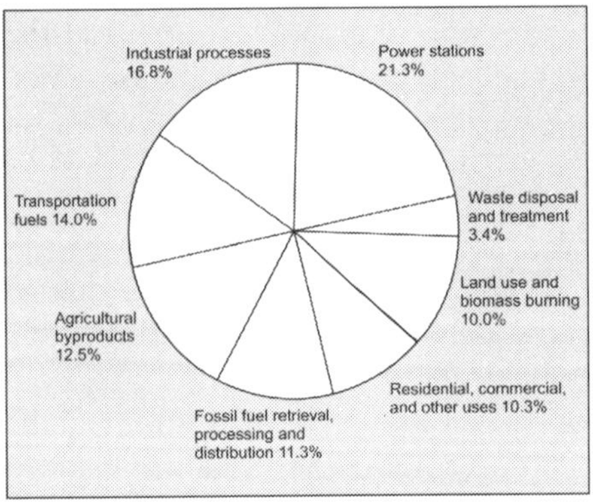

stations, cars, lorries, ships and aeroplanes, etc), cement production and deforestation. Figure 16.3 shows the percentage contribution from different sectors of human activity to overall GHG emissions, and to the three main individual gases (CO_2, methane and nitrous oxide).

These human-made CO_2 emissions account for 72 per cent of the added greenhouse effect to date. Once CO_2 is in the atmosphere, it stays there for thousands of years before dispersing. Indeed the world's most prestigious and influential body researching global warming, the Intergovernmental Panel on Climate Change (IPCC) (2007: 17), warns that

> ... both past and future anthropogenic carbon dioxide emissions will continue to contribute to warming and sea level rise for more than a millennium due to the time scales required for removal of this gas from the atmosphere.

The IPCC (2007:10) also tell us that:

> Most of the observed increase in global average temperature since the mid-20th century is very likely due to the observed increase in anthropogenic greenhouse gas concentrations.

Per Capita Greenhouse Index

National rankings reflect the contributions of individual countries to the atmosphere's growing store of greenhouse gases. But it is also of interest to compare countries by per capita contributions of these gases, since the citizens of our planet vary widely in their relative responsibility for these emissions.

Per capita figures also can be looked at as an indication of how much greenhouse gas emissions and global warming potential might grow if developing countries, in the process of development, significantly increase their per capita emissions. If just China and India for example, raised their per capita emissions to the current world average, 1.1 metric tonnes of carbon dioxide heating equivalent per annum, total worldwide additions to the atmosphere would increase 28 per cent; if raised per capita emissions to the level of France, worldwide additions to the atmosphere would increase 68 per cent.

Figure 16.4
Per Capita Net Greenhouse Gas Emissions

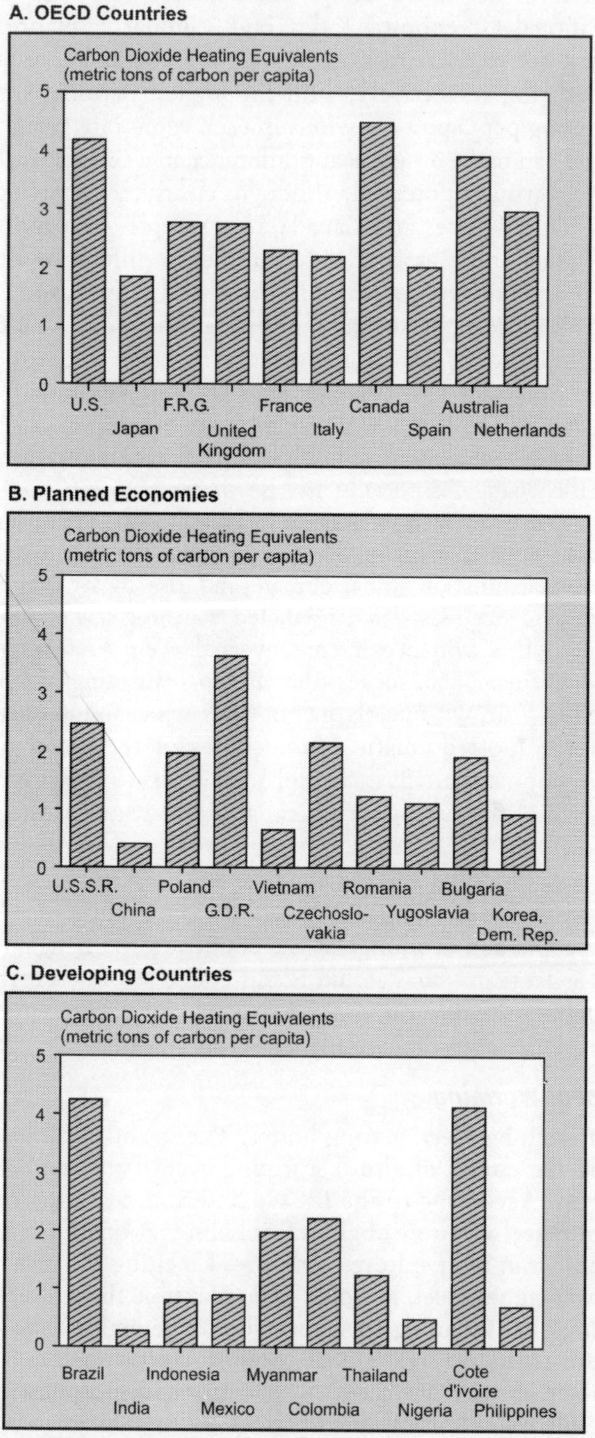

Figure 16.4-A compares the per capita emissions for the 10 countries of the Organization for Economic Cooperation and Development (OECD) with the highest national Greenhouse Index rank. Similarly, Figures 16.4-B and 16.4-C compare per capita emissions for the 10 planned economies and 10 developing countries, respectively, with the highest national emissions. The large ranges among per capita emissions in each economic group suggest that national policies can make a significant difference in a country's contributions to atmospheric warming potential, although, clearly, geography also plays a role. Both the United States and Canada, for example, emit more than twice the per capita greenhouse gases that Japan does, a difference attributable to both geography and policy. Mexico emits about half the per capita emissions of Colombia and about one quarter those of Brazil. At the same time, differences between economic groups are a reminder that any agreement to limit greenhouse gas emissions will have to take into account questions of individual equity and the legitimate aspirations of nations to develop.

The increase in atmospheric concentration of greenhouse gases are responsible for climatic change, especially in rise of temperature, in other words, responsible for global warming. Changes in greenhouse gas concentrations are closely correlated with changes in the earth's surface temperature. It has been established that warming of global climate and rise in CO_2 and CH_4 levels coincide. Some evidence exists that predicted warming has begun to manifest itself already, but this evidence is controversial. Temperature records taken around the world since 1860, suggest that a global warming of 0.5°–0.7° C has occurred since this year, with nearly half of this rise occurring since 1965.

Amongst the most dramatic consequences of the greenhouse climatic disruption is an unprecedented rise in global sea levels to the extent of 8 to 25.5 inches by the end of the century if gas emissions continue at the present rate. Melting of polar ice and glaciers, thermal expansion of seas would cause world-wide coastal flooding and rise in sea level will result in submergence of many coastal areas. It will not only effect the ecological balance but will be a threat to the people and economy of the countries like Bangladesh, Netherlands, as well as to many coastal and island countries. In fact, the impact of world-wide climatic disruption could have hazardous consequences for the survival of the planet both biologically and economically.

Effects of Global Warming

It is clear that earth has been getting hotter. The scientists all over the world have examined the effects of global warming even if the temperature rise is only few degrees. According to the IPCC (2007:5) ... warming of the climate system is unequivocal as is now obvious from observation of increases in global average air and ocean temperature, widespread melting of snow and ice, and rising global average sea level. It would seem plausible that if human activities have added to global warming then reversing these actions should, in a few years time, lead to its reverse, global cooling. Unfortunately, this will not happen. We noted above that once CO_2 gets into the atmosphere, it stays there for a considerable time. Carbon dioxide is, of course, absorbed by vegetation, especially forests, and by water, for example, seas. The issue is that the increase in GHGs being emitted into the atmosphere now outstrips the capacity of forests and seas to absorb them. The consequences of this are truly awesome ... the climate change that takes place due to increases in carbon dioxide concentration is largely irreversible ... atmospheric temperatures do not drop significantly for at least 1000 years. (Solomon et al., 2008: 1704)

The effects of the global warming will be as follows:

Rising of Sea Level

A significant proportion (up to 80%) of the temperature rise is absorbed by the oceans, expanding the water and contributing to increase in the sea level. Between 1880 and 2000, it rose by 200mm. The average annual sea level rise throughout the 20th century was 0.17mm; between 1962 and 2003, it rose to 1.8mm; and between 1993 and 2003, it reached 3.1 mm.

There has also been widespread melting of glaciers and ice caps particularly in Greenland and Antarctica, which has also contributed to sea level rise. Since 1978, Arctic sea ice has shrunk by 2.7 per cent per decade and the Arctic ice cap is now less than half the size it was 50 years ago. The situation in Greenland is perhaps even more troubling; although its contribution to the rise in sea level has so far been less than that of Antarctica, observations in 2008–09 have identified rapid melting in the Greenland glaciers and fears have been expressed of rapid acceleration leading to a 'tipping point' (Houghton, 2009: 177–9). If the whole Greenland ice sheet melted, this would trigger a globally catastrophic rise in sea levels of 7 metres (23 feet). Even if these specific catastrophes do not occur, sea levels are still set to rise by at least 18–38cm by 2100 even on the least pessimistic of IPCC's scenarios – and up to 50 cm on the most pessimistic scenario.

Change in Rainfall Patterns

Long-term trends in precipitation (that is, rain, sleet, snow, etc.) are much more variable than those in temperature – there have been increases in rainfall in eastern areas of North and South America, northern Europe and northern and central Asia. On the other hand, drying has occurred in the Sahel, Mediterranean and parts of southern Asia. More intense and longer droughts have been observed since the 1970s, particularly in the tropics and sub-tropics. Frequency of heavy precipitation has increased over most land areas (IPCC, 2007:5). The IPCC (2007:15) predicts a further 10–20 per cent reduction in precipitation in the Sahel by 2090. Rainfall pattern would be altered, with some areas getting more rainfall and others suffering more droughts.

Coastal Flooding

Coastal flooding will increasingly occur with rising sea level. People who are poor and overcrowded may be forced from homes in low-lying countries such as Bangladesh. Prosperous countries such as the Netherlands, long accustomed to fighting back the sea, may lose land too. A rising sea level can exert great damage even in less critically affected regions, e.g. beaches, beach-front property and may increase storm surges along coastlines. A higher sea level also means that, the salty water can infiltrate fresh groundwater in coastal areas, making it undrinkable. Such infiltration of ocean water can also destroy coastal ecosystems, and the species they support. Inland wetlands and ecosystems are also at risk.

Increased Likehood of Storms and Heat Waves

Storms of increased severity are expected. This happens because trapped heat energy drives atmospheric air circulation and oceanic water circulation. These circulations have been positive forces because they distribute heat energy more evenly around the world making Earth on average a more moderate place to live. However, hurricanes and typhoons feed on this warmth too, and as the energy in air and water increases, storms can be expected to grow to greater severity than in the past.

Another logical accompaniment of increased warming is more severe heat waves. Higher temperatures would worsen air quality – recall that photochemical smog increases on hot summer days.

Ocean Acidification and Bleaching of Coral Reefs

The oceans are significant absorbers of atmospheric CO_2. As the concentration of CO_2 increases in the atmosphere, so the amount absorbed into the oceans increases and this leads to changes in the acidity of the water. Even though the changes in ocean chemistry appear small (a decrease from pH 8.2 in pre-industrial times to pH 8.1 currently), the effects on marine life and ecosystems are huge, including the destruction of coral reefs and the reduced capacity for sea creatures to develop shells and skeletons. One-third of coral reefs now appear to have severely damaged by warming sea.

Spread of Diseases

Diseases currently restricted to existing hot regions, may move into the newly warming regions. Malaria is one such disease, spread by mosquito vectors; infecting and killing millions each year in warm climates. As temperature increases in currently temperate regions, mosquitoes and other insects are expected to move in, spreading diseases as they go. In addition, indigenous disease organisms previously killed by winter cold will be better able to survive milder winter.

Loss of Plankton Due to Warming Seas

The enormous (900 mile long) Aleution island ecosystem of orcas (killer whales), sea lions, sea otters, sea urchins, keep beds and fish populations, appear to have collapsed due to loss of plankton, leading to loss of sea lions, leading orcas to eat too many sea otters, leading to urchin explosions, leading to loss of keep beds and their associated fish populations.

Species Would Migrate Poleward

Perhaps the most severe disruption of the terrestrial carbon cycle could come as result of the rapidity of the expected global warming. One consequence of this warming will be a movement towards the poles of the range of many plant species. But will plants be able to migrate poleward as fast as their ranges shift? Species migration is usually a rather slow affair, especially for many long-lived tree species. However, a rapid rise of 2°–5° C in global temperatures in the next 100 years could well translate into a 500–1,000 kilometre range shift poleward – a distance that many species will not be able to travel. This may result in the gradual extinction of many forest communities and, perhaps, a concomitant decrease in carbon storage rates.

Cloud Behaviour

The behaviour of clouds in future greenhouse scenarios is also a major source of uncertainty in global climate models. Clouds play an important role in regulating the earth's energy budget. They routinely cover about half the earth and account for a good deal of the 30 per cent of incident sunlight that the earth reflects into spaces. But clouds do not merely reflect solar radiation from the earth; they also absorb some of the heat radiated by the earth and reflect it back to the surface below. In other words, they also exert a greenhouse effect of their own. Currently, averaged over the globe, the reflective property of clouds dominates their greenhouse effect and, thus, they exert a substantial, net cooling effect on the earth.

Both the reflective effect and the greenhouse effect of clouds are large compared with the greenhouse effect exercised by CO_2 and other trace gases in the atmosphere. Recent data from a satellite-borne experiment show that, on a scale where the warming effect from CO_2 doubling is 1, the cloud warming effect is currently 7. Therefore, the result is a net cooling effect four times larger than the expected warming that would be caused by a doubling of CO_2.

Fertilization Effect of Higher CO_2 Levels

Anticipating how this balance will change is complicated by the many competing influences that could pertain under greenhouse conditions. For instance, photosynthesis rates are known to be sensitive to atmospheric CO_2 level. The effect known as CO_2 fertilization can cause an average 30 per cent increase in growth rates in many, but not all, species when CO_2 concentration is doubled. Recent data suggest that this effect is temperature-dependant and could be increased up to threefold by a 3°C rise in air surface temperatures.

Further, there is evidence that CO_2 enrichment helps some plants minimize water loss, presumably making them more drought-tolerant – an important consideration if the drier conditions predicted by climate models for many midcontinent areas come to pass. If all these favourable effects of greenhouse conditions were realized, a net increase in plant matter might increase the carbon uptake of the biosphere and act as a negative climate feedback, helping to alleviate further CO_2 buildup. Indeed there is evidence that the annual carbon uptake in the Northern Hemisphere has increased during the growing season in recent years, perhaps indicating greater carbon storage.

This potential rise in methane production is particularly worrisome. Methane is 20–30 times more efficient than carbon dioxide at trapping heat, and its concentration in the atmosphere has been increasing more than 1 per cent per year. Within 50 years, it may be the most significant greenhouse gas. Large methane releases from a warmer tundra – or from other sources sensitive to warming, such as deep ocean deposits – could create a major climate feedback.

Biosphere Response

Predicting how living systems will respond to global warming and what climate feedbacks they will exercise may be just as difficult as solving the riddles of ocean dynamics or cloud behaviour and provide still another major source of uncertainty in climate modeling. Terrestrial ecosystems play an important role in the global carbon cycle, storing about 100 billion metric tons of carbon each year in plant materials through the process of photosynthesis and releasing about the same amount through the processes of respiration and decay. This annual flux is equivalent to almost 30 per cent of all the carbon held in the atmosphere, so even minor changes in the balance between photosynthesis, respiration, and decay could affect the level of atmospheric CO_2 quickly.

Although the overall effects of global warming are relatively clear (as noted above), it is not easy to be specific about the precise consequences or locations of the changes that are taking place. According to Lewis (2006:1),

> ... this is because the Earth is an integrated system consisting of a network of sub-systems ... each linked by various processes ... in complex ways.

He makes the important point that it is not just the complexity of the world's natural systems that renders it impossible to make robust predictions but it is also their interactions with global political and social systems. The European Union (EU) Commission is more specific than Lewis in its predictions for Europe, which include the following potential effects:

- *Agriculture and fisheries* increased risk of crop failure through extreme weather; reduction in soil fertility through depletion of organic matter; threats to fisheries from severe effects on coasts and marine ecosystems.
- *Energy* hydro-electric power will be affected both negatively and positively; negatively by the impact of drought in southern Europe leading possible to a 25 per cent reduction; and positively in northern Europe by melting glaciers leading to an increase.
- *Infrastructure* will come under severe threat, particularly in low-lying urban area through flooding.
- *Tourism* will suffer in Alpine regions through snow loss and in Mediterranean regions through excessive temperatures.
- *Water* as well as the flooding danger from too much water, the increased stress placed on water supplies from too little water, particularly for irrigation and drinking, will be a major issue in southern Europe with 'Europe's high water stress areas expected to increases from 19 per cent today to 35 per cent by the 2070s' (CEC, 2009:4–5).

In global warming, increased carbon dioxide and warmer temperatures can have a positive effect too. Agricultural crops, up to an unknown point are expected to respond with increased growth as the concentration of the nutrient carbon dioxide increases. However, the regions suitable for growing particular crops and trees will change, but, this minor positive impact is not in anyway justifying the climate change due to global warming.

Reducing the Risk of Global Warming

International Agreements and Efforts

International communities are well aware with the problem of global warming. In this direction, United Nations Organisation is doing commendable work, but still, it is a long way to go. Not only several international conferences, seminars, workshops have been organized but important agreements have also been signed for international cooperation in this direction. The following are the important conferences, agreements and declarations:

1. *First World Climate Summit*, 1979, Geneva (Switzerland)
2. *Conference on Industries and Climate,* 1980, Vienna (Austria)
3. *Vienna Convention, 1985*
4. *Agreement on Montreal Protocol,* 1987
5. *Toronto Summit, 1988*
6. Formation of *Intergovernmental Panel on Climate Change (IPCC)*, 1988
7. *Second World Climate Summit*, 1990
8. UN Conference on Environment and Development known as *First Earth Summit* or *Rio Summit*, 1992 held at Rio de Janeiro (Brazil)
9. Signed Climate Change Contention during *first Earth Summit*, which aimed at reduction of carbon emissions and to stabilize the emissions at 1990 level by 2000 AD, but it could not implemented.
10. *First Climate Change Convention* also known as *Berlin Summit,* 1995
11. *Second Climate Change Convention* held at Vienna, 1996
12. *UN Second Earth Summit*, New York, 1997
13. *Third Climate Change Summit* held at Kyoto city (Japan), 1998, in which Kyoto Agreement was signed regarding 5.2 per cent cut in the emission of carbon dioxide by the developed countries.

    Other UN Climate Change Conferences held are as follows:

    4th Climate Change Conference, Buenos Aires, 12 November 2000
    6th Climate Change Conference, Hague, 4 September 2001
    7th Climate Change Conference, Marrakesh, 11 November 2001
    8th Climate Change Conference, Delhi, 2 November 2002
    10th Climate Change Conference, Buenos Aires, 24 December 2004
    11th Climate Change Conference, Montreal, 29 December 2005
    13th Climate Change Conference, Bali, 1 January 2008
    14th Climate Change Conference, Poznam, 1 January 2009
    15th Climate Change Conference, Copenhagen, 30 December 2009
    16th Climate Change Conference, Cancun, 4 January 2011

17th Climate Change Conference, Durban, 4 January 2012

18th Climate Change Conference, Doha, 2 December 2012

Apart from above mentioned conferences, agreements and conventions, in all the international conferences related to environmental issues, the climate change and global warming remained an important point of discussion. Regarding reduction of greenhouse gas emissions, special mention of UN Framework Convention on Climate Change and Kyoto Protocol is necessary.

UN Framework Convention on Climate Change

In 1992, at the UN Earth Summit in Rio de Janeiro, many countries signed this convention. Its objective was to stabilize 'greenhouse gas concentrations ... at a level that would prevent dangerous anthropogenic interference with the climate system.' Anthropogenic interference means human interference. About 150 countries ratified the convention. By ratifying it, a country was agreeing to return its greenhouse gas emissions to 1990 levels by 2000. In the United States, then President Clinton developed a Climate Change Action Plan to lower greenhouse gas emissions. However, the steps of the plan were voluntary and US rate of emissions, far from decreasing, continued to climb. The same was true for many other countries. This result indicated that stronger measures were needed to reduce greenhouse gas emissions, a plan that clearly specified: (a) the exact reductions in greenhouse gas emissions; and (b) when these reductions were required.

The Kyoto Protocol

After strenuous negotiating sessions, the Kyoto Protocol (also under the UN Framework Convention on Climate Change) was adopted in late 1997. Under the protocol, industrialized countries agreed to reduce greenhouse gas emissions to an average of 5.2 per cent below 1990 levels between the years 2008 and 2012. Nations signing the protocol then laboured further to resolve three controversial points. (1) Find acceptable means of accomplishing greenhouse gas reductions. (2) Agree on an accounting system that gave nations credit for their reductions. (3) Agree on penalties for lack of full compliance with agreed-upon reductions. This third point was left until later to resolve.

In July 2001, 1,700 diplomats representing 178 countries resolved the first two of these points. Meanwhile, the United States had become only an observer to negotiations because the new President, George W. Bush, refused to consider this ratification 'fatally flawed' document. This made matters difficult for the protocol, because, for it to come into force, 55 countries (accounting for at least 55 per cent of the world's 1990 emissions) had to ratify it. With the United States – accounting for 25 per cent of the world's emissions – standing aside, it is difficult to reach 55 per cent although as of 2003, this may still happen. One US objection was that the protocol does not require less-developed countries to reduce their emissions. They were omitted because it is industrialized countries that are largely responsible for current levels of atmospheric greenhouse gases. And, industrialized nations are the ones able to develop the technology to lead the way in making reductions.

Measures and Policies to Reduce Greenhouse Gas Emissions

To reduce greenhouse gas emissions or even to slow their rate of growth on a global scale will require an extraordinary degree of political consensus. Any one country or even one region can prevent the buildup of greenhouse gases by itself, although leadership by individual countries will be important in achieving global consensus. To be broadly acceptable, ideally, prevention policies should confer other local benefits on their adopting countries in addition to the worldwide benefit of reducing the risk of a global warming.

Many polices have been proposed and are under study. It seems clear, however, that any successful prevention strategy will include five key elements:

- increasing the efficiency of energy production and use,
- switching from carbon-intensive fuels such as coal to hydrogen-intensive fuels such as natural gas, where possible,
- encouraging the rapid development and use of solar and other carbon-free energy sources,
- eliminating the production of most CFCs and developing the means to recapture those now in use, and
- reducing the rate of deforestation

Four approaches suggested to cut greenhouse gas emission and reduction of global warming are: *mitigation, adaptation, geo-engineering* and *conservation.*

Mitigation

The word 'mitigation' has come into common usage in relation to reducing GHGs, because it has several varying meanings. The most apt is, 'To moderate or reduce in strength (heat, cold, light, etc.) so as to make more bearable, to temper the severity of (a climate).' (*Oxford English Dictionary*).
Lewis (2006: 2) summarises mitigation as:

> The basic solution to climate changes is obvious but rarely articulated: most fossil carbon must not get into the atmosphere. Currently the only proven way to do this is to leave most fossil fuels in the ground. That is, no new oil fields, no new coal mines.

Governments are already responding to this challenge, for example, by returning to the construction of new nuclear power station and supporting the expansion of areas such as biofuel production. Funding is being put into the development of technologies to capture and store CO_2, for example, by adding technological interventions onto existing power stations and factories, or by diverting CO_2 from its point of production into storage facilities underground. Industry is also responding by increased investment in alternate energy sources such as wind and aiming to produce more energy efficient technology. Individual people are being asked to cut down on energy usage too. Groups have emerged that aim to tackle climate change and related problems; for example the Transitions Network (TN, 2009) aims to drastically reduce emissions through re-localisation and to address the depletion of fossil fuels through increasing social resilience.

Adaptation

The EU Commission's impact assessment identified three major options for adaptation.

- 'grey' infrastructure approaches based on engineering science including design of the built environment to withstand extreme events;
- 'green' structural approaches including increasing ecosystem resilience, halting biodiveristy loss and restoring water cycles;
- 'soft' non-structural approaches including land use controls and economic incentives to reduce or prevent disaster vulnerability.

Central to these 'soft' approaches are water and land use policies to maintain eco-systems that are optimally resistant to climate change and minimize adverse effects of potential disasters particularly in relation to human well-being and health and the protection of property.

Geo-engineering

Here the approach to reduce atmospheric CO_2 levels involves massive feats of engineering, on a global scale. Possible examples include removing gases from the atmosphere and solar radiation management. Several different proposals have been put forward for removing CO_2 from the atmosphere.

One is to find a way to remove it mechanically and store it either underground or in the oceans. Another is to find a way of removing it directly from the atmosphere, for example, via 'synthetic trees'. Solar radiation management – reducing the amount of sunlight getting to earth – can be undertaken chemically or mechanically. Chemically induced reduction in solar radiation take place naturally when sulphuric particles are emitted by volcanic eruptions. Mechanical approaches include the suggestion to use giant mirrors located in space to reflect the sun's rays away from the earth.

Conservation

The ultimate solution of global warming is the conservation. Conservation of natural resources, along with substantial contraction of consumption, is an essential requisite of the deep environmental approach associated with strong notions of sustainability. This approach sees an intrinsic value in nature that must be sustained at all costs.

The following measures may be adopted by industrialists, individuals as well as by government for the conservation and reduction of GHGs.

Energy Efficiency

Increased energy efficiency appears to be the single most promising option for reducing the risk of global warming. Higher efficiency in energy production and in a wide range of energy uses is the fastest and most cost-effective method for reducing emissions of carbon dioxide and other energy-related greenhouse gases. The replacement of a single 75-watt incandescent bulb with an 18-watt fluorescent bulb, for example, provides the same light but prevents the emission of over 100 kilograms of carbon as CO_2 over the life of the new bulb. Replacing a car that burns 10 litres of gasoline per 100 kilometres (km) with one that is twice as efficient would prevent the emission of more than five metric tons of carbon over a 10-year period. Experience during the past 15 years in the United States, Western Europe, Japan, the People's Republic of China, Korea, the Republic of China and many other countries has shown that large improvements in energy efficiency are possible without a detrimental effect on the rate of economic growth. The evidence suggests that economic growth is not tied inexorable to increased energy consumption, as was once thought. A recent study published by the World Resource Institute suggests that an energy strategy based on increased efficiency can sustain economic growth in both developing and industrialized countries.

More Efficient Power Generation

On the supply side, new power generation technologies offer significantly higher energy conversion efficiencies than have been available in the past. The spread of cogeneration offers some of the greatest potential savings. Cogeneration systems convert primary energy in the fuel to electricity and then use the otherwise-wasted heat to make steam or hot water for local processes requiring thermal energy. Such systems can deliver between 60–80 per cent of the heat in the fuel to useful purposes, compared with about 33 per cent for conventional steam-electricity technologies. Cogeneration has many important applications in industry. In the US paper industry, for example, almost all process steamed and heat requirements are met through the use of cogeneration systems. In Brazil, cogeneration systems in alcohol distilleries provide revenue from elecricity sales as well as heat for distillation and purification processes. In Germany and Scandinavia, cogeneration systems are used widely to supply steam or hot water to heat nearby buildings and provide power.

Carbon-Free Energy Sources

Energy sources and fuels that produce no CO_2 are ideally suited to reduce the risk of global warming. To the extent that biomass energy is produced in a renewable manner – with growth of new biomass equaling that which is burned – no net CO_2 is added to the atmosphere. Such technologies are in operation already in many applications around the world, although in many cases, they are not yet economically competitive with fossil fuels. Some carbon-free energy sources – such as hydroelectric dams and conventional nuclear power stations – raise serious environmental questions of their own. Nonetheless, hydroelectric, wind, solar thermal photovoltaic, and nuclear technologies deserve additional attention. As the eventual alternative to coal-based energy, carbon-free energy sources occupy a unique strategic role.

Solar photovoltaic systems are finding applications in remote area, but are still too costly for central station applications. Technical progress is rapid, however, and may lead to photovoltaic power becoming economically competitive with conventional electricity sources within a decade. Hydrogen may prove to be an important carbon-free fuel for transportation use. Hydrogen is an energy carrier and must be produced by primary energy source through, for example, the electrolysis of water. Hydrogen has the advantage that its combustion products are primarily water vapour, but as yet it is expensive to produce and difficult to store.

Slowing Deforestation and Altering Agricultural Practices

Activities related to agriculture and forestry contribute significantly to greenhouse gas emissions. The largest emissions arise from landuse changes, principally deforestation, which release large quantities of CO_2 and methane to atmosphere as well as N_2O and Ozone. Therefore, reducing and ending ultimately is an immediate priority that would reduces emissions of two principal greenhouse gases dramatically. By suspending tax credits that encouraged deforestation as a means of establishing title to land in the Amazon region, Brazil has decreased the rate of deforestation sharply in the past two years, although the rate remains high.

Some agricultural practices release CO_2 and N_2O through energy inputs, fertilizer inputs, and soil erosion. Enclosing manure piles or feedlots would capture and use methane as a commercial quality fuel, rather than allowing it to become an additional atmospheric burden. Revising agricultural practices to reduce fertilizer requirements would reduce N_2O emissions. No-till agriculture can stabilize carbon in the soil rather than releasing it to the atmosphere through erosion, and improved productivity reduces pressure for deforestation, which currently emanates from the need to rotate continually among poorly performing agricultural plots.

Policies and Politics after Kyoto Protocol

The Kyoto Protocol of 1997 was the beginning of the development of international policy and concerted action on global warming. Its purpose was to firm up the commitments made by virtually all the world's countries in 1992 at the United Nations (UN) Earth Summit which pledged to stabilize GHGs' at a level that would prevent dangerous anthropogenic interference with the climate system. There were two major problems with the Kyoto Protocol. The first is that it was only a very small beginning: the goal was for industrialized nations, on average, to cut their annual GHG emissions by 2012 to 5.2 per cent below their level in 1990. This works out at 29 per cent below that they would otherwise have been if they had continued rising.

The second problem is the continued reticence among many countries for actually doing anything about global warming. The Protocol required ratification by enough industrialized countries to represent 55 per cent of the CO_2 emissions of the developed world and it took a full seven years after the signing of the Protocol for it to be enacted (Kyoto Protocol Status of Ratification, 2006). Australia ratified it in 2008 and Kazakistan in 2009. Up to recent times the least enthusiastic country, and thus the biggest stumbling block to the full implementation of the Kyoto Protocol, has been the US, which at the time of writing had still not ratified Kyoto. Even though it has less than 5 per cent of the world's population, the US is responsible for about 25 per cent of total global warming.

The progress in controlling CO_2 is not satisfactory. Not a single country in the world is on target to meet its contribution to keeping the increase in CO_2 emissions at a maximum of 2°C above pre-industrial levels – every country in the world is going to overshoot.

Nevertheless, some countries, particularly Sweden and Germany, and to a lesser extent, the UK, are making some progress, Sweden has an effective carbon tax with a significant bite, so much so that when it was introduced, income tax was cut by half to compensate. Between 1990 and 2005 its economy grew by 44 per cent but its GHG emissions went down by 9 per cent. Germany lead the world in micro-generation of energy from renewable source: it has 'feed-in-tariffs' where householders and small businesses get paid for generating their own energy and inputing it to the national grid. The price is highly subsidized and fixed for 20 years so Germany is far ahead of any other country in micro generation.

Developing countries are caught in a dilemma in relation to global warming. It is clear that they are the countries which will suffer the most from the impacts of global warming already set in train by the developed countries, but they are also being warned that any attempt on their part to join the club of developed nations by expanding their economies must be constrained in order to minimize carbon emissions. The UN Climate Summit held in Copenhagen in December 2009 was anticipated to produce a step – change in how countries would address environmental concerns. The accord, drafted by the US, China, Brazil, India and South Africa, speaks only of reducing gas emission to limit additional temperature rise to 2°C above pre-industrial levels. There was no agreement on how to achieve these limits in concrete terms.

Several conclusions can be drawn from the preceding analysis of the scientific and political issues surrounding global warming:

- The threat posed by global climate change to human and natural systems is unparalleled in recorded history and means that the community of nations making up the planet must confront the likelihood of rapid and significant global warming.
- Technological and policy decisions made by nations in the relatively near term will affect substantially the timing and severity of any global warming that occurs.
- The technologies needed to prevent, reduce, or mitigate substantially further human impacts on the atmospheric concentrations of the gases contributing to global warming exist, or are likely to exist soon, although they are not disseminated widely.

References

CEC (2009), Adapting to Climate Change: Towards a European Framework for Action, White Paper , Brussels.

Hobes, J.E. (1980), *Applied Climatology*, Butterworths, London.

IPCC (Intergovernmental Panel on Climate Change) (2007), Climate Change, Synthesis Report, Geneva.

Houghton, J. (2009), *Global Warming: The Complete Briefing*, Cambridge University Press, Cambridge.

Kerski, J. and Ross Simon (2005), *The Essentials of the Environment,* Hodder Arnold, London.

Lewis, S. (2006), 'Climate change: in my view', *Royal Society: Science Issues: Climate Change*, http://royalsociety.org/page.asp?id=4689&tip=1.

Small, J. and Witherick, M. (1995), *A Modern Dictionary of Geography,* Arnold, London.

Hodgson M. Susan and Phillips, D. (2011), The Environmental Challenge, in Tony Fitzpartick (eds.) *Understanding the Environment and Social Policy,* The Policy Press, Bristol.

Solomon, S. Platter et al. (2008), Irreversible Climate Change due to Carbon dioxide Emissions, *Proceedings of the National Academy of Sciences of the USA,* Vol. 106, No. 6, pp. 1704–09.

17

Sustainable Development

Man has continually, and at an increasing rate, been changing the forms and modes of his interaction with the environment. Natural processes and factors that depend on the structure and other characters of the earth, and social processes and factors that depend on structure and characteristics of our society, have been closely interwoven. On one hand, population of the earth is growing rapidly and by the year 2010 AD, it might exceed the 7.5 billion mark, while on the other hand, with the technological and scientific revolution, developmental activities have been accelerated to such an extent that in certain areas they have become a threat to the environment. The consumption of natural resources is increasing at an alarming rate, causing great concern for their conservation. The degradation of the environment, due to industrial and other wastes discharged into the atmosphere and hydrosphere, has also caused great concern and there is a big question – what will happen next? All these environmental problems associated with development have raised several questions regarding the type and nature of development and this has given rise to the concept of 'sustainable development'.

Concept of Sustainable Development

The term 'sustainable development' was first used in the Cocoyoc Declaration on Environment and Development in the early 1970s. Since then, it has become a buzzword for international organizations dedicated to achieving environmentally benign or beneficial development. The term has served to catalyse debate over the relationship between economic change and the natural resource base in which it is grounded. Definitions vary, but the most widely adopted one is contained in *Our Common Future* – the 1987 report of the World Commission on Environment and Development: *'Sustainable development is the development that meets the needs of the present without compromising the ability of future generations to meet their own needs.'* This definition is far from adequate – it begs the critical question of what constitutes an acceptable 'need' in the context of present global situation – but it does encapsulate the essential point – that we all have to learn to live within our planetary means. Various UN publications, in an effort to define strategy for sustainable living and development, refer to it as 'improving the quality of human life while living within the carrying capacity of the supporting ecosystem'. Some definitions of sustainable development are:

— A strategy that meets the needs of the present generation without compromising the ability of future generations to achieve their own requirements.
— Improving the quality of human life while living within the carrying capacity of supporting ecosystems.
— Environmental care 'married' to development.
— An environmental 'handrail' to guide development.
— Balancing the fulfilment of human needs with the protection of the natural environment.

The field of sustainable development includes three constituent parts - environment, society and economy. These are interrelated each other as depicted in Figure 17.1.

Figure 17.1

Scheme of Sustainable Development at the Confluence of their Constituent Parts

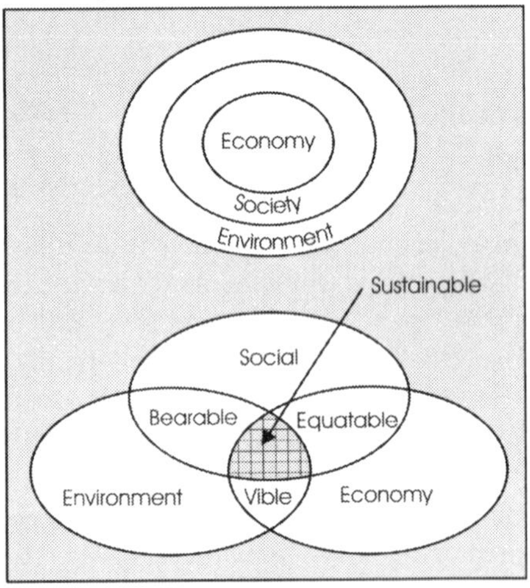

The environment, society and economy are integrated in such a way that each element plays a vital role. The entire concept of sustainable development can be understood through Figure 17.2.

Figure 17.2

The Concept of Sustainable Development

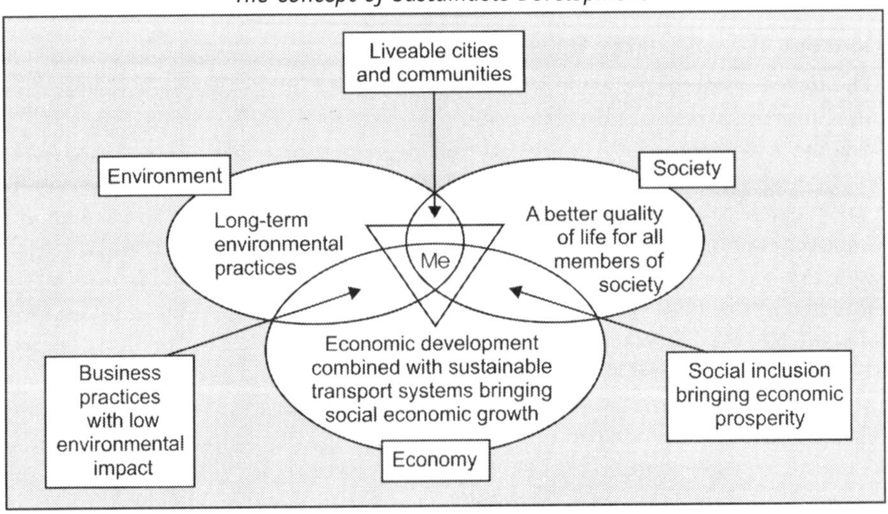

Source: www.uitp.org

Figure 17.3
Sustainable Development Matrix Growth of the Concept of Sustainable Development

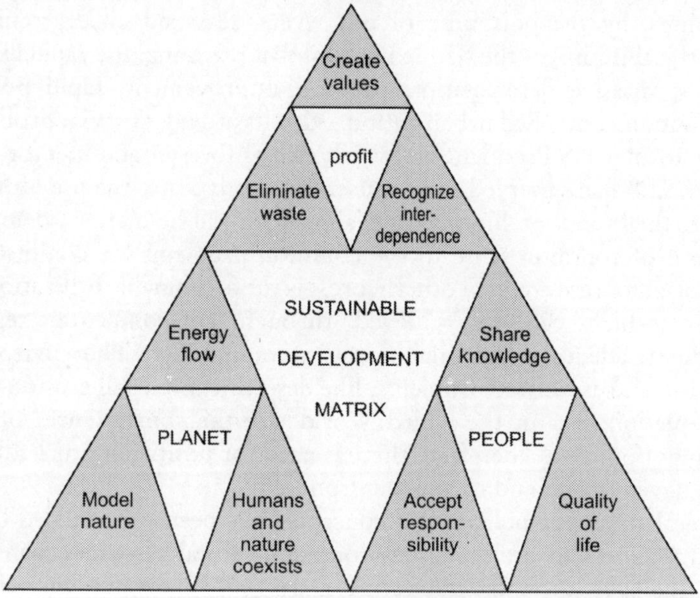

It becomes clear from the above figure that better quality of life in cities and communities can be developed through long term environmental planning based on sustainable economic development. The growth of economy should be directed towards the protection of environment and social well-being.

A matrix of sustainable development shown in Figure 17.3 indicated how planet Earth, its people engaged in various profit making activities can adopt an approach of sustainable development.

Growth of the Concept of Sustainable Development

The concept of sustainability, has a long historical background. In ancient Indian writings, 'nature' or environment has been considered as the controller of all human activities, including economic development. They have given the status of God to all the components of environment, such as air, water, land, natural vegetation, animals, etc. In spite of the fact that there was no need for their conservation, the concept of their protection has been put forward in all the religions of India. In Greek writings, there is a reference of *Ge* or *Gaia* as the Goddess of the Earth – the mother figure of natural replenishment. In fact, that was a time when population was limited and there was mutual adjustment between man and environment.

It was only after the industrial revolution, followed by transport revolution and urbanization, that degradation of the environment started and which has given rise to the concept of sustainable development. The sudden interest in sustainable development is, of course, a response to warnings that the world is facing an environmental and, therefore, social catastrophe in the near future unless mankind radically modifies certain practices and perspectives which have created the present crisis. This crisis is characterized, among other things, by the poisoning of our rivers, seas and underground water sources; the thinning of the ozone layer; global warming; the rapid extinction of species, massive deforestation and soil improvement; rapid population growth; and uncontrolled urbanization, with its attendant social problems.

The former US President Roosevelt's chief forester and mentor, Gifford Pinchot (1910) had observed that 'conservation advocates the use of foresight, prudence, thrift and intelligence in dealing with public matters, demands the application of common sense to the common problems for common good'. Conservation of resources, in other words, is the 'sustainable utilization' of the resources, which can be achieved through environmental regulation, management, adequate surveillance and accountability. The environmental degradation and associated problems, like desertification, soil erosion, tropical forest depletion, etc. in the Third World, are the consequences of present development policies. Therefore, there is need for positive mutual adjustment between development and natural environment.

Several strategies, policies and concepts have been propounded by ecologists and planners to deal with environment/natural resources management, such as the zero growth strategy of Daly (1977), organic agriculture for self-sustaining societies of Naess (1977), a narrow and overspecialized approach of environmental economics by Norgaard (1985) and the well-known concept, i.e. use of environment-friendly techniques and products.

The publication of *US Global 2000 Report* (1980) and other publications like *Resourceful Earth* (Simon and Kohn, 1984), *The Global Possible* (Repetto, 1985) and *Our Common Future* (World Commission on Environment and Development, 1987) have generated a debate highlighting the consequences of overexploitation of natural resources and neglect of global 'common interest'. The various programmes of the UNO, specially the programmes of UNEP, have emphasized the need for sustainable development, also referred to as 'eco-development'.

Despite its simplicity, the phrase 'sustainable development' is capable of carrying a wide range of meanings, sometimes allowing quite different ideas about what should be done to exist side by side without the conflict between them being clear. Both radical environmentalists and conventional development policy pragmatists have seized the phrase and used it to express and explain their ideas about development and environment. In the process they have created a theoretical maze of great complexity (LéLé, 1991; Redclift, 1996; Redclift and Benton, 1994; Adams, 2001). The superficial conformity of writing and thinking about sustainable development hides very real and fierce battles behind the scenes over the meaning.

Sustainable Development at Rio, 1992

The United Nations Conference on Environment and Development (UNCED) at Rio in 1992 (the Earth Summit) was attended by 128 heads of states and in total by the representatives of some 178 governments. Debate at the conference drew very directly on the mainstream ideas about the environment and development that had evolved during the 1980s. The Rio Declaration and the much larger *Agenda 21* were the fruits of endless negotiations at a series of Preparatory Commission meetings and at Rio itself, between teams of diplomats determined to surrender as little as possible of their national interest. A key feature of these debates was the gap between countries in the industrialized North and the underdeveloped South, which became steadily more glaring in the run up to, and during, the conference. The issues of climatic change and biodiversity that dominated the Rio Conference are vitally important to certain countries (especially those vulnerable to sea-level rise), but they are not the principal environmental problems faced by most countries of the South.

The Rio Declaration ended up as a bland list of 27 principles. It was long-winded, and in places, self-contradictory; even after the long debate, the US delegation released an 'interpretative statement' that effectively dissociated it from a number of the principles agreed. It dissented from principle 3 that there was a right to development, the Americans arguing that development was not a right but simply a shared goal, and from any interpretation of principle 7 that suggested that there was an international liability to make development sustainable (i.e. for rich countries to pay for it). Choices about development and environment were matters for individual countries to make their minds up about, and not an issue over which they should be subject to international opinion.

Agenda 21 was even more burdened by divergent opinions, becoming a bloated volume of more than 600 pages in 40 separate chapters. Its scope was enormous, covering issues from biodiversity and water quality to the role of women, children and organized labour in delivering sustainable development. It reflected previous mainstream thinking about sustainable development in several ways. First, it made the need for economic growth a central theme, as in the Brundtland Report. In the sustainable development mainstream, everything is predicated on economic growth, both globally and nationally. Second, Agenda 21 emphasized the familiar straightforward issues of environmental management: all the familiar environmental issues from the *World Conservation Strategy* appear developed but unmistakable. Third, Agenda 21 was technocentric. The first six key themes make this quite clear: growth with power and technology will direct the evolution of policy towards more efficient use of the environment and hence towards a more sustainable world economy. The 'essential means' to achieve sustainability also reflect this technocentrism, building on information, science and environmentally-sound technology. Fourth, Agenda 21 presumed a multilateral benefit to be derived from a sustainable development strategy, as did the Brundtland Report. It suggested that change would arise from the mutual interest of industrialized and non-industrialized countries, and from the concern of present generations about the future. It suggested that this shared interest would cause international financial resources and technology to flow, directed and promoted by international agencies and structured and regulated by international legal instruments. Fifth, like its predecessors, Agenda 21 called for sustainable development through participation. As in *Caring for the Earth*, women, children, young people, indigenous people, trade unionists, businessmen, industrialists, farmers, local authorities and scientists are all summoned to play a role – a rainbow coalition to put flesh on the endless skeleton on the text of Agenda 21.

After Rio

At the close of Earth Summit+5, the nineteenth special session of the United Nations General Assembly, held from 23 to 28 June 1997 in New York – delegates from over 165 countries adopted the *Programme for the Further Implementation of Agenda 21*. In the final document, governments acknowledged that the global environment has continued to deteriorate since 1992, with rising levels of greenhouse gas emissions, toxic pollution and solid waste. Renewable resources, notably fresh water, forests, topsoil and marine fish stocks, continue to be used at rates that are clearly unsustainable.

To address these concerns, governments took action on several fronts at the special session, as reflected in the final document. Among other decisions, they agreed to reconfirm the political commitment to sustainable development from all members of the international community, as well as from all major groups of civil society:

- reconfirm the financial commitments;
- establish an Inter-governmental Forum on Forests under the Commission on Sustainable Development to continue policy dialogue on this issue, including more focused consideration of elements for a possible legal instrument;
- open high-level inter-governmental dialogues on fresh water, and on energy and transport, which will be taken up by the Commission on Sustainable Development at upcoming sessions;
- make a stronger commitment at the global level to such issues as tourism, changing production and consumption patterns, and eco-efficiency; and
- set a more focused work programme for the Commission on Sustainable Development.

Since Rio, there have been three further milestones. First, the Kyoto Protocol was negotiated in 1997, recognizing the importance of a global reduction in GHG emissions in order to slow down the effects of climate change. Second, the Millennium Summit and *Millennium Development Goals* (MDGs). Third, in 2002 the Johannesburg Earth Summit was held and looked back at the progress of Agenda 21, and also set targets for sanitation, energy use, health, agriculture and biodiversity. In addition to meetings, reports and treaties mentioned earlier, the Millennium Declaration and MDGs are of great significance to global environment policy. These were built on a number of UN global conferences during the 1990s. The Millennium Declaration:

> ... marked a strong commitment to the right to development, peace and security, gender equality, the eradication of many dimensions of poverty and sustainable human development. (UN, 2005: v)

Table 17.1 outlines the main components of the Brundtland Report, Agenda 21 and the MDGs. The similarities between these are striking.

Table 17.1

Brundtland Report, Agenda 21 and the MDGs

Brundtland Report (1987)	*Agenda 21 (1992)*	*MDGs (2000)*
Common challenges	Key social and economic dimensions	Eradicate extreme poverty and hunger
Population and human resources	International cooperation	Primary universal education
Food security: sustaining and potential	Combating poverty	
Species and ecosystems resources for development	Changing consumption patterns	Promote gender equality and empower women
	Demographic dynamics	Reduce child mortality
	Human health	Improve maternal health
		Combat HIV/HIDS, malaria and other diseases
Energy: choices for environment and development	Human settlements	
	Strengthening the role of major groups:	Ensure environmental sustainability
Industry: producing more with less	Women	A global partnership for development
The urban challenge	Children and youth	
	Indigenous people	
Common endeavours		
Managing the commons	Farmers implementation	
Peace, security, development and the environment		
	Education and awareness	
	Capacity building in developing countries	
Towards common action: proposals for institutional and legal change	International institutions.	

The similarities in these documents indicate the close-knit relationship between international development and sustainable development. In all these documents, further commitment was asserted between developed and developing nations on sustainable development. There is a need for mutual understanding between developed and developing countries so that sustainable development can be possible, at least to a moderate limit; but still a long way to go.

Eco-development

The concept of eco-development implies that all developmental activities should be performed or designed in such a way that regional ecological balance can be maintained. This was originally advocated by UNEP for environmental planning and was defined as 'development at regional and local levels consistent with the potentials of the area involved, with attention given to the adequate and rational use of the natural resources, and to application of technological styles' (UNEP, 1975).

Dasman (1985) described eco-development as a triangle, one side of which is *basic needs*, the second *self-reliance,* and at the base *ecological sustainability.* Each of these variables needs to be considered, since development will not be sustainable unless poor people are involved in meeting their aspirations. In fact, 'the environment and development are means, not ends in themselves. The environment and development are for the people, not people for environment and development' (Chambers, 1986). There is thus need for 'environment thinking' with 'development thinking' and 'sustainable livelihood thinking'. In Third World countries, the main consideration should be given to 'basic needs' and 'self-reliance' which can be achieved through social planning norms. While designing eco-development for a region, consideration should be given to both short-term and long-term planning. For sustainable eco-development, it is also necessary to give priority for the livelihoods of the poor; this can be achieved through political changes at the local, national and international levels. The political, economic and social forces should be directed to fulfil the needs of population in such a way that environment remains protected.

Basic Aspects of Sustainability

The question of sustainable development has emanated from overexploitation of resources as well as mismanagement of technology. The aspects, which require monitoring of sustainability, include climatic change, biodiversity, disposal of hazardous and toxic wastes, disposal of pollution-generating industries and food and ecological security.

Swaminathan (1991) has identified nine principles for success in promoting ecologically-sound agriculture. These are: (i) land, (ii) water, (iii) energy, (iv) nutrient supply, (v) genetic diversity, (vi) pest management, (vii) post-harvest system, (viii) systems approach, and (ix) location-specific research and development.

The deterioration in the ecological base in various countries in spatio-temporal terms due to irrational management of the resource and environmental systems having damaging repercussions is reflective of unsustainable policy frame and planning strategies followed so far. Its observable signs can be listed as:

1. extensive deforestation accounting for loss of flora, fauna and some rare species;
2. drying up of drinking water resources and fall in the underground water levels;
3. intensifying rate and frequency of flood and droughts;
4. land degradation due to desertification, wastelands, salinity and waterlogging;

5. deterioration in quality of air and water;
6. pressure of population resulting in unemployment and mass migrations; and
7. unplanned urbanization and unprecedented growth of urban slums, etc.

The environmental problems are multidimensional and varied in nature in both developed and developing countries. On one hand, there are global problems which have had their impact throughout the world. On the other hand, every country has its own development as well as environmental problems. Apart from this, regional and local problems need immediate attention. The problems created by technology transfer from developed countries to Third World countries have become a cause for concern, because in the absence of proper management it has resulted in environmental degradation. Such alarms are embodied in the unsustainable activities which may include:

1. intensive cultivation of land without taking adequate care of soil fertility;
2. development of irrigation facilities without proper water management, which leads to waterlogging, alkaline or saline soil;
3. improper use of pesticides, fungicides, herbicides, etc. which causes soil damage and biological imbalance;
4. excessive trapping of underground water accounts for steep fall in underground water level;
5. replacement of high yielding hybrid varieties leads to spread of diseases capable of wiping out the entire crop as it happened with Irish potato crop in 1985 and Bengal rice famines in 1942;
6. excessive use of non-degradable material like plastic creates problems of waste management;
7. discharge of industrial and municipal waste in water bodies leads to problems like unmanageable water pollution; and
8. automobiles and industries have become a major cause of air pollution.

In fact, development without proper management has become a cause of eco-destruction, for which sustainable development is the only solution. In a report for a sustainable world, following steps have been suggested:

1. A production efficiency era of minimum environmental damage costs through *energy* transition.
2. For a stable world population, a *demographic* transition.
3. From non-renewable to renewable *resource* transition.
4. By a global mutually agreed objective between South and North, *political* transition.

While all these initiatives are necessary, the critical shift required is in the perception of man. For 'industrial man' has grown accustomed to thinking that (a) the world's resources are infinite, and (b) they exist primarily to be exploited for the benefit of mankind. These assumptions are leading us to disaster. What is required is that we replace this outmoded perspective with one that sees us all as crew members of spaceship earth and fragile life-support system spinning through space.

Guiding Principles of Sustainability

The guiding principles of sustainability cut across ecological, economic, social and cultural dimensions and there are obvious trade-offs. What needs to be addressed is sustainable development – for whom, in what context, and with what objectives?

The guiding principles of ecological sustainable resource management (Hare et al., 1990) are as under:

- Inter-generation equity providing for today while retaining resources and options for tomorrow.
- Conservation of cultural and biological diversity and ecological integrity.
- Constant natural capital and 'sustainable income'.
- Anticipatory and precautionary policy approach to resource use, erring on the side of caution.
- Use of resources in a manner that contributes to equity and social justice while avoiding social disruptions.
- Limits on natural resource use within the capacity of the environment to supply renewable resources and assimilate wastes.
- Qualitative rather than quantitative development of human well-being.
- Pricing of environmental values and natural resources to cover full environmental and social costs.
- Global, rather than regional or national, perspective of environmental issues.
- Efficiency of resource use by all societies.
- Strong community participation in policy and practice, during the process of transition to an ecologically sustainable society.
- The importance of individual, local and institutional sustainability to agricultural and environmental sustainability needs to be stressed.
- Poorly conceived globalization/trade liberalization threatens the sustainability and existence, economically and practically, of many developing countries (Dragun and Tisdell, 1999).

Temporal dimension of sustainability becomes obvious when we try to relate it to attributes that characterize ecosystems. A miniature ecosystem of microorganisms created in a culture flask may have a brief turnover time of less than an hour. On the other hand, a village pond, a grassland or a forest ecosystem has a turnover time that may have to be viewed in terms of decades or hundreds of years. And, the time frame becomes even greater for a landscape of interacting ecosystem.

Measuring Sustainability

It is well recognized that conservation and sustainable development represent two sides of the same coin, closely interlinked with one another, as one cannot be achieved at the expense of the other. From a human angle, such an integrated approach demands satisfying basic human needs in an equitable manner, maintenance and promotion of social, cultural and biological diversity, and indeed the ecological integrity of the system as a whole. It is in this context that one needs to view the whole question of evaluating sustainable development.

Indicators for Evaluation of Sustainability

Sustainability cannot be measured with a single indicator. A number of different indicators – ecological, economic and social – need to be reconciled in any evaluation of sustainable development of a landscape unit. The economic indicator is the most obvious one. Energy efficiency is an ecological indicator, which is also helpful in the decision-making process.

How does one monitor sustainable development? We need a variety of indicators. The indicators of sustainable development are varied; therefore, compromises are called for. Monitoring and evaluation have to be done using a number of diverse indicators that may be ecological, economic, social and even cultural. Basically, these indicators add up the 'value system' that a given society desires to have.

Ecological Indicators

Ecological indicators may be land use and land cover changes, biomass quality and quantity, water quality and quantity, and soil fertility. Energy management becomes the key measure for achieving the best in terms of the desired objectives.

Land Use and Cover Change: By sequential monitoring of land use and land cover changes, using a variety of technologies such as remote sensing, aerial photography and ground truth measurement, one can get a reasonable estimate of land quality. Any conclusions that suggest land degradation in the long run, obviously are not sustainable. Thus, for land use decisions, on an ecological plane, a plot level analysis, leading through ecosystem level, and eventually getting to the land-scale level analysis, becomes most obvious; a similar set of units, at social scale, would be a family, a village and a cluster of villages in a given ecological landscape. This would ensure that the end-product is sustainable livelihood as a short-term strategy, leading to sustainable development of the region as part of a long-term strategy, through a participatory and interactive analytical framework.

Biomass Quality and Quantity: The quality of biomass generated on the land (biodiversity), and the quantity of biomass generated through rehabilitation measures of a given ecosystem, are important indicators. We need quality and quantity of biomass for multiple use, for ecosystem integrity as well as for sustainable use by the humans.

Water Quality and Quantity: Any land use and cover change, and indeed any land-based human activity, if unsustainable, may get reflected in groundwater recharge and the water quality. Monitoring this through sequential recharge characteristics of the systems and/or monitoring water quality in rivers, ponds and lakes in the area immediately would give an idea about the sustainability of a given developmental pathway chosen for that landscape.

Soil Fertility: Sustainable management of soil fertility implies the effectiveness with which one is able to maintain soil fertility by strengthening internal biological processes within an ecosystem, both within the soil profile and above ground, and thus minimize energy subsides through inorganic fertilisers. The lesser the energy subsidy, the more sustainable the soil fertility management becomes.

Economic Indicators

Economic indicators are very much standardized in the already prevalent developmental paradigm. Monetary output/input analysis, capital savings or asset accumulation and dependency ratio are some of the measures adopted for economic evaluation. On a macro-scale, the gross national product (GNP) is a unified indicator of national developmental level.

Social Indicators

In a social sense, 'quality of life' is a composite indicator, which is difficult to monitor, and even more difficult to reconcile with the sub-units contained within. More easily measurable social indicators are health and hygiene, nutrition, food security, morbidity symptoms, educational status, etc. The difficult to quantify indicators such as societal empowerment, and the less tangible ones in the area of socio-cultural values are more difficult to evaluate.

The best way to approach sustainable development is an integrated approach through full understanding of linkages between the ecological and social processes. Figure 17.4 depicts indicators of sustainability of natural resources.

Figure 17.4

Indicators of Sustainability of Natural Resources

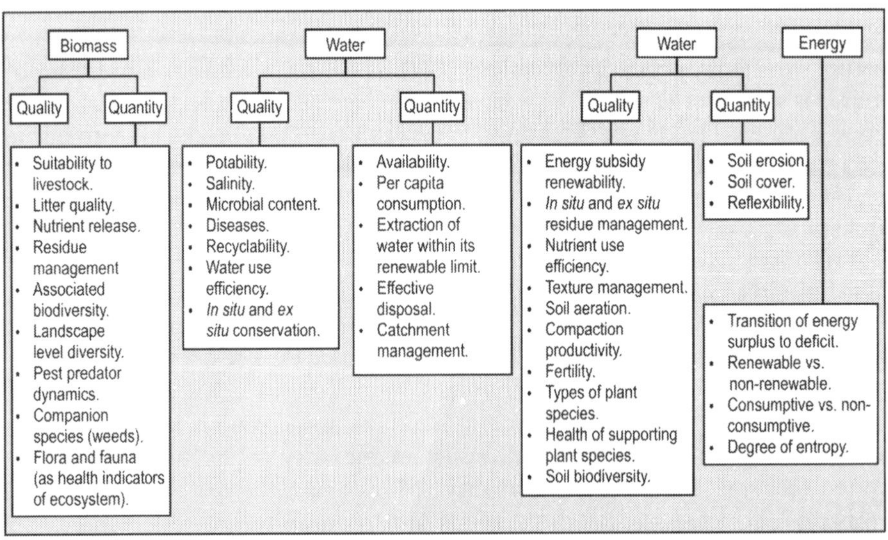

To cope with increased demand of the basic requirement of life and the limited supply of the natural resources, along with considerations of environmental degradation and ecological balance, we need to make adjustments and adaptations with these complex global phenomena, and move towards sustainability of the biosphere. These were the considerations that moved an international group of ecologists from 14 countries to come together in Cuernavace, Mexico, and propose an interdisciplinary cooperative research programme – the International Sustainable Biosphere Initiative – in the early 1990s. These were the considerations that prompted the US National Research Council to set an agenda for action, which will enable the humans to complete the transitional journey towards sustainability. The council has developed an 'Integrated and Location-Specific Approach for Protecting Sustainability' in the following form:

Reporting on the Transition

- Indicators are essential to inform society how, and to what extent, progress is being made in navigating a transition towards sustainability.
- It needs to be understood that multiple indicators are needed to chart progress towards the goals for meeting human needs and preserving life support systems, and to evaluate the efficacy of actions taken to attain these goals.

Integrated Knowledge and Action

- Since the pathway to sustainability cannot be charted in advance, it will have to be navigated through trial and error and conscious experimentation. The urgent need is to design strategies and institutions that can better integrated incomplete knowledge with experimental action into programmes of adaptive management and social learning.

Priorities for Research Sustainability Science

- Develop a research framework that integrates global and local perspectives to shape a place-based understanding of the interactions between environment and society.
- Initiate focused research programmes on a small set of understudied questions that are central to a deeper understanding of interactions between society and the environment.
- Promote better utilization of existing tools and processes for linking knowledge to action in pursuit of a transition to sustainability.

Priorities for Action: Knowledge-action Collaboratives

- Accelerate current trends in fertility reduction.
- Accommodate an expected doubling or tripling of the urban system in habitable, efficient and eco-friendly manner.
- Reverse declining trends in agricultural production in Africa.
- Accelerate improvements in the use of energy and materials.
- Restore degraded ecosystems while conserving biodiversity elsewhere.

References

Barrow, C.J. (1995), Sustainable Development: Concept, Value and Practice, *The Third World Planning Review*, 17(4), pp. 369–387.

Chambers, R. (1986), 'Sustainable Livelihoods: An Opportunity for the World Commission on Environment and Development', Institute of Development Studies, University of Sussex, Brighton, England.

Dasman, R.F. (1985), Achieving the Sustainable Use of Species and Eco-systems, *Landscape Planning*, 12, 211–219.

Dragun, A.K. and Tisdell, C. (1999), *Sustainable Agriculture and Environment: Globalization and the Impact of Trade Liberalization,* Edward Elgar, Cheltenham, UK, p. 308.

Hare, W.L. et al. (1990), *Ecologically Sustainable Development,* Australian Conservation Foundation, Fitzory, Victoria.

Redclift, Michael (1993) Sustainable Development, *Environmental Values*, Vol. 2, pp. 3–20.

Swaminathan, M.S. (1991), 'Sustainability: Beyond the Economic Factors', in The Hindu Survey of Agriculture, 1991.

U.N. Publication, *Earth Summit+5:* Programmes for the Further Implementation of *Agenda 21.*

UNEP (1992), *Agenda-21*: Programme of Action for Sustainable Development.

18

Towards Sustainability

Sustainable development is the need of the present time not only for the survival of mankind but also for its future protection. Unlike the other great revolutions in human history – the Green Revolution and the Industrial Revolution – the 'sustainable revolution' will have to take place rapidly, consciously and on many different levels and in many different spheres, simultaneously. On the technical level, for example, it will involve the sustainable technologies based upon the use of non-renewable, fossil fuels for technologies that take advantage of renewable energies like the sun, wind and biomass, the adoption of conservation and recycling practices on a wider scale, and the transfer of cleaner and more energy efficient technologies to countries in the developing world.

On the political and economic levels, it will involve, among other things, the overhauling of development and trade practices which tend to destroy the environment, and the improvement of indigenous peoples, a fairer distribution of wealth and resources within and between nations, the charging of true cost for products which exploit or pollute the environment, and the encouragement of sustainable practices through fiscal and legal controls and incentives. On the social plane, it will involve a renewed thrust towards universal primary education and health care, with particular emphasis on the education and social liberation of women. On the environmental level, we are talking about massive afforestation projects, renewed research into and assistance for organic farming practices and biopest control, and the vigorous protection of biodiversity. On the informational level, the need is for data that will allow the development of accurate social and environmental accountancy systems.

The aim of ecologically sustainable development is to maximize human well-being or quality of life without jeopardizing the life support system. The measures for sustainable development may be different in developed and developing countries according to their level of technological and economic development. But developing countries, like India, can focus attention on the following measures:

- ensure clean and hygienic living and working conditions for the people;
- sponsor research on environmental issues pertaining to the region;
- ensure safety against known and proven industrial hazards;
- find economical methods for salvaging hazardous industrial wastes;
- encourage afforestation;
- find out substitutes for proven hazardous materials based on local resources and needs instead of blindly depending on advanced nations to find solutions;
- ensuring environmental education as a part of school and college curriculum;
- encourage use of non-conventional sources of energy, specially solar energy;
- as far as possible, production of environment-friendly products should be encouraged;
- use of organic fertilisers and other biotechniques should be popularized;
- environmental management is the key for sustainable development, and it should include monitoring and accountability; and
- need for socialization and also humanization of all environmental issues.

The prime need for sustainable development is the conservation of natural resources. For conservation, the development policy should follow the following norms:

1. Make all attempts not to impair the natural regenerative capacity of renewable resources and simultaneously avoid excessive pollution hampering the biospherical capacity of waste assimilation and life support system.
2. All technological changes and planning strategy processes, as far as physically possible, must attempt switch from non-renewable to renewable resource uses.
3. Formulate a phase-out policy for the use of non-renewable resources in general.

Thus, for a worldwide sustainable growth, there is need for efficient and effective management of available resources. In this field, the production of 'environment-friendly products' (EFP) is a positive step. With the industrialisation and technological development, markets are flooded with products of daily consumption. They could however be a source of danger to health and damage to our environment. There is thus need to distinguish the more environmentally harmful consumer products from those which are less harmful, or have a more benign impact on the environment right from the stage of manufacture through packaging, distribution, use, disposal and reusability or recycling.

Throughout the world, emphasis is now being put on the production of EFP. In India, plans are afoot to market EFPs with combined efforts of Bureau of Indian Standards, Ministry of Environment and Forests and Central

Pollution Control Board. Since 1990, a scheme of labelling *ECOMARK* has also been started. In its first phase, the items included in this are soaps, plastics, papers, cosmetics, colours, lubricating oil, pesticides, drugs and various edible items. The objectives of the scheme are: (i) to provide an incentive for manufactures and to reduce adverse environmental impact of their products, (ii) to reward genuine initiatives by companies to reduce adverse environmental impact of their products, (iii) to assist consumers to become responsible in their daily lives by providing them information to take account of environmental factors in their purchase decisions, (iv) to encourage citizens to purchase products which have less harmful environmental impact, and (v) to improve the quality of the environment and to encourage the sustainable management of resources.

Not only in consumer goods production but in the field of energy production also, environment-friendly techniques of power generation can be used. For example, in power production from coal, PFBC (Pressurized Fluidized Bed Combined Cycle) technique is useful in which coal is burnt efficiently and cleanly in combined cycle plants.

To cope with increased demand of the basic requirement of life and the limited supply of the natural resources, along with consideration of environmental degradation and ecological balance, we need to emphasize on optimal management of land, water, minerals and other natural resources. There is also need to utilize the native wisdom of those people, who live close to nature and earth, for eco-restoration along with development.

In order to apply the principle of sustainable management in reality, a highly complex way of looking at the problem is required, involving various disciplines. Sustainability is first and foremost a mental question. Without a grasp of the need or the will to change awareness, we will not succeed in realizing the principle of sustainability in agriculture. It is upon the decision-makers in politics to create the right framework and the pre-conditions for a sustainable development in agriculture. Global involvement, on the other hand, must not be left out of account. Sustainability reflects our understanding of necessity and responsibility on the question for whom, for what and how production can be guided into the future in a way that is efficient, environmentally sound and sparing on resources.

Global change is an ecological phenomenon, whereas globalization is concerned with economic change. A recent analysis of sustainable agriculture in the context of trade liberalization and globalization raises equally significant concern for a more informed decision-making process at local, regional and international levels. The emerging issues related to the impact of globalization on sustainable agriculture are as follows:

- There are explicit problems with the conventional, theoretical and economic conditions for agricultural sustainability, especially when applied at the global level.

- The processes of trade liberalization and globalization will not be uniform given the ecological and institutional diversity of the nations of the world.
- There will be disparities in globalized impacts between rich and poor countries – for agriculture, industries, sustainability and environment as well as income and poverty.
- There is need for serious analysis of problems and policy initiatives, since the risk of disruption to agricultural systems and environmental deterioration, social disruption and dislocation in the poorer countries of the world is clearly very high.
- The type of production technology research, facilitated by private research, will not address the significant public good and externality issues facing developing countries.
- There is need to focus on local farming situations as a basis of dealing with global problems, especially in poor countries.
- There is need to understand local institutional situations so as to determine appropriate remedial economic policies based on institutional sustainability.
- Integrated approach is essential for research and action at the regional scale related to water, atmosphere and climate, and species and ecosystems.

The pursuit of sustainability demands choices about the distribution of costs and benefits in space and time. There is also need to take advantage of the 'traditional ecological knowledge' (TEK), which encompasses all issues related to ecology and natural resource management, both at local and regional levels. Along with political dimensions of environment-society relations, the TEK can be used for both eco-restoration and sustainable development.

Economics of Sustainable Development

The development and the environment have a two-way relationship: economic growth can have both good and bad impacts on environment. Environmental economics has sought to extend neo-classical theory, by encompassing the environment, and attaching monetary values to losses in natural capital (Redclift, 1993). The mainstream of sustainable development is based on a set of ideas that Low and Gleeson (1988) label 'market environmentalism'.

Environmental economists like Pearce (1985) have advanced thinking with economics in a number of important ways, such as:

1. They recognize that changes in natural capital stocks involve both costs and benefits.
2. There is also a recognition that both 'non-use' and 'use' values are important. Environmental values include 'existence values' which can be arrived at notionally, and a (theoretical) price attached to them.

3. Environmental economics also recognises uncertainty and irreversibility as principles in resource conservation.
4. It is also recognized that 'optimality' refers only to economic uses and efficiency, and effectively excludes and social goals for conserving resources.

Pearce (1989: 8) further declares that 'we know natural capital is valuable because people are willing to pay to preserve it ... .' He also observed, 'Whether there is an actual market in the asset or not is not of great relevance. We can still find out what people would pay if only there were a market.'

Using economic concept, sustainability can be defined as constant stocks of both man-made and natural capital maintained over time. Economic development has always implied increasing capital stocks, but insights of environmental economics suggests the need to account for stocks of both human-made and natural separately. An alternative approach allowed the possibility of trade offs between losses to natural capital in one project and gains elsewhere.

Financing Sustainable Development

The world is facing serious environmental problems, which can be solved and controlled only when proper finances are available. The question is – who will pay? Naturally, those who are responsible for it, but at the same time, it is the responsibility of those developed countries, which are capable both technologically as well as financially. Efforts have been made in this direction but still there is long way to go.

Accepting sustainable management as an objective has been easier to achieve than that of the contentious issue of financing it. The available financial estimates vary in quality, and should be treated only as indicative and order of magnitude estimates. They do, however, provide the starting point for analysing financial requirements for sustainable estimates. This lack of robustness of the estimates also makes firm international agreements difficult to achieve.

Financing sustainable development involves following three interrelated issues:

- the level or total amount of funding needed, depending on the scope of the overall programmes and targets to be achieved in specific areas;
- terms and conditions involved in the sharing of costs between countries specifically between the developed (high income) and the developing (low and middle income) countries; and
- countrywise mobilization of the requisite internal resources.

Funding necessary to provide sustainable development to the developing countries has been estimated in three different ways as discussed below:

Estimate 1

The two conventions of UNCED (Framework Convention of Climatic Change and Convention on Biological Diversity) established the principle that the developing countries should be refunded the full agreed incremental costs associated with measures taken under them with the Global Environment Facility (GEF) as the interim funding mechanism, However, as environmental protection and development are inextricably linked, the term 'full agreed incremental cost' has proved to be difficult to define and operationalise. The two conventions did not mention any specific cost.

The GEF was endowed with an initial funding of $1.3 billion for a three-year period from July 1991 to June 1994. It concentrates on four key areas: global warming, biological diversity, pollution of international waters, and stratospheric ozone depletion. Since March 1994, the GEF has become a permanent financial mechanism.

The GEF is managed through a collaborative tripartite arrangement involving the World Bank, UNEP and the UNDP. In practice, the World Bank, where the facilities are located, had been the dominant partner. Up to 80 per cent of the GEF funds can be attached to the Bank's sector loans. The GEF grants up to early 1996, totalled over $1 billion to fund 113 projects in 63 developing countries, with approximately 45 per cent of the total commitments funding biodiversity projects and a further 40 per cent funding projects dealing with climatic change. A group of 30 wealthy donor countries is already committed to tripling the GEF core fund to $4 billion. The actual GEF funding of around $1 billion is regarded as too small, especially in comparison to the official development assistance ($51.7 billion in 1993) and private financial flows ($157.7 billion in 1993) to make a significant difference. The supporters of the GEF argue that although the amount involved is small, it can act as a catalyst for global sustainability. For example, by providing only the incremental cost for financing more eco-friendly technology, the GEF can optimise its leverage from the limited resources, and have a much larger impact that its own resources would suggest.

Estimate 2

The UNCED secretariat estimated the average annual (over the 1993–2000 period) cost of implementing Agenda 21 programmes for the developing countries. The estimates were based on incomplete supporting documents and not reviewed by the national governments. The estimated annual cost was just below $600 billion, with $130 billion (22%) provided through grant or concessional requirements. This amount is put in perspective when the GDP of all low-income and middle-income countries is noted: $5,276 billion in 1994. The $470 billion ($600–$130 billion) cost to be borne by the developing countries is 8.9 per cent of their 1994 GDP.

The $130 billion concessional element to be funded by the high-income countries is 0.65 per cent of their 1994 GDP of $20,120 billion. However, the total official long-term net resource flows to developing countries in 1993 was only $51.7 billion, or 40 per cent of the estimated $130 billion. The private net flows of 1993 to the developing countries were $157.7 billion. Thus, the importance of official flows is now much less than that of private flows.

Even if the newly industrialized countries, such as Taiwan, South Korea and Singapore, begin making overdue contributions to net official flows of assistance to the developing countries, the requisite amount is unlikely to be realised. The fiscal stringency being faced by the high-income countries, along with the ideological shift towards the private sector as opposed to the public sector solutions, makes it unlikely that the estimated concessional resource flows would be forthcoming.

The UNCED secretariat estimates were carried out according to the chapter/programme areas of Agenda 21. The largest sums were for the following areas: human settlements ($215 billion), fresh water ($55 billion), human health ($51 billion), agriculture ($32 billion), deforestation ($31 billion), poverty ($30 billion), biotechnology ($20 billion), and education, public awareness and training ($15 billion). It should be stressed that Agenda 21 takes a broader view of the environment and therefore poverty reduction, trade liberalization, health care, shelter, etc. are included as part of the environmental management effort.

Estimate 3

The 1992 World Development Report was devoted to the development and environment. It is estimated that in 2000 AD the additional investment in selected environmental programmes would amount to $75 billion or about 1.4 per cent of the GDP of developing countries for that year. The major items of additional investments are soil conservation and afforestation ($15–20 billion), reducing emissions, effluents and wastes from industry ($10–15 billion), controlling main pollutants from vehicles ($10 billion) and family planning ($7 billion). The report also recommended that such investments should be made on a self-financing basis to the extent consistent with equity and other objectives.

The above three examples of financial estimates suggest that clarity in this area is much below the desired level, in spite of its crucial importance. The methods used to arrive at these estimates not always being explicit, their robustness cannot be determined. It is not clear whether in kind services are included in the estimates, and what is the level of efficiency assumed in carrying out environmental projects.

Financial resources and mechanism play a key role in the implementation of Agenda 21. In general, the financing for the implementation of Agenda 21 will come from a country's own public and private sectors. For developing countries, ODA is the main source of external funding and substantial new and additional funding for sustainable development. Renewed efforts are essential to ensure that all sources of funding contribute to economic growth, social development and environmental protection.

Agenda 21: Programme of Action for Sustainable Development

Agenda 21 is a programme of action for sustainable development worldwide. This is also known as the Rio Declaration on Environment and Development, adopted by more than 178 governments at the United Nations Conference on Environment and Development, known as the Earth Summit, held in Rio de Janeiro, Brazil, from 3 to 14 June 1992.

Agenda 21 is intended to set out an international programme of action for achieving sustainable development in the 21st century. It stands as a comprehensive blueprint for action to be taken globally by governments, United Nations Organisation, development agencies, non-governmental organizations and independent sector groups, in every area in which human activity impacts the environment. It provides a broad review of issues pertaining to sustainable development, including statements on the basis for action, objectives, recommended activities and means of implementation.

Principles

Agenda 21 is based on the following principles:

1. Human beings are at the centre of concerns for sustainable development. They are entitled to a healthy and productive life in harmony with nature.
2. States have, in accordance with Charter of the United Nations should pursue their own environmental and developmental polices and make sure that they do not damage the environment of other states or of areas beyond the limits of national jurisdiction.
3. The right to development must be fulfilled so as to equitably meet development and environmental needs of present and future generations.
4. In order to achieve sustainable development, environmental protection shall constitute an integral part of the development process and cannot be considered in isolation from it.
5. All states and all people shall cooperate in the essential task of eradicating poverty as an indispensable requirement for sustainable development.
6. The special situation and needs of developing countries, particularly the least developed and those most environmentally vulnerable, shall be given special priority.
7. States shall cooperate in a sprit of global partnership to conserve, protect and restore the health and integrity of the earth's ecosystem.
8. To achieve sustainable development and a higher quality of life for all people, states should reduce and eliminate unsustainable patterns of production and consumption and promote appropriate demographic policies.
9. States should cooperate to strengthen endogenous capacity-building for sustainable development by improving scientific understanding through exchanges of scientific and technological knowledge.
10. Environmental issues are best handled with the participation of all concerned citizens at the relevant level.
11. States shall enact effective environmental legislations.
12. States should cooperate to promote a supportive and open international economic system that would lead to economic growth and sustainable development in all countries.

13. States shall develop national law regarding liability and compensation for the victims of pollution and other environmental damages.
14. States should effectively cooperate to discourage or prevent the relocation and transfer to other states of any activities and substances that cause severe environmental degradation or are found to be harmful to human health.
15. In order to protect the environment, the precautionary approach should be widely applied by states according to their capabilities.
16. National authorities should endeavour to promote the internalization of environmental costs and the use of economic instruments, taking into account the approach that the polluter should, in principle, bear the cost of pollution.
17. Environmental impact assessment, as a national instrument, shall be undertaken for proposed activities that are likely to have a significant adverse impact on the environment.
18. States shall immediately notify other states of any natural disasters or other emergencies that are likely to produce sudden harmful effects on the environment of those states.
19. States shall provide prior and timely notification and relevant information to potentially affected states on activities that may have a significant adverse transboundary environmental effects.
20. Women have a vital role in environmental management and development. Their full participation is therefore essential to achieve sustainable development.
21. The creativity, ideals and courage of the youth of the world should be mobilized to forge a global partnership in order to achieve sustainable development and ensure a better future for all.
22. Indigenous people and their communities and other local communities have a vital role in environmental management and development because of their native knowledge and traditional practices. States should recognize and duly support their identity, culture and interests and enable their effective participation in the achievement of sustainable development.
23. The environment and natural resources of people under oppression, domination and occupation shall be protected.
24. Warfare is inherently destructive to sustainable development. States shall, therefore, respect international law providing protection for the environment in times of armed conflicts and cooperate in its further development, as necessary.
25. Peace, development and environmental protection are interdependent and indivisible.
26. States shall resolve all their environmental disputes peacefully and by appropriate means in accordance with the Charter of the United Nations.

27. States and people shall cooperate in good faith and in a spirit of partnership in the fulfilment of the principles embodied in this Declaration and in the future development of international law in the field of sustainable development.

Structure

Agenda 21 is a report of over 500 pages, comprising 40 chapters, organized into four sections, which address the major areas of political actions as follow:

Section I: Social and Economic Development

1. International cooperation to accelerate sustainable development in developing countries and related domestic policies
2. Combating poverty
3. Changing consumption pattern
4. Demographic dynamics and sustainability
5. Protecting and promoting human health
6. Promoting sustainable human settlement development
7. Integrating environment and development in decision-making

Section II: Conservation and Management of Resources for Development

1. Protection of the atmosphere
2. Integrated approach to the planning and management of land resources
3. Combating deforestation
4. Managing fragile ecosystems: combating desertification and droughts
5. Promoting sustainable agriculture and rural development
6. Conservation of biological diversity
7. Environmentally sound management of biotechnology
8. Protection of oceans and seas
9. Protection of the quality and supply of fresh water resources
10. Environmentally sound management of toxic chemicals
11. Environmentally sound management to hazardous wastes
12. Environmentally sound management of solid wastes and sewage-related issues
13. Safe and environmentally sound management of radioactive wastes

Section III: Strengthening the Role of Major Groups

1. Preamble
2. Global action for women towards sustainable and equitable development
3. Children and youth in sustainable development
4. Recognizing and strengthening the role of indigenous people and their communities
5. Strengthening the role of non-governmental organizations
6. Local authorities' initiatives in support of Agenda 21

7. Strengthening the role of workers and their trade unions
8. Strengthening the role of business and industry
9. Scientific and technological community
10. Strengthening the role of farmers

Section IV: Measures of Implementation

1. Financial resources and mechanisms
2. Transfer of environmentally sound technology, cooperation and capacity building
3. Science for sustainable development
4. Promoting education, public awareness and training
5. National mechanism and international cooperation for capacity-building in developing countries
6. International institutional arrangements
7. International legal instruments and mechanism
8. Information for decision-making

The programme areas are generally structured in four heads as under:

- The basis for action
- The objectives
- The activities
- The means of implementation

In fact, Agenda 21 is a guideline for sustainable development. A programme for the further implementation of Agenda 21 was also adopted at the ninetieth special session of the United Nations General Assembly (Earth Summit +5), held from 23 to 28 June 1997 in New York by over 165 countries. This programme was designed to reconfirm the political commitment to sustainable development from all members of the international community, as well as from all major groups of civil society.

Millennium Development Goals

Another major step in sustainable development which covers the whole range of human activity and social development is the 'millennium development goals' set by the UN in 2000 for achievement by 2015. These are :

1. Eradicate extreme poverty and hunger: reduce by half the 1.3 billion people living on less than $1 a day and the 815 million without enough food to meet daily energy needs.
2. Achieve universal primary education: 115 million children; most in sub-Saharan Africa and Southern Asia, do not go to school. Reduce this number.
3. Promote gender quality and empower women: in three main areas – education, employment and political decision-making.

4. Reduce child mortality: 11 million children die before the age of 5, most from treatable diseases. The aim is to cut this by two-thirds.
5. Improve maternal health: particularly in sub-Saharan Africa, by investing in health facilities, birth attendants and education.
6. Combat HIV/AIDS, malaria and other diseases: An estimated 39.4 million people are affected by these diseases each year, The aim is to reduce these numbers significantly.
7. Ensure environmental sustainability: reverse the loss of environmental resources – forests, biological diversity, the ozone layer – and provide safe water, adequate sanitation and affordable housing.
8. Develop a global partnership for development: encourage the North to increase aid to developing countries, to move to fairer trade and cancel the debts of struggling developing countries.

Environmental Issues and Sustainable Development

Man, with his technological and scientific skills, has made rapid developments in agriculture, mining, industries, transportation, forestry, land management and other areas. Unfortunately, in doing so, there has been much disruption of the functioning of natural environment. Now, people have begun to realize the side-effects of the developmental activities and have started their efforts in minimizing these effects by adopting methods and technology for sustainable development. Some of the important environmental issues associated with development, and the role of ecology in possible solution of these problems, i.e. sustainable development, are as follows:

Food Production and Organic Farming

The increasing demand of food for growing population has become a serious problem in many parts of the world, specially in developing countries like India, China, Indonesia, Egypt, Brazil and many other Afro-Asian countries. Agricultural scientists have been successful in increasing the quantity and quality of food and other crops. This has been possible due to application of chemical fertilisers, high yielding seeds, crop rotation practices, weed control, pesticides as well as due to mechanization in agriculture. This resulted in the Green Revolution beginning in the late 1960s with an attempt to increase food production in developing countries. New varieties of wheat, rice and corn were developed that responded better than the traditional varieties to fertilisers, irrigation and chemicals for pest and weed control.

But, along with the increase in food production, there came an increased dependence on energy-based expensive technology and several ecological problems have emerged due to the excessive use of chemicals either in the form of fertilisers or as pesticides. Now, it has been realized throughout the world that use of chemical fertilisers and other chemicals is harmful to soil productivity and also a cause of water and air pollution. The Green Revolution has not only disturbed the ecosystem but is also a cause of social revolution. In countries like USA, farmers are now shifting from routine use of herbicides, pesticides, fertilizers, etc. to organic farming in which no chemicals are used.

The only fertilisers used are organic matter, such as manure, crushed limestone, etc. This system is similar to that of ancient Indian system which is more eco-friendly. The adoption of organic farming is a way for sustainable development. By adopting proper management practices, one can develop high yielding varieties even in this system. Along with agriculture, *aquaculture* can also be developed. The best way is to develop fish farming, which fulfils the increasing demand of foods, specially that of protein.

Figure 18.1
Fundamentals of Sustainability in Agriculture

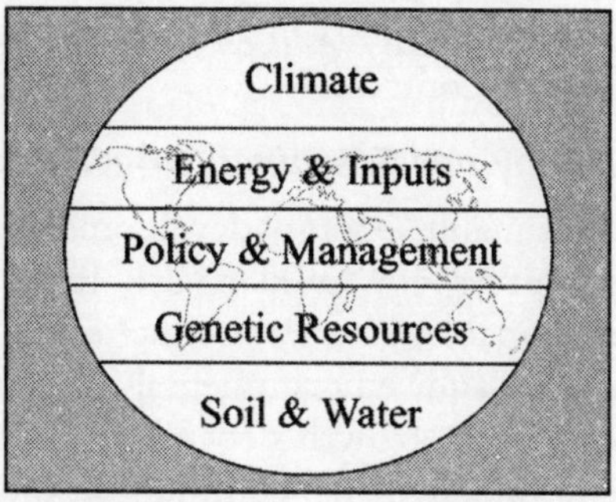

The fundamentals of sustainability in agricultural production systems could be summarised in five major elements as displayed below in Figure 18.1:

1. *Climate*: constraints and impacts.
2. *Energy and inputs*: energy resources, fertilisers, plant protection, ecological farming, techniques and technology.
3. *Policy and management*: political dimensions including economical, cultural and social issues, research and development as well as population control policy.
4. *Genetic resources*: identification, evaluation and utilization of genetic resources.
5. *Soil and water*: resources and requirements.

By adopting certain techniques agriculture can become sustainable. Some of the agricultural technologies with high potential sustainability are as follows: (A. Elliot, 2010)

Intercropping – The growing of two or more crops simultaneously on the same piece of land. Benefits arise because crops exploit different resources, or interact with one another. This interactions may also serve to control pests and weeds.

Rotations – The growing of two or more crops on the same piece of land. Benefits are similar to those arising from inter cropping.

Agro-forestry – A form of inter cropping in which annual herbaceous crops are grown with perennial trees or shrubs.

Sylvo-pasture – Similar to agro-forestry, but combining trees with grassland and other fodder species on which livestock graze.

Green manuring – The growing of legumes and other plants in order to fix nitrogen and then incorporating them in the soil for the following crops. Commonly used green manures are *Sesbania* and the fern *Azolla* which contains nitrogen fixing, blue green algae.

Conservation tillage – Systems of minimum tillage or no tillage, in which the seed is placed directly in the soil with little or no preparatory cultivation.

Biological control – The use of natural enemies, parasites or predators, to control pests.

Integrated pest management – The use of appropriate techniques of controlling pests in an integrated manner that enhances rather than destroys natural controls.

Energy and Sustainable Development

Sustainable energy development (SED) involves the harnessing of energy resources for human use in a manner that makes development more sustainable. Energy production and use have strong links with the economic, social and environmental dimensions of sustainable development. *First*, economic growth depends on energy services such as heating, refrigeration, cooking, lighting, communications, motive power and electricity.

Second, energy is a basic social need that affects human well-being. Wide disparities in access to affordable commercial energy and energy services in both urban and rural areas threaten social stability. Access to decentralized small-scale energy technologies is an important element of successful poverty alleviation. Women and children, who are relatively more dependent on traditional fuels, suffer disproportionately.

Third, there are many environmental links (mainly negative), since energy production and use continue to be a primary source of local, transnational and global pollution. Specific impacts include: groundwater and air contamination; land degradation and changes in use; marine and coastal pollution; ecosystems destruction and loss of biodiversity; damage to health, structures and natural systems from SO_2 and *Nox* and ash particulates which degrade air quality; and, finally, greenhouse gas emissions which have harmed the global environment.

Energy for sustainable development will require: the more efficient use of energy, especially at the point of end use in buildings, transportation and production processes; increased reliance on renewable energy sources; and accelerated development and deployment of new energy technologies. Some strategies and policies for globally achieving SED include: encouraging greater international cooperation in areas such as technology procurement, harmonization of environmental taxes and emissions trading, and energy efficiency standards for equipment and products; adopting policies and mechanisms to increase access to energy services through modern fuels and electricity; building capacity among all stakeholders to improve energy decisions; advancing innovation, with balanced emphasis on all steps of the innovation chain, encouraging competitiveness in energy markets to reduce total costs of energy services to end user; cost-based prices, including phasing out all forms of subsides for fossil fuels and nuclear power and internalizing external environmental and health impacts; encouraging greater energy efficiency; and developing and/or diffusing new technologies to wider markets.

In the ecosystem, energy plays a very important role. The sun is the original source of energy on our planet. All forms of energy, i.e. nuclear, wind, coal, electric, water, gasoline, etc. originate from the sun. The total energy received by the earth's surface from the sun is enormous and much more than required by human beings. But, in modern times, the use of fossil fuels, hydroelectricity and to some extent nuclear fission are the sources of energy. The following table indicates the production and consumption of various sources of energy in the world:

Table 18.1

Production and Consumption of Energy in the World

Sources	*Production*	*Consumption*
Petroleum	126,900	127,222
Natural gas	72,791	68,290
Coal	102,322	101,660
Nuclear energy	18,373	18,373
Hydroelectricity	22,493	22,493
Total	342,872	338,038

The rapid exploitation of these energy sources has resulted in an energy crisis. In coming times fossil fuels such as oil, gas and coal will become scarce. The impact of big dams and thermal as well as atomic power stations is a cause for concern to environmentalists. Therefore, for sustainable development, it is necessary to adopt non-conventional sources of energy like solar, wind, tidal, etc. Of these, the best is solar energy, provided that we are able to harness it with cheapest possible technology. In tropical countries like India, solar energy is the only alternative, which will provide an energy without any adverse impact on environment. Steps have already been taken for the development of photovoltaic cells, solar-energized water pumps, air conditioning units, solar water heaters, solar home-heating units, solar cookers, etc. Research in this field is also going on. What is needed is the development of cheaper technology to harness the solar and other non-conventional sources of energy.

Energy is directly related to the most critical social issues which affect sustainable development. These are poverty, jobs, income levels, access to social services, gender disparity, population growth, agricultural production, climate change and environmental quality, and economic and security issues. Without adequate attention to the critical importance of energy and all of these aspects, the global social, economical and environmental goals of sustainability cannot be achieved. Indeed, the magnitude of change needed is immense, fundamental and directly related to the energy produced and consumed internationally.

World primary energy demand is projected to expand by more than one-half between 2005 and 2030, or 1.6 per cent annually (IEA, 2005). Over two-thirds of this growth will come from DCs (Developing Countries), where per capita use is low. At present, DCs comprise over three-quarters of the world's population, but utilize only one-quarter of the world's energy. Most of this population live in rural area, have little or no access to commercial energy, and rely heavily on traditional biomass fuels such as wood, crop waste and animal dung.

Energy demand has been growing rapidly in DCs in recent decades, creating supply shortages. Making energy development more sustainable in the developing world will require reconciliation of economic growth (including increased use of energy) and poverty alleviation efforts, with responsible stewardship of the environment. This has to be done without overburdening already weak economies and without diverting funds from other key goals such as education and healthcare.

The importance of energy in agricultural production, food preparation and consumption is evident and essential (UNDP, 1997). The world still continues to seek energy to satisfy its needs without giving due consideration to the social, environmental, economical and security impacts of its uses. It is now clear that current approaches to energy are unsustainable. It is the responsibility of political institutions to ensure that research and development of technologies supporting sustainable systems should be transferred to the end-users. Scientists must bear the responsibility to understand the earth as an integrated whole and the impact of our actions on the global environment in order to ensure sustainability and to avoid disorder in the natural life cycle.

The key challenge in realizing these targets is to overcome the lack of commitment and to develop the political will to protect people and the natural resource base. Failure to take action will lead to continuing degradation of natural resources, increasing conflicts over scarce resources and widening gaps between the rich and poor. We must act while we still have choices. Implementing sustainable energy strategies is one of the most important levers the humankind has for creating a sustainable world. Most present trends in energy indicate a deteriorating situation (UNDP, 1997). Furthermore, current energy patterns are aggravating this process by an over-preoccupation with centralized energy supply and fossil fuels to the detriment of energy efficiency, decentralized supply and renewable energy.

The use of non-convential and renewable energy is necessary for sustainable development. These are as follows:

Small Hydro

Mini- and micro-hydro plants are popular for providing local power supply and irrigation without the massive environmental and social impacts arising from large hydro schemes. Micro hydro-projects are currently underway in many DCs, which have the advantage of providing electricity to rural areas beyond the central grid.

Solar Energy

Advances in solar thermal technologies and direct electricity generation by photovoltaic (PVs) have reduced costs significantly since the late 1990s. While these techniques pose no significant occupational and health risks at the generation stage, some environmental impacts may arise from the loss of land use resulting from its high land intensity. Solar energy is constrained by its limited applicability.

Geothermal Energy

Geothermal energy uses the earth's heat to drive steam turbines, Cost-competitive exploitation of geothermal energy with current technology is largely limited to the volcanically active Pacific 'Ring of Fire' and the Mediterranean, where suitable steam reservoirs are located within one mile of the surface. Geothermal steam carries with it a number of atmospheric pollutants, including carbon dioxide, mercury and radon. Currently, the toxins are commonly re-injected into the reservoir. With recent advances, hot dry rock technology will be competitive with conventional geothermal technology and fossil-fuelled plants.

Biomass Power Generation

A wide range of options exists for converting biomass into electric power. Most potential applications are for remote power supply in DCs. Concerns about the effects of dedicated fuelwood plantations on the local environment and competing land uses impose constraints on the widespread adoption of dendro-thermal power supply schemes. It is economical to use biomass for energy purposes only where it is available as a byproduct of other processes. Another source of biomass for power generation is municipal solid waste, whose use is limited both by public concern over emissions and the shortage of concentrated sources of solid waste. In tropical areas, the water content of solid wastes reduces it attractiveness as a fuel.

Wind Power

Large grid-connected wind generators (75KW–450KW) in multi turbine 'wind farms' now supplement power grids in many countries. The total installed wind-power capacity was about 47300 MW in 2004, with Germany, Spain and the USA accounting for 35 per cent and 14 per cent respectively (MIT, 2005). However, wind energy provided only 0.3 per cent of the total electricity supply,

Ocean Energy

Pilot plant experience with ocean thermal energy conversion (OTEC) indicates that temperature differences of 20°C between surface waters and waters about 1000m below are sufficient for economic power generation. There is a vast potential for this energy source in the tropical oceans. Due to high capital requirements and the unproven technology, no large commercial plants are functioning.

Pollution and Environmental Degradation

Pollution is the result of human activities associated with industrial and automobile revolution. The forms, causes, impact and management aspects of pollution have been discussed in detail in earlier chapters. Pollution is a man-made problem, mainly of affluent countries and now a serious problem of the developing countries as well.. The developed countries used to exploit the natural resources of developing countries and also export dump such materials and machines which have become a source of pollution.

The common forms of pollution are air, water, soil, noise and radioactive pollution, which can be divided into (i) non-degradable, and (ii) biodegradable pollutants from ecosystem viewpoint. The main pollutants are deposited matter, carbon monoxide, sulfur dioxide, sulfuric acid droplets, hydrogen sulfide, fluoride and other halogens, metals like lead, iron, zinc, bromine, iodine, agricultural pollutants like pesticides, insecticides, herbicides, fungicides and fertilisers, complex organic pollutants, photo-chemical pollutants from automobiles, radioactive minerals, unwanted noise, municipal wastes, etc. All these pollutants have become a threat to mankind, as they not only result in several diseases, but also environmental degradation and immediate steps should be taken to bring about a pollution-free environment. For sustainable development pollution should be checked first at source. The location of industries should be selected in such a way that their pollutants could not affect the nearby localities, and there must also be treatment plants. The various measures to check pollution have already been discussed in detail in preceding chapters. There is now need for environmental education, monitoring, environmental management and also to fix accountability and responsibility for damages, if any. The production of eco-friendly products is an innovative concept in this direction (Discussed in detail in Chapters 7, 8 and 9).

Problem of Waste Disposal and its Management

In developed countries enormous use of commodities in daily life is causing environmental problems. This problem is now not limited to developed countries but growing day by day in developing countries also. Wastes and toxic substances are accumulating at a rapid rate in our surroundings. These substances enter the biogeochemical cycles. In man-made ecosystems, various types of wastes accumulate in the environment which upset the growth of microorganisms. A variety of synthetic matter like plastic, rubber, glass, etc. are not degradable, thus, causing environmental problems. Wastes include domestic wastes, industrial and mining wastes and agricultural wastes.

Various types of wastes, if properly managed, can be used and steps have already been taken in this direction. Through gene engineering, new strains of bacteria and other microorganisms are being bred, which can help to decompose various harmful industrial effluents into non-toxic agents.

Scientists have also developed methods and technology to utilize sewage waste for wood, cattle feed, fertiliser and fuel production. Municipal sewage treatment plants remove large objects, floating matter and settleble solids and also through physico-chemical and biological processes break down the organic matter and remove harmful microbes before release of the treated sewage into water bodies.

Due to the lack of proper facilities, sewage is allowed to enter nearby water bodies without any treatment. The untreated sewage causes serious water pollution and also damages the local ecology. It is, therefore, high time that sewage is recycled in an economically feasible and environmentally safe manner to meet the various needs of mankind. Several techniques have been developed for sewage treatment and to extract water for irrigation, fodder, fuel and fertiliser from it. Efforts are also being made to use sewage on a large scale to improve farm lands, reclaim spoilt areas and create new recreational facilities. Use of sewage water for fish farming is common in countries like Germany, Poland, Russia, Hungary, Israel, China, USA, Indonesia, etc. Sewage is being also exploited for fuel, the methane gas released due to bacterial growth can be used as fuel. There is thus an urgent need to develop an efficient waste management system.

Irrigation and Problem of Salinity and Waterlogging

During the last fifty years, vast areas have been provided irrigation facilities. This has definitely increased the agricultural production throughout the world. But, due to the lack of water management and also due to illiteracy of the farmers and the geological structure of a particular region, vast areas are now suffering from the problem of salinity and waterlogging.

Due to the gradual depletion of organic matter and microbial population involved in decomposition, there is increase in salinity and waterlogging, which have adverse effects on soil fertility. Waterlogging and increased salinity are mostly the outcome of poor drainage and mismanagement of water in areas covered by large irrigation projects. The recent example is the Ganganagar and Bikaner districts of Rajasthan, where problem of waterlogging has become serious and several villages have been evacuated and there is no possibility of crop production in these areas. This has happened due to the irrigation water of the Indira Gandhi Canal. In Punjab, the seepage from unlined canals as well as due to irrigation has raised the water table by 7–10 metres above the previous levels. Similar is the problem of salinity, which has damaged the irrigated land. These problems can be checked by proper water management system.

Deforestation and Desertification

The problem of deforestation has arose mainly due to destruction of forests for various developmental or commercial purposes. The growth of population, rapid expansion of cities and other settlements, demand for land for new agricultural farms, mining, construction of roads, dams, etc. have all resulted in mass cutting of natural vegetation. Many useful plants have been made extinct and several thousands are nearing extinction. In the absence of reforestation, the forest cover throughout the world is declining at a rapid pace. The dense tropical forests too are suffering from the problem of mass cutting for commercial purposes.

In order to protect the environment and ecosystem, action plans should be implemented not only to protect the present forest cover but also for large-scale afforestation, so that 30 to 33 per cent of the land area can be developed as natural and/or planted forests. There is also need for the enforcement of present laws strictly to avoid illegal and indiscriminate felling of trees by businessmen, industrialists and rural masses. Both social and agro-forestry should be encouraged and waste lands be developed as forest lands. Besides, public participation should also be encouraged at all levels.

The threat of desertification and expansion of desert area is also due to overgrazing, loss of natural vegetation and other human activities. UNESCO studies have shown that approximately 43 per cent of total land area in the world has been affected by desertification. The suggested measures for the check of desertification are improvement in landuse patterns, stabilization of sand dunes, development of natural vegetation suitable to arid climate and reduction of population pressure, etc.

Sustainable development should essentially be based on maintaining the fragile balance between productivity functions and ecosystem. For sustainable development, our approach should be holistic on every resource sector and at all levels of local, regional and global scale. Unless we adopt an approach for sustainable development, the problem of environmental degradation cannot be solved.

References

Elliott, Jennifer A. (2010), *An Introduction to Sustainable Development,* Routledge, Abingdon, UK, p. 166.

Kayastha, S.L. (1997), Geography and Sustainable Development: Environmental Scenario and Perspectives, *Annals of the NAGI,* 1997, Vol. XVII, No. 1–2.

Pearce, D. (1985), Sustainable Futures: Economics and the Environment, Paper Department of Economics, University College, London.

Ramakrishnan, P.S. (2001), *Ecology and Sustainable Development*, NBT, New Delhi.

Redclift, M. (1993), Sustainable Development, *Environmental Values,* 1993, Vol. 2, pp. 3–20.

UN Publication, *AGENDA-21, Rio Declaration on Environment and Development,* 1992.

UN General Assembly (1997), '*Earth Summit+5*', New York.

19

Governance and Geo-politics of the Environment

Environmental issues now have become too much important that not only state required special efforts for its governance but also needs global governance because most of the environmental issues are 'trans-boundary' in nature. In the words of the Commission on Global Governance (1995:2):

'Governance is the sum of the many ways individuals and institutions, public and private, manage their common affairs. It is a continuing process through which conflicting or diverse interests may be accommodated and cooperative action may be taken. It includes formal institutions and regimes empowered to enforce compliance, as well as informal arrangements that people and institutions either have agreed to or perceive to be in their interest.'

Thus governance relates to both formal government institutions as well as the less formal arrangements of social movements and non-governmental organizations. The question is what form of governance is required to tackle environmental problems properly. Tony Fitzpatrick (2011) has suggested a number of relevant categories.

Levels: Global, trans-regional, regional, sub-regional, national, sub-national, local and municipal.

Actors: International organizations, nation states, networks and alliances, philanthropic and research foundations, firms and multinational corporations, cities, NGOs, civil society organizations, public-private partnerships, communities, pressure groups and campaigns, media, social movements, administrators and practitioners.

Sectors: Economy, polity, private sector, voluntary and independent sectors, public sectors, welfare systems.

These categories intersect one another in complex and dynamic way. The governing agencies often change, originate and flow across boundaries. Governance across boundaries has become important in drawing conclusions about how environmental policies are currently organized. Table 19.1 shows the format of governance across boundaries:

Table 19.1

Governance Across Boundaries (in context with environment)

	Global	*Regional*	*National*	*Local*
Economic incentives	Carbon markets	EU Emissions Trading Schemes	Subsidies (such as reduced sales tax) and/or low interest loans to facilitate green choices, energy efficiency	Council Tax rebates
Regulation	International conferences, treaties and conventions since 1972 Millennium Development Goals (MDGs) IPCC	Polluter pays principle EU Directions	Huproved growth system to local authorities Changes to transper and planning infrastructure	Recycling and waste management Adaptations to buildings and energy use, e.g. combined heat and power
Education, knowledge and information	Earth Days to promote public awareness	Directive on freedom of access to information	Engagement with NGOs and other relevant actors within society	Growth of sustainable communities, e.g. UK eco-towns Local Agenda 21

Source: Tony Fitzpartick, *Understanding the Environment and Social Policy*, p., 165.

The environment polices include three types – economic instruments, regulation, education and information. *Economic instruments* are concerned with financial incentives, such as carbon trading, higher or lower taxes, subsides, in order to steer behaviour in green directions.

Regulation means 'command-and control', which is a dominant approach of many governments. Regulation includes establishment of environmental goals and objectives, quality criteria, quality standard, level of pollution, emissions standards etc. and enforce the standards and other requirements through negotiations, penalties and other sanctions. While the final policy instrument is the *education and information*, through which actual causes and effects of the whole system is not only explained to public but also their active participation can be obtained.

The state plays a vital role in forming policy for the various environmental issues. Every country of the world is having its own department of environment and also have rules and regulations to protect its resources and environment. Organized conservation and preservation groups have undoubtedly had a measure success in regulating the impact of modern societies on the natural environment. *Green politics* is also developing. It is different from the mainstream of formal politics because it takes society's treatment of the natural environment as a key political issue.

The promise of Green politics is to introduce the condition of the natural environment into all areas of political decision making in order to create societies that will be sustainable. Green parties have also developed Global

Greens, a network of Green parties from across the world, who produced a statement of principles in their 2001 Canberra Charter. These principles are:

- ecological wisdom;
- social justice;
- participatory democracy;
- non-violence;
- sustainability; and
- respect for diversity.

International or Global Governance and Geo-politics of Environment

The dilemma is that the environment is international, but the power remains with the nation-states, therefore, a system of global governance is required, which can be possible only through mutual international cooperation. Like many issues, the environment has also become an issue of international politics, especially between the North and the South. But, at the same time, a hierarchical structure is emerging for planning and overseeing environmental issues at the global level and also for linking them with the development. This arrangement operates in two ways. First, the establishment of international organizations, conferences and agreements such as the Montreal Protocol, CSD, Agenda 21 and UNCED leads to periodic reviews of the state of the world's environment. Second, environmental degradation often occurs with a development project, such as power plant, construction of dams and reservoirs, or planning a new town. In most cases, environmental impact assessment (EIA) is required. The UNDP and the World Bank are two organizations, which usually provide such assistance, and both are also very much aware of the adverse effects of project-related environmental degradation. But, the main problem is the ineffectiveness of international systems over state sovereignty and also lack of enforcement power. In fact, there was no real response at the international level to transboundary environmental problems before the United Nations Conference on the Human Environment (UNCHE) was held at Stockholm in 1972.

United Nations and the Environment

UN is playing a key role in international issues such as the environment. Its actual beginning, as stated above, has started from Stockholm in 1972. The Stockholm Conference originated from the North's concern about pollution. At an early stage, it was already clear that the South was suspicious about the northern motives. At the meeting in 1971 of the Preparatory Committee (PrepCom) at Founex, Switzerland, the northern delegations agreed that lack of development was in itself an agent of pollution and had accepted that the southern states had legitimate concerns in seeking to emulate the pattern set by the North. In the *Stockholm Declaration on the Human Environment*, accordingly, the 26 principles set out included specific recognition that, given 'rational planning' and/or 'integrated development', development could be achieved without harm to the environment. However, neither the Declaration nor the accompanying *Action Plan for the Human Environment* was specific about what would constitute either of these two desirable goals.

The more important consequence of the Stockholm Conference, therefore, was the establishment of the United Nations Environment Programme (UNEP) - the first UN agency to have its headquarters in a southern state - Kenya. In its first 25 years of existence, UNEP has some important achievements to its credit, including raising environmental awareness and helping to foster national environmental protection, even if mainly in the AICs.

The *World Conservation Strategy* (WCS), published by the IUCN in 1980 with the support and assistance of UNEP, was the first major statement to employ the term 'sustainable development' and to try to clothe that concept in specifics.

The Brandt Commission's first report, North-South: A Programme for Survival, accepted that development based on simple considerations of economic growth had failed to secure 'human dignity, security, justice and equity' for much of the world. After its second report, in 1983, the Brandt Commission was dissolved, and in the same year the General Assembly set up a new World Commission on Environment and Development (WCED), chaired by the Prime Minister of Norway, Gro Harlem Brundtland. The numerical majority of the southern states in the UN was reflected in the selection of the other 22 members.

The new Commission recognized that the twin concept of environment and development had to be addressed together. It found the solution to the apparent contradiction between them in the concept of 'sustainable development'. The Brundtland Report, *Our Common Future* (WCED, 1987) defined sustainable development as: 'development that meets the needs of the present without compromising the ability of future generation to meet their own needs'. Sustainable development does not use non-renewable resources faster than substitutes can be found, or renewable resources faster than they can be replaced, nor does it emit pollutants faster than natural processes can render them harmless.

It was the Stockholm Conference on the Human Environment, with its slogan 'Only One Earth', the first of many large international conferences, that added the environment as a huge global policy problem to the UN agenda and set the scene for the signing of a series of key international treaties. Previously, single environmental issues had been dealt with by different UN agencies. Just as Stockholm led to the creation of UNEP, so its concurrent Environmental Forum brought (accredited) NGOs formally into the process of environmental policy making for the first time, and set the pattern that Global Forum would follow at Rio.

UNCED was a consequence of the deepening environmental crisis, which was increasingly subject to investigation and thus more understood as a result of scientific progress. It was this understanding that led to the Brundtland Commission publishing *Our Common Future,* which provided the backdrop to the preliminary meetings that were to culminate in the UN Conference on

Environment and Development in Rio de Janeiro: UNCED, also known as ECO 92 or the 'Earth Summit'. It was the largest gathering of world leaders ever held, bringing together representatives of 179 states, including 118 heads of states in person, and it put ten related issues squarely on to the global environmental agenda.

UNCED recognized the interconnectedness of the environmental and development crises. Nevertheless, the stress placed upon these two aspects of the conference by varied participants. Developing countries emphasized their crisis of development and international economic inequality as the main source for the environmental crisis. They also expressed their fears of environmental imperialism.

The United Nations Development Programme (UNDP) organizes and administers most of the technical assistance provided by UN system to individual countries in their social and economic development. At the same time, it also has concern about environmental degradation. Planning and implementation of development projects necessarily requires environmental consideration, which is carried out within the UNDP. Under the present system, each project proposal should be reviewed for possible negative impacts on the environment. The UNDP is also involved in building country-level capacity for sustainable development as formulated in Agenda 21: information gathering, project evaluation and project monitoring. In 1992, in its 31st meeting, the Governing Council of the UNDP urged the UNDP administrator 'to ensure the environmental concerns are taken into account in all United Nations Development Programme programmes and projects to the maximum extent possible'. The summary of the annual report of the Administrator for 1992 states:

> Faced with the mounting recent evidence of accelerating environmental degradation and the serious obstacles it presents to poverty eradication, UNDP has intensified its efforts to integrate environmental concerns into the mainstream of its development activities. Since the Earth Summit, UNDP has also moved quickly to launch a major programme in support of Agenda 21. This programme includes, among other components, reviewing UNDP's own capacities for supporting sustainable development activities; reviewing the fifth cycle country programmes and preparing for a more detailed review to be carried out during the cycle to ensure that they fully support a sustainable development strategy; and launching a major capacity-building programme – Capacity 21 – to support recipient countries in their implementation of Agenda 21.

The UNDP has put in place following three major environment-related activities:

- environmental overviews for country programmes (EOP);
- environmental overviews of specific projects/programmes and a formulation of management strategies (EOPMS); and
- Capacity 21.

The UNDP programmes for specific countries are planned on the basis of five-year units. At the end of each five-year period, a new series of programmes are designed for each specific country, which are now called Country Cooperative Frameworks (CCF). It has been recommended that the basic geographic conditions and environmental signatures of a country are considered and incorporated within the programme when a new country's programme is set up. This is the environmental overview for the country programme (EOP) whose objective is to highlight the critical environmental conditions so that any development project or programme may utilize the existing strengths such as technical facilities, and at the same time, avoid accelerating possible major environmental degradation, such as soil erosion. An EOP will include the following:

- a brief description of the natural environment of the country;
- identification of the main environmental issues;
- an assessment of the relationship between economic development and the environment;
- an evaluation of the ability of the country to achieve appropriate environmental management and sustainable development;
- the identification of likely environmental impacts associated with the implementation of the UNDP country programme; and
- a comparison of the alternatives to the proposed country programme.

The United Nations Conference on Environment and Development (UNCED), held at Rio de Janeiro (3–14 June 1992), was a remarkable event in the direction of global governance for environment. As many as 178 governments actively participated in the conference. The highlight of the conference was signing of several agreements. A number of international agreements have come into force in the twenty years between Stockholm and Rio. Examples include treaties and conventions on World Heritage (1972), Convention on International Trade of Endangered Species of Wild Flora and Fauna (CITES), the Antarctic (1980), the Law of the Sea (1982), the Montreal Protocol on CFCs (1987), and the Movement of Hazardous Waste (1989).

A number of agreements regarding aid and trade practices between the developed and developing countries could be added to the list. Besides the Brundtland Commission, two other UN commissions have submitted their reports: the Brandt Commission on North-South issues and the Palme Commission on security and disarmament. The recommendations of the Bruntland Commission were followed by a UNEP report, *Environmental Perspectives to the Year 2000 and Beyond* and the IUCN publication, *World Conservation Strategy*. The time seemed appropriate to call a global meeting to translate the recommendations into action and usher in the age of sustainable development. In December 1989, the UN General Assembly convened the United Nations Conference on Environment and Development (UN Resolution 44/228), and the Brazilian Government offered to host the meeting. The participatory governments affirmed the following five agreements:

Framework Convention on Climate Change

It is a legally binding treaty aimed a lowering the emission of CO_2 and other greenhouse gases. The treaty was expected to come into operation immediately. The developed countries should lead, and the developing countries should be compensated for costs incurred for taking necessary measures for controlling climatic change. Governments should submit reports on their activities and meet regularly to review progress.

Convention on Biological Diversity

It is another legally binding treaty aimed at halting the destruction of biological diversity, protection of species, and also establishing standards of conduct in the sharing of research information, profits and technology in the area of genetic research. It was agreed that states have sovereign rights over their biological resources but fair and equitable sharing should take place.

Rio Declaration

The declaration is a set of 27 non-binding principles which emphasize the link between environment and development.

Agenda 21

The agenda for the 21st century is a document of over 500 pages that contains a non-binding action programme for sustainable development which was adopted by consensus by more than 170 countries. It covers different aspects of environment, resources, population and development. A participatory and community-based approach to problem-solving is preferred.

Forest Principles

It is a shortened version of a more comprehensive plan for forest protection and management which was originally discussed but failed to secure world-wide affirmation. It also recognizes the sovereign rights of states to utilize forest resources.

Apart from the above five agreements, other environmental issues and management arrangements were discussed, including (1) desertification and droughts at the insistence of a number of African nations; (2) institutional arrangements for environmental management including the agreement to create a high-level commission on sustainable development, financial resources and mechanisms; and (3) the role of women related to environment and development.

Certain very positive results came out of Rio. Tremendous enthusiasm and publicity were generated regarding environment and development. The concept of sustainable development was globally recognized. The roles of unofficial organizations and common people were recognized. Various post-Rio arrangements were made to supervise the global environmental scene as discussed earlier. However, a number of shortcomings remained, to the disappointment of various environmental groups and member countries of G77. Two of the shortcomings are disturbing. First, the financial arrangements for governing the environment of the world fell short of the target. The UNCED secretariat in Agenda 21 estimated the need for $600 billion a year, most of which had to come from the developed countries. A minuscule portion of this was forthcoming. Second, a number of countries felt that the Convention did not give enough importance to the environmental problems of the developing countries. The real test of UNCED will, of course, be the extent to which the governments ratify and follow up the commitments they have made in Rio, especially regarding the climatic change and biodiversity conventions. Finally, the role of the United Nations was strengthened as the legitimate leading organization for advancing sustainable development and also for supervising global environmental agreements and conditions.

On 12 February 1993, the United Nations Economic and Social Council (ECOSOC) established a high-level monitoring agency, the United Nations Commission on Sustainable Development (UNCSD) to encourage and oversee progress in implementing Agenda 21 at various levels - nation, regional and global. In brief, the UN is doing a commendable job in global governance for environment.

The World Bank and the Environment

The International Bank for Reconstruction and Development (IBRD), also known as the World Bank, is one of the two economic institutions set up at the Bretton Woods Conference at the end of the Second World War. The other institution is the International Monetary Fund (IMF). The World Bank initially concentrated on the reconstruction of Europe, and subsequently on development, essentially through projects, programmes and sector loans.

The linking on Bank-assisted projects with environmental degradation, primarily in the 1970s, prompted structural and policy changes. In 1970, the Bank established the post of Environmental Advisor. In May 1984, the Operational Manual Statement 2.36, *Environmental Aspects of World Bank*, was issued, requiring the introduction of environmental considerations at the time of project identification and preparation. This was to be followed by modifications, if necessary, at later stages of the project. The speed of the integration of environmental concerns into the normal working procedure of the Bank is partly due to the realization that environmental degradation is incompatible with promoting development. It was further stipulated that projects with severe environmental impacts would not be financed without proper mitigatory measures. However, incidences of environmental degradation, caused by Bank-financed development projects continued, with some of these developing into well-publicized cases and confrontations with local NGOs. Projects that contributed to the destruction of tropical rainforests with associated loss of biodiversity and resettlement of indigenous people are clear examples.

In May 1987, a series of structural changes were implemented to strengthen the Bank's environmental policies, procedures and resources. These included the creation of an Environment Department along with Environment Divisions in each of the four regional technical departments.

Currently, the Environment Department includes the Land, Water and Natural Habitats Division (ENVLW), the Pollution and Environmental Economics Division (ENVE), and the Social Policy and Resettlement Division (ENVSP). October 1989 saw the introduction to the Operational Directive on Environmental Assessment, which required an environmental assessment (EA) for all projects with possible environmental connotations. All prospective loans needed to be screened and classified into categories depending on the nature and magnitude of probable environmental impact. The original four categories (A-D) were later reduced to three (A-C). The determination of the screening category depends on a combination of factors: project location, sensitivity of environmental issues, nature of impact and magnitude of impact. For example, the location of a project near sensitive and valuable ecosystems, archaeological sites, cultural and social institutions, a high density of population, water courses, etc. may give it an 'A' rating, which is the highest potential class for adverse environmental impact.

- *Category A*: A full EA is required for projects that fall into category 'A' as such projects are expected to have adverse environmental effects that may be sensitive, irreversible and diverse. Such projects may involve large-scale pollution discharge directly into soil, water or air, large-scale physical disturbances, replacement of an existing forest, modification of hydrological cycle components, storage of hazardous material, or displacement of people.
- *Category B:* A full EA is not required but a limited environmental analysis is necessary. The impact is seen as less significant and not irreversible. Examples include rehabilitation, agro-industries, aquaculture, small-scale irrigation and drainage, and tourism.
- *Category C*: Category 'C' includes projects that do not require an EA or other environmental analysis. Examples include general projects dealing with education, family planning, health matter, and human resource development.

Globalization, Trade and Environment

Globalization can be defined as the process by which events, decisions and activities in one part of the world can come to have significant consequences for individuals and communities in quite distant parts of the globe. With the growing globalization and international trade, its impact is also significant on environment.

The impression of globalization has been reinforced by the strengthening of international economic organizations. These include the summit meetings of the Group of Seven, the IMF, the World Bank and, most recently, the WTO. With the end of the Cold War, G7 (and G8) summits have gained a new significance. The media correctly recognize that these are more important than sessions of the UN. Henry Kissinger originally proposed a 'summit' to consider international economic issues in 1971. However, it was at the time of the first oil shock in 1974 that President Giscard of France, whose constitution gave him exceptional power among western leaders, took up the idea as a way of concerting action among the leading capitalist economies, and it was at his invitation that the first meeting was held at Rambouillet in November 1975. The first essential feature of the summit was that it should be small, select and personal. It should be limited to countries which carried weight and influence. It should bring together those directly responsible for policy. They should be able to talk together frankly and without inhibitions. Southern state qualifies for membership, but over time the agenda for the summit meetings has taken on board more and more global issues.

The summits of the G7 are just one sign of the increasing control by the First World of trade relationships between themselves and the South. The fact that it has recently been enlarged to include Russia says less about Russia's ambitions to be recognized as a First World country than about the fact that, as a result, it no longer poses a threat as an alternative centre of economic power. In the meantime, the conclusion of the Uruguay Round of the GATT has culminated in the formation of the World Trade Organization (WTO), which large companies are already using to eliminate competition to their interests.

Trade and environment are linked as follows (UNCTAD, 1994):

- Multilateral trade obligations, such as those under the WTO, may come into conflict with domestic environmental protection arrangements consistent with national priorities and preferences.
- Trade agreements should permit transborder and global environmental concerns to be addressed in a manner that allows for equitable distribution of responsibilities between countries at different levels of income and development.

The high and growing level of world trade and the globalization of economic activities have increased the need to tackle the link between trade and environment. The preamble to the agreement establishing the WTO recognizes the need to protect and preserve the environment. A number of international agreements on environmental issues also deal with trade. Such agreements include the Montreal Protocol on Ozone Depleting Substances, the Basle Convention on Hazardous Substances, and the Convention on International Trade in Endangered Species (CITES). It is essential that these provisions and the WTO rules and practices regarding environment be consistent with each other. One possible approach would be to use the WTO dispute resolution mechanism to enforce the trade-related aspects of the international environmental agreements.

WTO established a Committee on Trade and Environment (CTE) in 1994, which is expected to

- analyze the trend towards unilateral action in addressing extra-jurisdictional environmental problems; and
- find a balance between multilateral solutions to environmental problems and ensure that the WTO members are not unduly affected by either disguised protectionism or the imposition of unacceptable environmental values and standards through trade policy actions.

The CTE is also expected to examine the implementation of the 'polluter pays principle' in an economically efficient manner.

Specific WTO rules and provisions related to trade and environment are listed below:

- *Agreement on Technical Barriers to Trade (TBT)*: This agreement governs the use of product standards and brings technical regulations adopted for environmental objectives more explicitly within its scope. Countries are encouraged to follow international standards, and are required to notify their standards or regulations when they are likely to have significant trade effects. The technical regulations should not be more restrictive than necessary. Participation of the developing countries is necessary for these regulations to be effective in protecting the environment and promoting trade. Any perception that these could be used by the developed countries in a protective manner could create a backlash against multilateral trading arrangements and environmental agreements.
- *Agreement on the Application of Sanitary and Phytosanitary Measures (SPS)*: This agreement deals with measures applied to protect human or animal life or health within the territory of the importing country from risks arising from additives, contaminants, toxins, or disease-causing organisms in food, beverages or feedstuffs, and to prevent the establishment or spread of pests. Special provisions exist for developing countries, including technical assistance.
- *Agreement on Subsidies and Countervailing Measures (SCM):* This agreement deals with subsidies that a country may provide to its exporters and the countervailing duties that an importing country may levy. Subsidies may be allowed when designed to help businesses enact higher environmental standards.
- *Agreement on Trade-Related Intellectual Property Rights (TRIPS):* Under such conditions, countries may exclude from patentability inventions which may cause serious harm to the environment. The protection of indigenous community rights and biodiversity is organized by a combination of patent and *sui generis* arrangements. The developing countries have a particular interest in preserving biodiversity and the traditional use of medicinal plants.

Important and complex linkages occur between trade and environment. This requires more research, both to identify the impact of trade on environment and in order to translate the 'polluter pays principle' into actual measures. This is unlikely to be achieved soon.

The North and South Debate

The question of environment, its protection and conservation has also become an issue of international debate. The South charges the North with 'eco-imperialism' arguing that because of northern colonial attitudes the problem of the degradation of environment at global and regional levels has been growing day-by-day. The South feels that the North is responsible for most of the environmental problems as well as for the exploitation of the resources of South, but not contributing their share towards the protection of environment. On the other hand, the North is of the opinion that the responsibility of environmental protection is 'common' to both.

These different interpretations are reflected in the two different uses of the term 'commons'. The basis idea is that there are certain parts of the land or sea that are not the private property of any person, company or state. However:

1. Pro-southern perspectives tend to use the term as those areas and resources upon which the poor who own nothing, rely.
2. The term has also been used by environmentalists to express the common inheritance stressed by the North at Rio and rejected by many southern leaders, though not by all. Southern interests have argued that the use of the term 'global commons' is effectively the moral hijacking of their resources, another example of eco-imperialism.

Perhaps, the real dichotomy in the use of the term 'commons' is illustrated by the reluctance of the USA to sign an accord (biodiversity) that might have constrained American-based TNCs from continuing to enclose (through patents) precisely those 'commons' on which the world's poor depend. At UNCED the global commons (and in particular rainforests) were amongst the main problems identified by the North. The others were climatic change the ozone layer and population. On the other hand, the countries of the South sought to emphasize a different agenda, i.e. poverty hunger and desertification.

Three new conventions were agreed as a result of the Rio Summit – on climate, biodiversity and desert – along with a new set of guidelines for forests. The Framework Convention on Climate Change was signed by 155 countries, However, its commitments were inadequate, especially those made (or not made) by developed nations to limit greenhouse gas emissions. The IPCC estimated that 60 per cent reductions were necessary to stabilize carbon dioxide concentrations at their 1992 level. The word 'aim', however, was chosen, as US negotiators freely admit, so as not to be a commitment, and the deadline of the year 2000 was also dropped from the sentence at the Bush administration's request. In the event, the wording was much weaker than the Alliance of Small Island States wanted or the EU was then prepared to accept.

North-South divisions focused on transfers of technology and funds that were needed if developing nations were to meet any of the Convention agreements. The dependency on those transfers was included in the Convention, but no commitment was made to make the transfers. The North agreed to provide financial assistance in compiling national inventories of 'sources' of carbon dioxide and the 'sinks' that remove it from the atmosphere, but did not make any arrangement to meet the costs of actually doing so. The problem is that such agreements are consensus documents and therefore tend to be minimal in terms of commitment. What was needed was consensus not just between different perceived interests within developed countries, but between the rich industrialized world and the developing states, not to mention between the hugely varied interests within those developing states. The consensus was very limited, and current levels of commitments will not slow global warming.

Despite its weaknesses, however, the Framework Convention provided evidence of what is needed (and known to be needed) to slow climatic change. It provided a framework for future discussions and an institutional structure for them, leading up to the Kyoto Conference in 1997. Since this, in turn, failed to make any real progress, the entire problem was again remitted to Buenos Aires in November 1998 and this in turn agreed to put off any action until long after any of those present was likely to be in the office. But very small climatic change would produce changes that are generally seen as likely to impact particularly heavily on the developing world. For example, drought risk in not equally spread, and very small changes in drought risk will produce big impacts on agriculture, leading to significant agricultural losses in Brazil, Peru, the Sahel, South-East Asia the Central Asian Republic and China. Sea level rises could potentially cause massive people movements. Relatively small rises in mean sea level could displace 30 million people in Bangladesh, for example.

Worse still, the completion of the Uruguay Round of the General Agreement on Tariffs and Trade (GATT) resulted in the creation of the new World Trade Organization (WTO), and the limited perspective already shown by it has gone far to destroy any trust that Rio might have engendered in the South. The stress on markets has intensified competition and put further pressure on the environment. It is not just globalization working against the Rio aspirations, but the nature of the fragmentation of international and national agendas. The UN is increasingly seen as unimportant by the wealthy developed nations. It is in internal turmoil and confused about its role. In key respects it has been successfully by-passed by the G8 – the club of rich nations.

In June 1997, the representatives of national governments reconvened at the UN General Assembly in New York for a review of progress since UNCED. This United Nations General Assembly special session, or 'Rio-plus 5', lacked the sense of purpose of the original. Only a few heads of states or governments attended, and it was hard not to feel disillusioned. But, Rio was not a complete waste of time and resources. It has in some respects at least helped to change public perception in the developed world. 'Green' investment, for example, now amounts to some US$ 600 billion worldwide, and the potential is hardly being tapped. As part of this development, the impact of Rio on business has been the increasing awareness of the negative images among consumers which environmental protests can inspire.

Arms and the Environment

A special environmental problem is presented by the widespread distribution of arms throughout the South, and the use of them in armed conflicts. Where either or both of the superpowers intervened to maintain the peace, in Korea, Vietnam and Somalia, new environmental problems were created and existing ones became much worse.

The collapse of communism ended the moral justification for much US intervention. The change was not immediate: the new public vocabulary included phrases such as 'ensuring international stability' or 'the worldwide crusade for democracy'. But, the then new US Administration of George Bush did not show much respect either for national sovereignty or for democracy. Its intervention in Panama in December 1989, claimed to be enhancing the cause of democracy resulted in several days of fighting in Panama City, in which hundreds died and many buildings were damaged. It was carried out in breach of the charters of both the UN and the Organization of American States (OAS), and was censured by the OAS. Then, in 1991, the decision after a long period of delay to launch Operation Desert Storm in Iraq was justified as defending what Bush claimed was the 'legitimate government' of Kuwait. USA and EU have used heavy bombing on Iraq in 2003, which resulted in several environmental problems in the entire Gulf region. All these conflicts will, of course, damage local environments, as much interference with everyday life as by deliberate damage.

Using weather as a weapon and manipulating the environment for military purposes is common and has come into focus in the Gulf war. The US was the first to use weather modification as a weapon in Indo-China region during Vietnam war. Artificial triggering of earthquakes and tidal waves, modifying hurricanes or hail storms to change their course, creating artificial fog to impede enemy traffic, etc. The use of chemical weapons are not only harmful to people and other organisms but to the ecosystem also. The emergence of cruise and ballistic missiles and other types of bombs are responsible not only for atmospheric pollution but also damaging for the entire ecosystem. Military experts are of opinion that every state has its environmental or geophysical vulnerabilities that can be exploited by an enemy in the event of war. All this can be checked through an international awareness and mutual understanding.

Non-Governmental Organizations

The campaign against landmines reflects the importance of non-governmental organizations (NGOs), through their capacity to sway public opinion in national and hence international politics. A multiplicity of NGOs is characteristic of the democratic government of AICs. The environmental movement began among NGOs, acting in the first instance as pressure groups on their own governments. Since then, they have assumed a wider international role, working across formal governmental boundaries and getting involved with international organizations on global issues requiring global solutions, in particular, with the UN and its proliferating special agencies.

By the early 1990s, NGOs had developed into a diverse and growing set of institutions operating at various levels. Globally, they numbered in thousands, and the total number of members and associates of individual organizations showed a remarkable increase. This recent proliferation is indicated by the age of the NGOs. Only about a half of all the NGOs from the developing countries are more than 10 years old. They still carry out the primary function of the environmental NGOs by publicizing instances of environmental degradation, although most modern NGOs will also be competent enough to suggest an alternative to the proposed project which they consider to be environmentally unfriendly. It has probably become easier to function as the NGOs have a network of their own and are recognized as an integral part of environmental management by UN institutions, the World Bank and a number of national governments. The Friends of the Earth (FoE) network, which covers USA, most western European countries, Malaysia and South Africa, is a good example. The funding situation is also better as they may receive financial assistance from a number of sources. NGOs have a variable geographical distribution, being scarce in certain regions such as China or North Africa, due to a combination of political, economic and cultural reasons. In certain countries (e.g. India, the Philippines and Brazil), well-organized groups operate as autonomous NGOs, whereas in other parts of the world, support from international NGO networks is crucial for their success.

At Stockholm, in 1972, 134 organizations were represented. Since then, there has been an explosion of activity. The problem is that not all groups are of equal weight. It is clear that a small number of well-funded transnational organizations enjoy special access to and influence with the UN. They act not only in public but in the continuing lobbying process that the international system requires. The main ones active at the global level are Greenpeace with 6 million members in 30 countries, the WWF, spawned by the IUCN in 1961 (now with 3 million members in 28 countries) and Friends of the Earth (FoE), with 0.5 million members in 46 countries.

The problem for all environmentalist groups is that they need to be constantly active, as threats to the environment are continuous. The groups therefore work on a variety of fronts at each level. They work with think-tanks such as the World Resources Institute (WRI) in Washington, the Worldwatch Institute and the International Institute for Environment and Development in London. They also make use of networks such as the Earth Council (Maurice Strong and UNCED), the Centre for Our Common Future, Geneva (Gro Harlem Brundtland and the former WCED) and Third World Network.

Environmental groups make up only some 10 per cent of the 1,000 NGOs with permanent UN affiliation. UNCED accredited 1,400 itself and 1,600 groups took part in Global Forum '92. NGOs had access to PrepCom meetings; they could attend (though not always simultaneously) all public sessions and they had the opportunity in the confined space of the Convention Centre to lobby national delegations. More controversially, 14 of the national delegations included some NGO representation, potentially an important precedent for other activities within the UN system. On the other hand, there was not so much evidence of NGO impact on existing specialized agencies, and particular disappointment at its effect (or lack of it) on the World Bank. By 1992, there was some concern at the emergence 'fake NGOs' – bodies set up by governments to give them the sort of advice they wanted to hear.

International organizations want to deal directly with NGOs for five reasons. Publicity brings increased funding. The ideological preferences of such organizations are for consultation and deliberation. Friendly criticism and the ability to generate support both enhance programme effectiveness and sustainability. External pressure is more easily handled by dealing with organizations rather than with the public at large. Lastly, the creation of constituencies brings them support in the internal battle against other agencies and organizations.

From the 1980s onwards, the strength of the NGO movement was enhanced by the establishment of direct contact with international organizations. The NGO World Bank Committee was founded in 1982 to provide a forum for policy discussion between senior World Bank officials and NGO leaders from different parts of the world. In recent years, subsequent meetings have discussed popular participation in development-related decision-making. A number of NGOs are currently kept informed of Bank-related projects with a possible environmental impact, or those projects where the potential for possible NGO involvement exists. In the 1980s a group of US environmental NGOs campaigned successfully against Brazil's Polonoreste road-building and colonization project. Another group of NGOs has created the Development Bank Assessment Network which urges the Bank to examine project impact on the poor in general and on the environment in particular. The World Bank has progressively encouraged the involvement of NGOs in projects financed by it. Over time, the profile of the NGO partners of the Bank has shifted from a majority of international NGOs to a majority of grassroot groups, the people who are directly affected by development projects. Both the Stockholm and the Rio conferences have contributed to the linkages between the UN institutions and the NGOs. Hundreds of NGOs have a consultative status with the Economic and Social Council (ECOSOC) of the United Nations. This gives the NGOs the right to attend ECOSOC meetings, to submit statements, to testify at meetings, and to some extent, propose items for consideration. Certain UN agencies work in collaboration with the NGOs. Small grants ($25,000 per country) are made to grassroot NGOs in support of innovative projects through UNDP's 'Partners in Development Programme'. The Sustainable Development Network Programme (SDN) of the UNDP identifies and links organizations (including NGOs) that could contribute to sustainable development. The UNEP has organized the Environmental Liaison Centre (Nairobi), the African NGOs Environmental Network and the Green Belt Movement – all of which require NGO networking. The United Nations Development Fund for Women (UNIFEM) is another institution that relates to NGOs in supporting the women's movement. UNIFEM provides funding for supporting poor women in the developing world in their attempt towards building a better life in an enhanced environment. For example, it has been involved in a sericulture project for village women near Udaipur, India; in introducing biodigesters as replacement for firewood gathering to the Yemeni women at Al-Habeel; and in training the women in building low-cost water pumps in Tempoal County, Mexico.

International Environmental Problem

The most important lesson from our experience with ecology, energy and land use has yet to be learned by mankind as a whole: that the planet earth and its atmosphere constitutes one unified ecosystem and that damage to any part of it inevitably results in damage to the whole. Everything is connected to everything else. Statesmen rarely give much thought to the long-range environmental consequences of their decisions and actions, and politicians do so even more rarely. Even though, as we pointed out earlier, an environmental consciousness has begun to spread around the world, it has not spread far enough or fast enough. Pollution, for example, knows no boundaries, yet there is still no adequate mechanism for preventing, containing or reversing pollution at the international level.

The nearest thing to a global environmental agency is the United Nations Environment Programme (UNEP), with headquarters in Nairobi. Its Oceans and Costal Areas Programme has generated and coordinated environmental protection clean up activities in designated portions of the global sea. Its Mediterranean Action Plan has had considerable success in cleaning up the Mediterranean Sea – one of the most polluted large bodies of water in the world. Progress has also been made in the Baltic, where the problem is at least as serious. Other international and regional organizations have expressed interest in environmental matters, but seldom do they rank very high on their priority lists. Even the World Bank has come under heavy fire for ignoring ecological considerations when funding huge infrastructure projects in developing countries, and it is beginning to reorder its priorities as a result of it.

Meanwhile, agricultural pesticides applied in the rich countries of the Northern Hemisphere wash down rivers into the global sea and eventually end up in the bodies of penguins in Antarctica. Industrialized countries in Western Europe are already sending 10,000 to 20,000 shipments of hazardous wastes every year to Eastern Europe for disposal and are seeking disposal sites in developing countries. Acid precipitation caused by the mixing of atmospheric water with airborne industrial and automotive pollutants is devastating lakes and forests in North America and Northern and Central Europe and is threatening farms and monuments in India, Mexico, Indonesia, Zambia, Brazil, and other developing countries.

The threat of major worldwide climatic change, usually expressed as 'global warming', is growing, with all of its ramifications, such as desiccation of agricultural areas and rising sea level as glaciers and icecaps melt. The depletion of the ozone layer in the upper atmosphere, which protects the planet from excessive ultraviolet radiation from outer space, is another global problem. So is deforestation – the rapid disappearance of the forests that once blanketed much of the earth and that provide much of our atmospheric oxygen as well as wildlife habitat, watershed protection, sustenance for soil, and regenerative forest products. This is one of the factors contributing to desertification, the spreading of deserts, particularly in Africa, over land that had been at least steppe and even savanna.

All of these intensify the pressure on finite global supplies of fresh water, which is rapidly becoming so scarce a commodity in much of the world that serious consideration is being given to towing icebergs from the Southern Ocean to costal deserts, as has already been done experimentally. Another result of both natural forces and especially man's activities is the rapid extinction of species of plants and animals. Who can tell how many potential sources of food, medicine, pest control, beauty and spiritual joy have been lost in this way?

The political and social consequences of this degradation of the global environment are incalculable. We are already experiencing growing, not diminishing, hunger and malnutrition in the world and huge waves of environmental refugees fleeing droughts, floods, famines, destruction of the land, and other ecological problems. The possibility looms of resource wars, water wars, food wars, even land wars, just as we used to have before ideology became the chief rationalization for war.

All of this, of course, both stems from and reinforces the maldistribution of both wealth and power in the world, generated in part by the Industrial Revolution and in part by the staggering growth of the world's population. All these points are discussed in greater details elsewhere in this book, but here they are seen in a broader context. Another point that bears repetition: Although at the present stage of world history populations are growing most rapidly on the poorest countries, it is likely that the greatest environmental damage is caused by the rich countries and their agents in the poor countries. Furthermore, it is in the rich countries that industrial and urban pollution is greatest.

Western Europe, for example, which can boast of being rich, sophisticated and proud, was caught completely off guard in November 1986 when a series of accidental spills of toxic chemicals in Switzerland flowed down the Rhine river, killing hundreds of thousands of fish, contaminating drinking water for millions of people, and otherwise damaging the environment of four countries. This calamity called to mind the words of the English Romantic poet, Samuel Taylor Coleridge:

The river Rhine, it is well known,
Doth wash your city of Cologne;
But tell me, Nymphs, what power divine
Shall henceforth wash the river Rhine?

What power indeed? Perhaps some global plan of action – backed by a worldwide consensus, determination, sustained effort, and plenty of cash – will emerge as a result of the 1987 report of the World Commission on Environment and Development. This Commission, composed of 23 people from all parts of the world and headed by Gro Harlem Brundtland, Prime Minister of Norway, was created by the United Nations, but was independent of it. It functioned for three years before issuing its massive report on an enormous range of contemporary world problems. Its recommendations, if followed urgently and faithfully, would certainly result in better world for all of us.

Since the release of the Brundtland Report, a number of specialized international conferences have been held to discuss some of the global environmental problems listed above. Each has generated serious activity, including international conventions that bind their parties to specific actions to help reverse the destruction of our environment. It is doubtful, however, if even after the end of the Cold War, there is yet the political will outside a few small countries to take the difficult and costly actions necessary to achieve the goal laid out by the Brundtland Commission: *sustainable* economic and social progress in an environmentally healthy world.

There are some signs of hope, other than the governmental efforts described already. One sign is the growing effectiveness of environmental NGOs around the world. The moratorium on whaling, the ban on large-scale drift-net fishing, the outlawing of chloroflurocarbons discharged into the atmosphere, and many other small victories can be credited at last in part to their efforts. A creative and potentially important arrangement advocated by environmental NGOs is that of forgiving portions of a developing country's foreign debt if it dedicates areas in the country as biosphere reserves or other types of protected parks or reserves. This 'debt-for-nature swap' has great potential to accomplish a number of environmental and development objectives at once – if it is administered wisely, on a large enough scale, and over a long enough period of time.

Early experience with this arrangement, however, is not encouraging. A recent GAO report says:

> From 1987 through 1990, 13 countries completed 26 debt swaps. These swaps retired debts totaling about $126 million (less than one-twentieth of 1 per cent of the countries' external debt and less than one-fifth of 1 per cent of their commercial debt.) Of the $126 million, $86.4 million was exchanged by Costa Rica. Fifteen swaps for nature accounted for nearly 90 per cent of the $126 million, while 11 swaps for development accounted for about 10 per cent. (United States General Accounting Office, Developing Country Debt: Debt Swaps for Development and Nature Provide Little Debt Relief, Washington, December 1991, pp. 1–2)

Another sign of hope is the 'greening' of politics at the local and national levels, which may eventually spread to the international level. 'Green' parties were organized in a number of European countries in the 1970s and 1980 to contest elections on environmentalist platforms. They have had some success in some countries, but so far have not really caught on elsewhere. Single-issue parties such as this can only be successful in electoral systems based on proportional representation, and even there the system places limits on what they can accomplish. If they try to broaden their appeal and win more votes, either by incorporating other planks in their platform or forging alliances with other parties with different programmes, then they automatically dilute both their message and their energy, possibly to the detriment of both the parties and the environmental movement. It is a classic dilemma in politics, and it has no simple solution. Nevertheless, formal, legal political activity has both publicized and legitimated the notion of environmental protection, and this has demonstrably influenced some national policies.

Perhaps, the crucial test of not only the concept of linking economic development and environmental protection but also of international action as a means to bring it about is the United Nations Conference on Environment and Development (Rio de Janeiro, 1–2 June 1992). This conference, scheduled just 20 years after the United Nations Conference on the Human Environment in Stockholm, was voted by the General Assembly in 1989 and preparation for it began almost immediately. Countries are to be represented by heads of states or governments; it will be the largest 'summit' meeting held to date. Another unique feature of the conference is the full and active participation of both non-governmental organizations and private sector interests. The scope, magnitude and pioneering nature of this conference are breathtaking. It will certainly be the most important conference since UNCLOS III and perhaps the most important since the San Francisco conference in 1945 that established the United Nations. Its true significance, however, will not be measurable for years – perhaps even decades – for the real measure is not what is said or done in Rio but how the inevitable resolutions, declarations, plans, treaties and other documents that emanate from it are interpreted and implemented afterward. The global environment issues have been discussed in detail in Chapter 16.

Conflicts in Ecology, Energy and Land Use

In this chapter, we have pointed out a number of problems relating to environmental politics from the urban neighbourhood to the global one. Each is difficult enough in itself, but their interrelatedness compounds the complexity. Here, we suggest a few interrelationships only to illustrate the point.

1. In the rush to 'energy independence', we are considering a return to coal as a primary fuel because it is domestically abundant, but we forget the reasons that we converted from coal to oil and gas in the first place: coal is bulky and dirty, and except for expensive anthracite, not very efficient. Underground coal mining, moreover, is hazardous and unhealthful, while surface mining is ecologically ruinous.
2. Expansion of the land under cultivation to increase food production means in most of the world using marginal or sub-marginal land, which would require a great deal of energy and can have most unfortunate ecological consequences.
3. Dispersal of industry to improve the urban environment and provide rural employment can simply accelerate urban sprawl, increase transport costs and energy consumption, and transfer the less desirable features of industry to rural areas, making these areas less attractive than they are now.
4. An increase in energy consumption means additional pipelines through fragile environments, oil spills, and blowouts; unsightly and perhaps dangerous electricity transmission lines; more rail, road and water transport, requiring more energy and raw materials to produce and operate; uncertainties about nuclear energy; and change in the composition of the earth's atmosphere.

5. Increased raising of livestock for food means increased competition with wildlife for forage and destruction of their habitats.
6. Landfills in urban areas or even for offshore facilities, such as airports and oil terminals, can gave adverse effects on the ecology of the coastal zone.
7. Economic development, considered a desirable political goal, inevitably requires intensified land use, ever-increasing energy supplies, and ecological damage.
8. Hydroelectric dams, which generate non-polluting energy at relatively low cost, can interfere with the migrations of anadromous and catadramous fish, cause siltation behind the dam and desiccate wetlands below it, and devastate the land and culture of nearby indigenous peoples.
9. The readily accessible fossil fuels have already been found; new deposits are likely to be found in environmentally fragile areas where the risks and costs are far greater than in the older ones.
10. Orderly economic growth today requires broad policies and plans for land use energy and ecology that can be developed and administered by government only at the expense of some limitations on free enterprise, private ownership of land and individual behaviour.
11. Successes already achieved in environmental protection are threatened by reversal as economic interests organize to achieve 'wise use' or 'balanced use' or 'multiple use' of land and resources, thereby obliterating wilderness altogether.
12. A growing population would require economic growth merely to maintain the present unsatisfactory levels of consumption of goods and services for the great majority of the people of the world, so that cutbacks in production can only make a bad situation worse.
13. Political democracy can be achieved and maintained in a society that is reasonably prosperous and in which wealth is reasonably equitably distributed, but these conditions can only be developed over a period of time at some environmental cost.

In fact, environmental management has come a long way since Stockholm conference. The success of the future environmental management of our planet depends on a greater cooperation. Otherwise, in the middle of a distressed world, we will perpetually remember the words of the Brundtland Commission: 'The earth is one but the world is not.'

References

Fitzpatrick, Tony (eds.) (2011), *Understanding the Environment and Social Justice*, Policy Press, Bristol.

Johnston, R.J. et al. (2013), *Geographies of Global Change*, Blackwell, Oxford, pp. 429–439.

20

Global Environmental Issues

During the 1960s and 1970s people began to think that human existence would be threatened unless they adjusted their relationship with their environment. Now it has been realized that environmental problems are not limited to local, regional and national level but there are several global environmental problems. The scientific and technological revolution has given multiple facilities to mankind but at the same time it is responsible for climate change, global warming, greenhouse effect, ozone depletion, depletion of resources, loss of biodiversity, deforestation, desertification, acid rain, environmental degradation, environmental health, overpopulation, loss of the fertility of soil, hazardous waste disposal, changes in land use, energy crisis, nuclear waste and radiation issues etc. All these and many other problems have a global dimension and the world community has already realized this fact that only cooperative efforts can, if not solve, at least, minimize these environmental problems.

Some of the global environmental problems have already been discussed in detail, such as climate change, global warming and green house effect (Chapter 16), biodiversity (Chapter 15), degradation of environment (Chapter 6), population and ecological crisis (Chapter 11), air, water, noise and land pollution (Chapter 7, 8, 9). In this chapter some other environmental problems are discussed in brief.

Thinning of Ozone Layer (Ozone Depletion)

Ozone is a form of oxygen with three atoms instead of the normal two, and forms a fragile shield scattered in the stratosphere absorbing the sun's ultra violet radiation, away from the earth's surface. It forms a three millimetre thick layer, but if it disappears or thins, all terrestrial life will be annihilated. Near the earth's surface, ozone is an increasingly troublesome pollutant, but safely up at a height of 20 to 30 kilometres from the earth in the atmosphere, it is as important to life as oxygen itself.

The thinning of this layer has generated global concern. This is due to several chemical pollutants discharged by industries or produced through other chemical reactions. Chlorofluorocarbons (CFCs) are substances that are inert, inexpensive, stable, non-inflammable, non-toxic, easy to store and produce. The population explosion is one of the causes of the increase of CFCs due to use of airconditioners, insulation, solvents and fire-killers. In 1986, three US agencies, viz. National Aeronautics and Space Administration (NASA), National Oceanic and Atmospheric Administration (NOAA) and Federal Aviation Administration (FAA) started reporting on ground and satellite-based data, on ozone depletion in stratosphere, by the use of chemicals like halon 1301, methyl chloroform, carbon tetrachloride, solvents, FC–113, aerosols, foams and refrigeration CFC–11, and air conditioning CFC–12 that are depleting the ozone layer 4, 5, 8, 12 and 26 per cent, respectively, on global scale, because of human activities. As far as CFC regional consumption is concerned USA is on top consuming 29 per cent, Russia and Europe 14 per cent, while in China and India it is merely 2 per cent. The total ozone amounts in the northern hemisphere, middle and high latitudes have decreased, and sudden and unexpected decrease during spring time in Antarctic zone is confirmed on 15 to 20 km in the stratosphere ozone layer.

The increase in ozone layer depletion will invite the lethal ultraviolet rays from the sun which increases cancer, eye damage, injury to plants and marine life and even reduces our immunity to diseases. The UV-B damages the genetic material DNA and also causes skin cancer. The ultraviolet radiation even reduces the efficiency of the immune system, thus impairing the body's resistance. Experts are of the opinion that animal and plant world would face a grim challenge if radiations are not checked. Certain crops have been found vulnerable to such radiations. Under the impact of radiations, proteins in plants, microbes and animals would be affected and such changes in spatial orientation would be critical in the formed part of the reaction centre of an enzyme. Near cities where population concentration is more, ozone could be formed near the earth's surface, there by having disastrous effects on human health, crops and ecosystems. It would also increase the incidence of smog in areas of industrial concentration. Ozone has its effect on the earth's climate by adding to the greenhouse effect.

The world has now been serious about the most dramatic evidence of ozone depletion, a 'hole' in the ozone layer which appears every spring over the Antarctica. This hole has grown over the last ten years. Till now, no clear explanation has been found to explain this phenomenon and scientific research in this field is in progress. But general opinion regarding the cause of this depletion is the CFC. The countries of the world now have started serious thinking about this problem. In fact, this is the duty of the developed countries, who have technology and infrastructure, to check the ozone depletion. Because this problem is not a problem of individual nations, therefore, combined efforts can only check the ozone depletion.

The concern about ozone layer protection began by the late 1970s and early 1980s. In 1978, the United States banned the principal CFCs for most uses; in 1980, European Community countries placed a limit on production of CFCs and cut their use in aerosol products by 30 per cent. Sweden, Norway and Canada also placed limits on CFC use. In 1975, UNEP has also expressed serious concern about ozone depletion and also developed a World Plan of Action on the Ozone Layer. Some of the important international actions taken on ozone protection are listed below:

1977	UNEP coordinating committee on the ozone layer established
1985	Vienna Convention for the protection of ozone layer
1987	Montreal Protocol on substances that deplete the ozone layer
1988	Entry into force of the Vienna Convention.
1989	Entry into force of the Montreal Protocol.
1990	Montreal Protocol amended in London. The amendment required signatory governments to regulate consumption and production of CFCs.
Jan.1991	Establishment of interim multilateral fund to provide financial assistance to developing countries in order to facilitate control measures.
May 1992	112 states have ratified the amendment to the Montreal Protocol.
Aug. 1992	Entry into force of the amendment.
Nov. 1992	At Copenhagen, parties agreed to accelerate, phase out schedule, to Jan.1996.

Ozone Protection Regime

Both the state and non-state factors played major roles in the regime building process. To a large extent, their roles were determined by their position in the global economic system.

At the beginning of the problem definition process, there was a scientific uncertainty about the gravity of the ozone depletion problem. The definition process reduced this uncertainty by accomplishing three major tasks: it confirmed that ozone depletion was occurring; it confirmed the connection between the use of CFCs and halons and the damage to the ozone layer; and it estimated the extent and rate of depletion.

The ecological epistemic community, which played a major role in defining the problem, was a transnational one. It was made up of atmospheric scientist and policymakers, including officials of UNEP, the US Environmental Protection Agency, and the US State Department's Bureau of Oceans and International Environmental and Scientific Affairs (OES). The epistemic community used a variety of channels to affect the decision making process. In the process of drafting documents and reports, gathering data, organizing scientific panels, and lobbying delegates, it was able to substantially define the issue and modify the agenda for the negotiations. Significant components of the epistemic community participated in UNEP's Coordinating Committee on the Ozone Layer, in which they were able to share the information.

The bargaining process over the ozone protection regime actually began before the issue was clearly defined. In the beginning, the Third World countries played a minor role in the negotiations, primarily because they did not have clearly defined interests. Nations with economic interests, however as well as industries involved in CFC production, did have a clear understanding of the interests at stake.

Nations brought dissimilar viewpoints to the negotiations. Some favoured a global ban on CFCs 11 and 12 in aerosols (except for essential purposes) and a limit on CFC emissions in nonaerosol uses. The United States, which accounted for 30 per cent of worldwide production, was prepared to be the lead state. However, other nations were reluctant to undertake such sweeping actions until they felt the scientific evidence against CFCs, compelling. The latter group would have been satisfied with a framework convention.

The European Community accounted for 45 per cent of world CFC output, and by the mid–1980s, it was exporting a third of its production to the developing states. Because of the major producer states' interests in continuing production, the EC member states constituted a veto coalition. Since participation of these states was necessary for a strong ozone protection regime, their cooperation was important. The large developing states such as India, China. Indonesia, Brazil, and Mexico also appear to have had some potential as a veto coalition. Their bargaining leverage was based not on their consumption of CFCs but on their potential as producers and consumers of CFCs. These countries were already producing on a small scale; although they were providing less than 5 per cent of the world's CFC production, their production was increasing by 7 to 10 per cent each year. However, in the negotiations on the Vienna Convention, these states did not take advantage of their veto power. At this stage, no Third World country played an active role.

The Toronto Group took the lead, pushing for simultaneous negotiation of a framework convention and associated protocols that would obligate states to reduce CFC use. However, the veto coalition would not support the negotiation of regulatory protocols, taking the position that the state of scientific knowledge was insufficient to support such a protocol. The lead state remained on the offensive, proposing a worldwide ban on nonessential uses of CFCs in spray cans in 1983. But, the veto coalition offered a counter-proposal: Total CFC production by 30 per cent beyond the anticipated 1985 level.

Over a three-year period, there were eight negotiating sessions, but the groups failed to achieve consensus on the ozone issue. In 1985, after three years of negotiation, the United States began to define ozone depletion as an urgent issue, but still the veto coalition would not yield. As a result, the parties had to be satisfied with the Vienna Convention, which was a framework convention.

The Montreal Protocol began the process of strengthening the ozone layer protection regime. Because of new findings, most nations accepted the idea of some controls. However, the extent and the timing of the controls were contentious issues. Lead states argued for a freeze, followed by a gradual reduction in the production of ozone-depleting substances over a period of 10 to 14 years, whereas the veto coalition wanted a production cap. The Toronto Group still pressed for a cut, but as a compromise, its members suggested a 50 per cent cut. As late as April 1987, the EC insisted it would agree to a reduction of not more than 20 per cent, but at the Montreal Conference later that year, the EC representative agreed to a 50 per cent reduction.

Developing countries successfully pressed for some concessions; they obtained a derogation that would allow them to increase their CFC use for the first decade of the agreement. They were also to receive technical assistance as well as information and advice on CFC recycling and conservation. The fear was that without some concessions, Third World countries would set up their own CFC production plants.

But these concessions did not address all of the developing countries' major concerns regarding the evolving regime. Of particular concern was the fact that there were no provisions for a fund to help them make the switch to CFC substitutes. The fund was not established because the United States, Japan, and the EC countries opposed it. Because there were no provision for financial assistance, three major developing countries – China, India and Brazil refused to sign the protocol. These were important holdouts, because they had the potential to become major producers and consumers of CFCs. They refused to participate in the regime until the industrialized countries agreed to provide financial and technological assistance. UNEP's Tolba supported them in their calls for the establishment of the fund.

Scientific revelations about ozone depletion continued to unfold, and they spurred the regime strengthening process. The Montreal Protocol came into force on 1 January 1989, and the negotiations on its amendment began promptly. At a meeting to amend the protocol in Helsinki in May 1989, the veto coalition began to shift its position dramatically. The EC members were among 80 nations voting for a complete CFC phaseout by the year 2000.

At the London meeting in June 1990, a new coalition of 13 industrialized states pushed for a 1997 deadline for the final elimination of CFCs. However, the four leading CFC-producing states – the United States, the United Kingdom, France and Italy – joined by the Soviet Union offered an alternate phaseout date of the year 2000. The push by some industrialized states for a 1997 deadline caused the United States to change from the lead to a veto role.

Developing countries were usually at a disadvantage with respect to the size and makeup of their delegations. A developing country was likely to be represented by a single generalist, whereas the industrialized countries were more likely to have a team of specialists. Countries that had a strong economic interest in the process were usually represented by large delegations. But, most delegations feel between these two extremes. They had, on average, two to four members, usually representing scientific, industrial, economic, legal, and political disciplines. The delegations that had specialists had an advantage, since these people usually had special knowledge of, or experience, regarding the ozone issue.

The potential for further regime transformation still exists. With the developing countries now on board, the regime can be further strengthened by speeding up the timetable for the phaseout.

Ozone Depletion Pollutants and Effects

The source of ozone depleting chemicals in stratosphere is the manufacture of chlorine-containing and bromine-containing chemicals by humans, especially CFCs and halons. With the ozone depletion, Ultra-Violet radiation becomes more which can harm any living creature exposed to it. This includes phytoplankton (microscopic algae) at the bottom of the food chain on which all animals, including humans, depend for food, directly or indirectly. UV light also affects the skin, eyes and immune system of the living organisms.

Figure 20.1
Ozone Depletion and its Effects

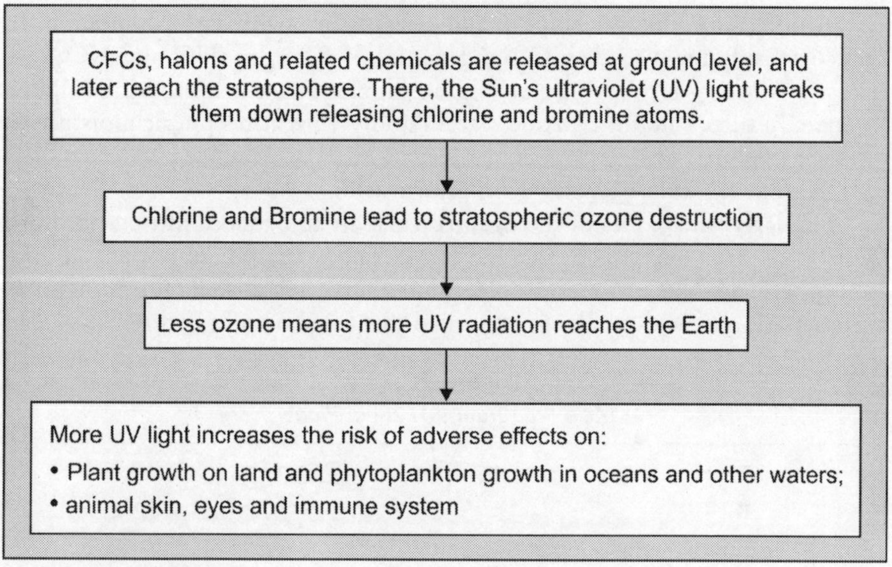

El Nino and Climate

The climatological event known as El Nino is a periodic rise in temperature of the Eastern Pacific, around Peru and Galapagos Islands. There was a strong El Nino in 1982 and 1983 and a lesser version of the phenomenon was observed in late 1990s. In 1997 we have seen its impact on a worldwide scale. It was the most powerful El Nino on record, resulting in heavy rain in the Eastern Pacific and unusual dryness in Southern Africa Central and South America.

In South-East Asia, there was a severe drought, causing serious food shortage in New Guinea and contributing to forest fires in Indonesia, which enveloped the whole region in a choking smog. A subsequent warming of the Indian Ocean has lead to devastating floods in Somalia and Southern Ethiopia. But far from being confined to the tropics, El Nino also had pronounced effects in Europe.

Scientists have been trying to come up with conclusive answers on the extent of damage El Nino would cause on the entire world. It has occurred twice in the last 15 years. Latin America, Cambodia and China and now the entire world is witnessing serious weather anomalies as an impact of El Nino, literally meaning 'little Christ' because, generally it has occurred during Christmas.

Acid Rain

Acid rain is a matter of great global concern and has become one of the major environmental problems. The term 'acid rain', used first by Robert Angus Smith, the Chief Alkali Inspector of UK in 1872, described the acidic nature of rain falling around Manchester. Acid rain is commonly defined as 'a condition in which natural precipitation becomes acidic after reacting chemically with pollutants in the air'.

Among air pollutants, sulphur and nitrogen are significant compounds. In particular, sulphur dioxide (SO_2) and oxides of nitrogen (NO and NO_2) are released from a variety of sources. Sulphur dioxide reacts in air to form sulphur trioxide. In turn, sulphur trioxide reacts rapidly with atmospheric moisture to produce sulphuric acid.

Sulphur trioxide + Water = Sulphuric acid.

Similarly, the oxides of nitrogen react in air to produce nitric acid. Both these acids are very strong acids. These materials, when present in the atmosphere, dissolve in water droplets and fall to earth as acid rain. A pictorial representation of acid rain from a thermal plant is shown in Figure 20.2.

Figure 20.2

Acid Rain from a Thermal Plant

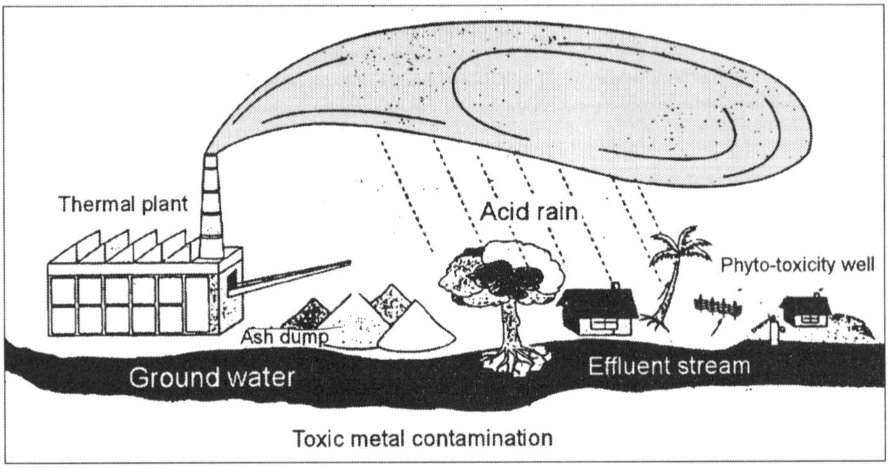

Under normal conditions, rainwater is slightly acidic because carbon dioxide dissolves in it and reacts to form a weak acid, which is not harmful. But man's contribution through fossil fuel, power plants, industries, automobiles has disturbed this natural acidic balance and converted natural rain into precipitated acid rain with several environmental implications.

The ill-effects of acid rain can be found on vegetation, soil, marine resources, monuments as well as on man. Air pollutants and acids generated by industrial activities are now entering forests at an unprecedented scale. Many forests in Europe and North America now receive as much as 30 times more acidity than they would if rain and snow fell without pollutants. Its reaction includes change in colour of leaves, premature drops of leaves and flowers, tree crowns progressively thin, and, ultimately, trees die. Marine resources can be totally devastated with loss of aquatic life, fishes can die and entire lakes and streams can be destroyed. The affected fishes, if consumed by humans and birds, cause dangerous effects. In soil, the rate of decomposition of organic matter and formation of nitrogen fixing organisms is reduced by acids. Human health may also be affected increasing respiratory and skin problems.

Acid rain is an increasing problem in the industrial areas of America, Europe and Asia, where large amount of fossil fuels are burned. The ill effects of acid rains have no national limits as nature knows no geographical boundaries. Only developed technology can save the world from the ill effects of acid rain.

Desertification

The UN Conference on Desertification (UNCOD) has defined desertification as 'the diminution or destruction of the biological potential of the land, which can ultimately lead to desert-like conditions'. There is no environmental problem in the world that affects poor people as extensively as land degradation or desertification. According to the UN Environmental Programme (UNEP), about 900 million people are threatened by desertification, which affects more than 6.1 billion hectares, about 35 per cent of the earth's land area (Figure 20.3). Each of the 16 years that have passed since the UN Conference on Desertification (UNCOD) in 1977, has seen approximately 6 million hectares of previously productive land rendered infertile. In terms of income, on an average, this is an annual loss of about 45 billion dollars.

Figure 20.3
Threat from Desertification

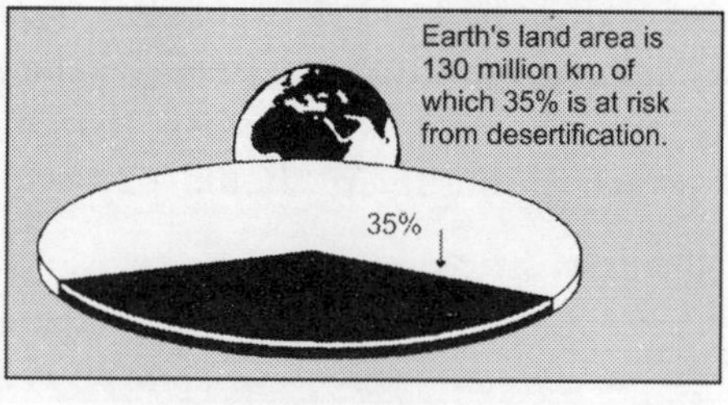

Desertification is the end-product of a vicious mixture of economic, political and technological forces, which leaves people economically, socially and politically marginalized. As the land gets further degraded, sheer survival takes up more and more human efforts, labour and time, till finally the men are forced to migrate and swell the ranks of the urban unemployed, while the women are left behind to struggle on.

Desertification is a slow process in which land productivity and resilience decline steadily. UNEP's latest study lists overgrazing, deforestation and unsustainable agricultural practices as the major causes of desertification. But these practices may be motivated by a whole chain of economic practices, which result in poor prices for agricultural and livestock products, or political compulsions like-debt, which force a country to promote adverse land use practices in order to earn foreign exchange.

What makes dryland soils specially vulnerable to degradation is the slowness of their recovery from disturbance. Because water is often not available or is available only in limited amounts, new soil is formed slowly, and salts, once accumulated, tend to remain *in situ*. Build up of organic matter, which is usually consumed very quickly under warm conditions, takes a long time. Moisture deficiency discourages recolonization by plants that have been removed or damaged.

Asia, with 1.3 billion hectares of degraded drylands, suffers the most desertification. But in terms of severity, Africa and North America are worse off, because nearly three-quarters of drylands are affected. In fact, desertification, is a major environmental problem and has worsened by the persistent droughts reported in the northern hemisphere during 1988. In Africa, drought and desertification rank as the foremost environmental concerns. During the last two decades, there have been at least 10 million environmental refugees from the drylands. Desertification is a matter of global concern.

Figure 20.4 shows existing deserts and areas into which deserts may spread. The risk of desertification is great in areas where rainfall is low and evaporation is high. However, the politics on desertification have turned full circle after the Intergovernmental Negotiating Committee for Convention to Combat Desertification (INC-D), held in May 1993 in Nairobi. To check desertification it is necessary that nations make combined efforts and if necessary, consumers should pay ecological costs of the products exported

Figure 20.4
Earth's Spreading Deserts

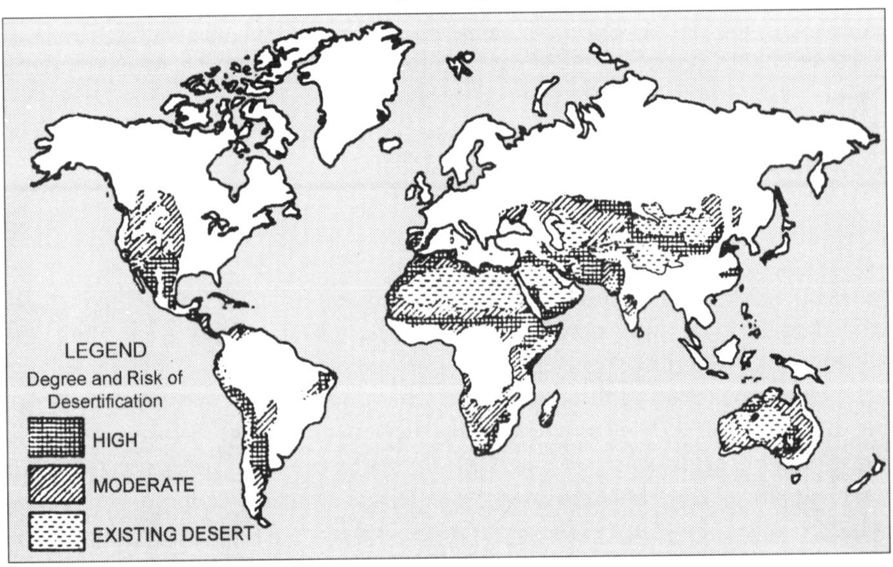

from developing countries. An international effort can only check or minimize the process of desertification.

Deforestation

Forests are not merely a natural resource to supplement multiple human requirements but are a very vital aspect of the environmental system. The forest plays a pivotal role in balancing the ecosystem. Growing industrialization, urbanization and ruthless exploitation of forests has created chaotic conditions and severe phyto-geographical and environmental imbalances. The use of forests has been linked with mankind from times immemorial basically for food, shelter and clothing, but population exodus and stresses on forest resources created a condition where natural growth of forests and their maintenance is incapable of coping with the demand.

Tropical forests are the richest forest resources of the world, their distribution has been depicted in Figure 20.5. But tropical deforestation has become an environmental issue over the past five years. Recent studies covering several key countries suggest that deforestation in the tropics may be much worse than was previously thought. The annual deforestation in the tropics was estimated to 11.4 million hectares in 1980 but now it has been estimated to be around 20.4 million hectares. The rate of deforestation is much higher in countries like Brazil, Cameroon, Costa Rica, India, Indonesia, Myanmar, Philippines, Thailand and Vietnam. Since pre-agricultural times, the world's forests have declined by one-fifth, from 5 to 4 billion hectares. Temperate forests have lost the highest percentage of their area (32–35%), followed by sub-tropical woody savannahs and deciduous forests (24–25%) and old grown tropical forests (15–20%). Tropical evergreen forests are now under the most pressure. Although most of the tropical forests are located in inaccessible areas and sparsely populated zones, now these forests are facing the tragedy of destruction.

Figure 20.5

Global Distribution of Tropical Forests

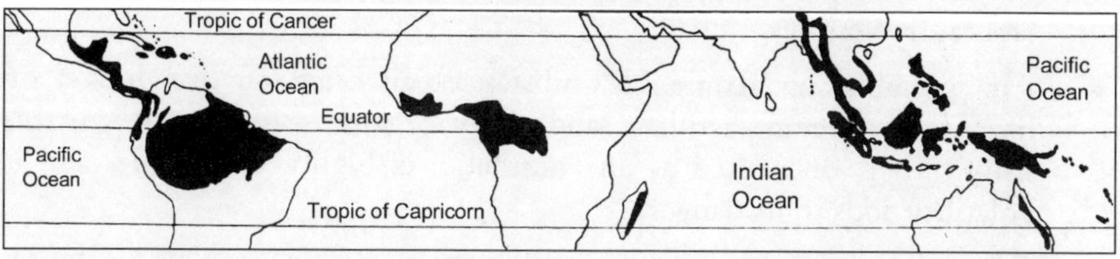

Tropical deforestation has three direct causes, which often act together in the same area. The first cause is the permanent conversion of a forest to an agricultural land. Logging is the another cause, although deforestation depends on the methods used and on what follows the logging. A third cause of deforestation is the demand for fuel wood, fodder, and other forest products where the resources cannot meet the demands. These direct causes as well as grazing, forest fire and drought are often interrelated and are exacerbated by government's economic policies, population pressure and poverty. The main cause of deforestation is the commercial exploitation of a forest.

There has been growing concern among professional foresters during 1970s and 1980s about the rate of tropical deforestation. With the introduction of the Tropical Forestry Action Plan (TFAP) in 1985, FAO, UNDP, World Bank and other governmental and non-governmental organizations have expressed the opinion that efforts to improve forest management had failed, because many causes of deforestation lie outside the domain of national forest departments; the problem that must be addressed at higher political and economic levels. TFAP was designed to deal not only with deforestation but also with forest degradation, with emphasis on agro-forestry, fuelwood, industrial forestry, and other methods by which forestry can provide goods and services to rural people and national economy. More than 60 countries have decided to prepare national forestry action plans to guide management of their forests. As the role of tropical forests in tempering global warming and harbouring a diverse range of species, has become clear, environmental groups have pressured international developmental agencies to halt projects that cause deforestation.

Hazardous Waste

As defined by US Resource Conservation and Recovery Act (RCRA) a hazardous waste has one or more of the following characteristics - it is ignitable, corrosive, reactive or toxic.

- A 'toxic' substance can adversely affect the health of living organisms exposed to it. Examples are arsenic and cyanide, pesticides and many metals.
- A 'corrosive' substance can cause grievous injury at the point of contact, i.e. the skin, eyes, lungs, or mouth. Strong acids and alkalis are corrosive. So are chlorine and hydrogen peroxide. The concentration of chlorine (as hypochlorite) or hydrogen peroxide found in household products is not corrosive, but much higher concentrations are often used in industrial and laboratory settings.
- An 'ignitable' or 'flammable' substance can catch on fire. It is a fire hazard. Petroleum distillates and many organic solvents are ignitable. Ethyl ether, once used as an anesthetic, is highly flammable, and its volatility adds to its danger.
- A 'reactive' substance reacts violently with water, air, or other substances. Ethers (chemicals used in industry and laboratories) can form explosive peroxides if left sitting in their containers for years. More familiar reactive substances are dynamite, gun ammunition, and firecrackers. In the household, chlorine, bleach and ammonia if mixed together will react to form toxic fumes.

Hazardous Waste Sources

Large petroleum refineries and chemical manufactures generate large quantity of hazardous waste. Manufacturers of metal products are another large generators, producing wastewater with high-enough concentrations of metals to be hazardous. Military operations generate hazardous waste, sometimes in large amounts; so do government agencies, especially the Department of Energy. Hospitals, many universities, and commercial laboratories generate smaller quantities of hazardous wastes. Other small-quantity generators include small businesses such as gas stations, photo developers, dry cleaners, sometimes even beauty parlours. Although the definition of hazardous waste varies, the UN Environmental Program estimates that 300 to 500 million tons (272 to 454 million tonnes) are produced worldwide each year.

Paint residue is the household hazardous waste produced in the largest amount. Others are paint thinners and strippers, pesticides, cleaners and polishes that contain petroleum distillates, and alkaline drain and oven cleaners. Even some discarded cosmetics, such as acetone nail-polish remover are considered hazardous wastes. MSW incinerator ash, although its origin was household trash, may be deemed hazardous waste. This happens if the ash does not pass an EPA-specified leaching test. A familiar hazardous waste is the used oil.

Health Threats of Hazardous Waste Site

The most often found chemicals are the four metals viz., lead, aresenic, mercury and cadmium, which are hazardous to health. Another four are volatile organic chemicals namely, vinyl chloride, chloroform, and trichloroethylene.

Human exposure to chemicals at hazardous-waste sites may occur through air, drinking water, soil or food. Following are the possible exposure pathways.

- *Air:* If the waste spilled onto the soil includes volatile organic chemicals (VOCs), these can evaporate. Direct inhalation by humans would ordinarily be very small unless they live extremely near the site (or unless on-site workers remain unprotected).
- *Food:* Evaporated VOCs could settle onto vegetation (such as the trees and corn plants). Animals may eat the contaminated vegetation or grain. Later, humans may eat contaminated meat, vegetation or grain.
- *Soil:* This is not usually a direct exposure source for humans, but could be for plants and animals on site.
- *Water:* Spilled waste could percolate into groundwater or run off into surface water. Groundwater that is used for drinking is considered the most likely exposure route for those living near to the sites.

Treatments of Hazardous Wastes

Once hazardous waste is generated, it is too late to treat, unless the waste treatment system is developed. Recycling, reuse, reducing volume, reducing toxicity, incineration, stabilization, neutralization, precipitation, filtration, distillation, etc. are the methods of hazardous waste treatment. The examples of these methods are as follows:

Table 20.2
Treatments of Hazardous Wastes

Treatment	*Examples*
Thermal	*Incineration* can destroy organic substances. If metals are in the ash, the ash may still be hazardous waste, but its volume, much reduced.
	Stabilization is a process in which a metal-containing waste is heated to high temperatures and melted. After it solidifies into a hard mass, it is difficult to leach metal from it. Thus, in many cases it may safely be placed in a landfill.
Chemical	*Neutralization* takes an alkaline or acidic substance, and converts it to a neutral form. An acid is neutralized with an alkali, or an alkali with an acid. After neutralization, the waste is no longer hazardous unless it has other hazardous constituents.
	Precipitation is often used to treat a metal solution. A substance is added that leads to the metals becoming insoluble and precipitating out. The precipitate too may be hazardous, but waste volume is much reduced. If enough metals precipitate from the solution and are carefully removed, the solution may no longer be hazardous.
Physical	*In filtration*, a liquid is separated from a solid using a membrane. The liquid goes through membrane pores. The solid particles are retained on its surface and its volume is much reduced. If enough hazardous material is filtered out, the liquid coming through the membrane (the filtrate) may no longer be hazardous.
	Distillation is sometimes used to separate a mixture of liquids. The mixture is heated. Lower temperature drives off the more volatile substance, leaving behind those with higher boiling points. Depending on the chemicals and the success of the distillation, the waste remaining may or may not be hazardous. Volumes of some of the components can be reduced too.
Bioremeiological (bioremediation)	Microorganisms degrade certain organic hazardous wastes using them as nutrients. Certain plants take up and concentrate metal from soil or water (phytoremediation). Some microbes consume concentrate metal or change the metals to less noxious forms.

International Transport and Trade of Hazardous Waste

In the late 1980s, the media published a spate of news stories about the dumping of hazardous wastes in the developing world. In the wake of these incidents, hazardous waste trade moved near the top of the international agenda. UNEP accelerated talks addressing the regulation of this trade, and the world's nations debated the issue. These negotiations resulted in the 1989 Basle Convention, which established a 'prior informed consent' system. The convention, adopted in March 1989, entered into force in May 1992. It was strengthened in March 1994, when the contracting parties agreed to a complete ban on shipping of hazardous wastes.

The term hazardous waste refers to material that poses a threat to human health and the environment, especially if it is not properly managed. Such material includes radioactive wastes, polychorinated biphenyl pesticides, toxic incinerator ash, chemical sludge and organic solvents. An estimated 90 per cent of the world's hazardous wastes is generated by industrialized market economies. The United States produces approximately 80 per cent of the world's total output. The main exporters of hazardous wastes in western Europe have been the Netherlands, Belgium, France, Italy and Germany. All these countries handle the major proportion of the wastes they generate. The United States handles the major proportion of its wastes domestically, exporting only one per cent. About one-fifth of the total annual global trade in wastes goes from industrialized countries to developing countries. Most of the third world countries lack the technology or administrative capacity to dispose of the wastes safely. The only attraction of developing countries has been the low cost.

To address such problems. the United Nations passed a treaty, The Basel Convention on the Control of Transboundary Movements of Hazardous Wastes and their Disposal. Its objectives are to: (1) minimize hazardous waste generation; (2) dispose of hazardous waste close to where they are generated; and (3) reduce the international movement of hazardous waste. The convention prohibits transboundary movements of hazardous or other wastes without *prior written notification* by the nation exporting the waste to authorities in the country that is importing it. Shipments must also be accompanied by appropriate documents. There are also outright bans on the export of hazardous wastes to certain countries. The Basel Convention is a global agreement ratified, as of April 2000, by 135 member countries and the European Union. The United States has not ratified the treaty, the only industrialized country not to do so.

Pesticides and their Effects on Environment and Humans

Pesticide use is very common throughout the world to control the enemies of agriculture and health. But with the excessive use of pesticide, several ecological and environmental problems have emerged and now agricultural scientists are agreed that their proper use is necessary. Pesticides are used

— to grow crops at times of year when they could not otherwise grow.
— to grow crops in ecologically inappropriate regions.
— for killing disease carrying organisms mosquitoes, flies, ticks and rats.
— for monocultures, in which large tracks of land are devoted to a single crop such as wheat, cotton, corn, etc.
— businessman, foresters, homeowners, utility owns also use pesticides for pest control etc.

The excessive use of chemical pesticides and insecticides has resulted in degradation of environment which has become a threat to human health. Throughout the ages, farmers have had difficulty in protecting their crops from small animals and disease organisms. At times, pest species have boomed and brought disease and famine. But most of the time, natural balance has been maintained, and humans have lived together with insects in reasonable harmony. Now-a-days, a great debate over the use and misuse of insecticides in agriculture is going on; on the one hand, scientists claim that more production can only be possible through control over pests and weeds, while on the other hand, they disagree and feel that chemical pesticides are harmful to the environment as well as human health.

Adverse Effects of Pesticides

Almost all pesticides can become pollutants because of the way they are applied, often by aerial spraying. There are some cases in which contamination is not a problem; the pesticide may swiftly degrade into innocuous products, it may not be very mobile and thus not reach water, and it may have low toxicity to non-target organisms. Unfortunately, in most cases, pesticide contamination is a matter of concern.

Herbicides and insecticides are applied over larger areas of agricultural fields and forests. Farmers may apply them a dozen times or more during a growing season. Less than half of the applied pesticide actually reaches the insect, weed, or any other pest. Most of it becomes a pollutant. Pesticides applied by spraying drift with air currents from the point of application. The largest amounts settle onto the land and water close to the point of application, but smaller amounts, sweep higher into the atmosphere with the winds, can be carried to many miles. Pesticides are detected in rainwater. The heavily used herbicide *atrazine* has been found at levels above 1 ppb in rainwater. Certain polychlorinated pesticides detected in wilderness lakes in the northern United States or Canada are not used in these countries and are assumed to have been blown north from Mexico or other Latin American countries. Soil or water, once contaminated with persistent pesticides may remain so for many years, especially in northern locations, where cold weather and lack of intense sunlight prevent them from degrading.

Pesticide applicators and those around them are the ones most likely to have high exposures. Birds, other animals and beneficial insects are also more highly exposed, and often suffer adverse effects or die.

- *Pollinating insects:* Consider the importance of pollinating insects for successful crops. Yet, in California, an estimated 11 per cent of honeybee colonies are lost each year to pesticide exposure. Responsible applicators take care to minimize exposure to desirable insects by carefully choosing the pesticides used, not applying them during the blossoming period, choosing to spray in the early morning or the evening when pollinating insects are absent, and using ground applications not aerial spraying, to lessen the pesticide drift that could harm nearby colonies.
- *Birds:* Cornell University has estimated that in the United States alone, at least 67 million birds die each year from pesticide poisoning. Birds die as a direct result from pesticide exposure or from eating pesticide-treated seeds, foliage, or insects; or if they are raptors through eating poisoned prey. On the US east coast, the deaths of as many as 2 million birds in one incident were associated with carbofuran, an insecticide, which is also extremely toxic to humans. When applied in a granular form, birds mistake it for dietary grit, eat it, and die. Carbofuran is now restricted or banned in many locales. In a tragic incident on the Argentine Pampas, nearly 6000 migrating Swainson's hawks died from exposure to the pesticide monocrotophos. The Argentine government later banned this pesticide. A great many, reported bird kills result from homeowners' misuse of pesticides. On an average, US homeowners use more than three times as much pesticide per amount of land as do farmers. Poisonings associated with pesticide use continue to occur.

- *Fish:* have died by the thousands or, in some cases by the millions, along with other aquatic species. Deaths are attributed to pesticide runoff or to pesticide drift settling onto water.
- *Amphibians:* An alarming trend is the disappearance, or reduced populations worldwide, of frogs and other amphibians.

Pesticides also contaminate groundwater. The organochlorine pesticides are much less water soluble. Groundwater contamination presents special concerns because even pesticides that are short lived in surface water may degrade very slowly in groundwater and harmful for health. Thus, long-term global contamination continues. What is needed is the use of organic fertilizers and natural system of control through certain bacteria. Pests can be controlled without killing them directly. There is a need for integrated pest management and to evolve scientific methods to control pests and weeds without adverse effects on ecosystem. In addition, government restrictions may serve as a drive to develop other, more environmentally sound methods of pest control. Organic farming is the best way to minimize the use of pesticides. Farmers can adopt pests by intercropping, crop rotation, and other traditional methods. Along with Integrated Pest Management (IPM), other tactics are - selection of pesticides with low toxicity to humans and other non-target organisms, application of pesticides in the lowest quantity and change in farming methods that could reduce pesticide use.

Pollution of the Oceans

For many years, people were relatively unconcerned about the pollution of the oceans because the great mass of the sea can dilute a huge volume of foreign matter to the point where it has little effect. In recent years, this attitude has changed because pollutants are often added in such relatively high concentrations that environmental disruption does occur before the pollutants have a chance to disperse. In addition, the quantity of pollutants dumped into oceans has grown so large that some scientists fear that global effects may be significant.

Oil spilling in oceans is a major problem of the pollution of the oceans. Oil spilling in ocean water has been occurred due to accidents of oil carrying tankers and also due to seepage of oil from these tankers. Figure 20.6 depicts the main tanker routes and major spills.

Some of the major oil spills are listed in Table 20.5 during 1967–99.

Table 20.5
Oil Spills

Date	*Oil-tanker*	*Oil lost (tonnes)*
1967	The *Torrey Canyon*: UK	119,000
1978	The *Amoco Cadiz:* Brittany, France	223,000
1979	The *Atlantic Empress:* Tobago	160,000
1989	The *Exxon Valdez*: Alaska	38,800
1993	The *Braer:* Shetland, UK	85,000
1996	The *Sea Empress:* Milford Haven, *UK*	72,000
1999	The *Erika:* Brittany, France	13,000

Figure 20.6
Main Tanker Routes and Major Spills

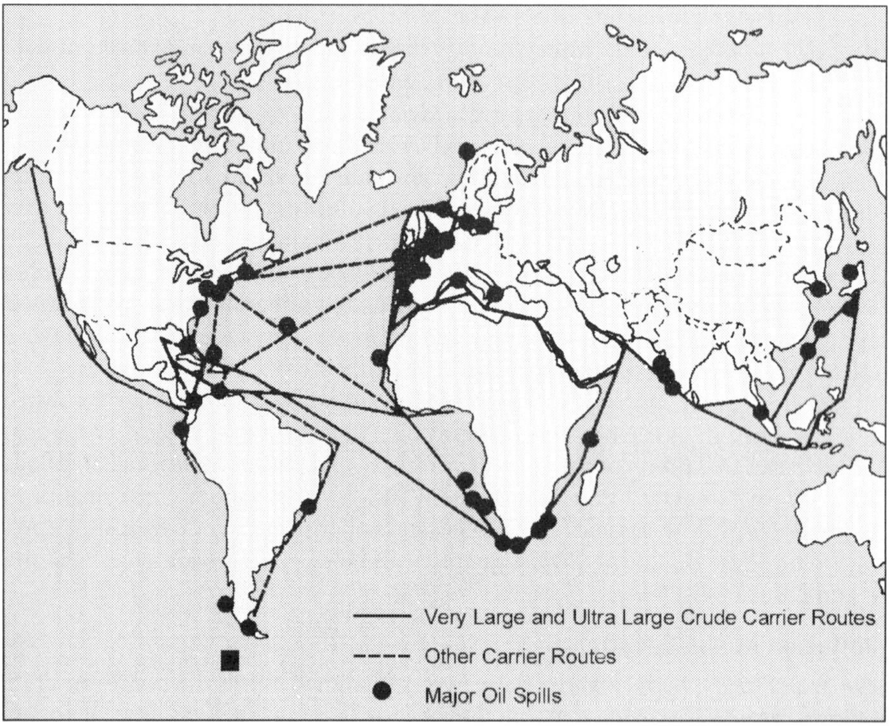

These spills (and the ones listed in table 20.5 are just a few of the most well known) represent the waste of millions of barrels of oil and, perhaps more importantly, create major pollution scares and potentially disastrous effects on local and regional ecosystems. One contributory factor has been the transportation of huge quantities of oil under so-called 'flags of convenience'. Ship-owners may register vessels more cheaply in a foreign country which also imposes lower taxes and has fewer regulations and restrictions. A typical example of this was the *Prestige*, an oil-tanker that sank off the coast of Spain in 2002 with 70,000 tonnes of heavy fuel oil on board. The ship was registered in the Bahamas, Greek-operated and Liberaian-owned, but it had been chartered by a Swiss-based Russaian oil company. Once registered, tanker owners insure against pollution, but only to a maximum amount, with any compensation payments over that amount coming from International Oil Pollution Compensation Funds. The ecological impact of oil spills depends in large measure on the type of oil spilled. Lighter and more volatile oil can evaporate, while heavier oils can be much more difficult to clean up. Other factors include the weather conditions at the time, which affect whether the oil makes land or remains in the sea, and the local environment into which the oil is spilled, which may be highly sensitive to disruption or more robust and resilient. The *Erika* spill of just 13,000 tonnes may not seem of the same order as that of the

Amoco Cadiz with 2,23,000 tonnes, but the *Erika's* cargo was heavy diesel oil which does not evaporate in the same way as the lighter crude oils, and therefore could have been more difficult to remove and damaging to wildlife. Environmental impacts also depend on the organized clean-up efforts. When some nine million barrels of oil were released into the Arabian Gulf by Iraqi forces in Kuwait in 1991, clean-up work was seriously hampered by the ongoing conflict.

Apart from oil spills, there is a tendency to dispose off high poisonous chemical wastes including chemical warfare agents, pesticide residues and radioactive substances into the oceans. The sealed drums are usually dumped into the sea. All these dumping activities have had adverse effect on the entire ecology of the oceans. During the 1972 Stockholm Conference on Human Environment, a study entitled, 'The Health of Oceans' was released, which has attracted the attention of world leaders but till now nothing substantial has been done in this field. The blasting of atom bombs under ocean waters is the greatest threat to the ocean environment, which must be banned immediately by all nations.

Nuclear Waste and Radiation Issue

Nuclear technology is relatively recent, with most nuclear power stations being built in the 1960s and 70s. Of course, nuclear weapons had been used against Japanese cities in 1945 by the USA, and many countries have carried out nuclear weapons testing both at sea and over and below ground. The connection between the peaceful civil use and military deployment of nuclear weapons is one of the key reasons why people have not taken up the nuclear option for energy production more enthusiastically. Nuclear power offers some unique advantages over many other energy sources. Nuclear power stations are costly to build but relatively cheap to run once in use. They also produce no carbon dioxide, the main greenhouse gas that contributes to global warming of the planet, thus making them potentially attractive to environmentalists as well as governments, looking to maintain power supplies, as fossil fuels run out. But nuclear power also has some unique disadvantages. It produces radioactive waste that remains harmful for thousands of years and, as yet, cannot be disposed of completely safely. Nuclear plants are potentially key targets for terrorist attacks and if more are built, they would need to be defended, militarily if necessary, on a much larger scale. Some people are concerned that this would probably mean some additional restrictions on people's civil liberties. Nuclear power generation also produces material for use in making nuclear weapons and, loosening international agreements on nuclear power production may lead to a proliferation of nuclear weapons, something that has been resisted until now. Finally, although accidents in coal mines and other energy production sites have in the past had some devastating consequences and losses of life, accidents inside nuclear power stations carry the potential to affect people at large distances from the source if radioactive material is released. Politicians therefore face a difficult assessment of the pros and cons of nuclear power that is only made more difficult with the advent of increased terrorist activity.

More than 60 incidents have been recorded in nuclear power plants since 1945, the vast majority in the USA and the former Soviet Union. Some of the more widely publicized accidents have led to nuclear power being closely associated with risk. The three worst accidents were at Windscale (now renamed Sellafield) in the UK (1957), Three Mile Island in Pennsylvania, USA (1979) and, the most serious of all, Chernobyl in the Ukraine (1986). A fire at the Windscale plant led to the release of radioactive material which forced the authorities to destroy the agricultural produce from farms in the surrounding area. The partial meltdown of a reactor core in the episode at Three Mile Island has a place in common folklore even though only small amounts of radioactivity was externally released and there were no deaths or injuries from it. It was the potential for disaster and the exposure of poor safety standards in the nuclear industry that brought home the risks of nuclear energy generation. The plant never did produce any usable energy.

The Chernobyl accident, however, was a different matter, It became a key reference point in debates on the future of civil nuclear power and was a significant symbolic factor influencing the break-up of East European communism. An unexpected power increase caused an explosion at the Chernobyl plant on 26 April 1986. The reactor was destroyed, releasing large quantities of radioactivity which spread over a very large area. Although direct fatalities from the accident numbered just 31, birth deformities and early deaths are thought to have significantly raised the numbers of people affected. Many neighbouring towns were badly affected and some, such as Pripyat had to be evacuated and effectively abandoned because of the risks of radioactive pollution.

Given the length of time it takes for radioactive material to degrade, there is no prospect of nearby towns becoming inhabited again in the foreseeable future. It would no be an exaggeration to say that he Chernobyl accident has had the biggest impact of any single pollution event. It reinforced an existing negative image of nuclear power and, in effect, negated the arguments of nuclear use advocated for almost two decades. It also led to a growing awareness of environmental issues which boosted support for environmental organizations.

Overpopulation

In Chapter 11, various aspects of population and ecological crisis have been discussed. Since overpopulation has become as matter of global concern, a note on overpopulation is necessary in context with its impact on environment.

Overpopulation occurs when human carrying capacity of an area is exceeded by the number of people living and using the resources in the area. Overpopulation puts enormous pressures on all types of environmental resources. It's main features are as follows:

The human population's birth rate has been surpassing the death rate at an alarming rate causing rapid population growth.

- The earth's current population is now more than 6 billion.

- Higher birth rates tend to be associated with poverty, low level of education, rural life styles and lesser opportunities for women.
- All of these factors tend to be typical of less developed countries (LDCs)
- Urbanization is a huge problem of LDCs because governments often are not able to provide basic services such as sanitation, education, water or food to city residents.
- Urbanization also leads to -

1. growth of slums;
2. industrial pollution;
3. air and noise pollution by vehicular traffic;
4. lower quality of life;
5. problem of safe drinking water
6. problem of waste disposal, etc.

Both the population and environment are complex systems and their interactions are also dynamic, complex and inconstant with time and space, therefore, planning for each country/region should be based on realistic research.

Environment and Human Health

It is an accepted fact that, environmental degradation is contributing to human health threats worldwide. The climate change, ozone depletion, agrosystem, population, land degradation, biodiversity, decline of several ecosystem services, etc. are directly influencing the human health.

Figure 20.7 depicts the relationship between degradation of global environment and their effects on human health.

Clean water, air and soil are prerequisites for human health, and pollution of these threatens human health on a global scale. Human health effects from environmental degradation of any sort are often cumulative, long term and complex (WHO, 2009). Environmental degradation of water, land and air leads to cancer, respiratory, cardiovascular, neuropsychiatric and communicable diseases (OECD, 2008).

Despite international efforts to mitigate major emissions, it is projected that there will be a six-fold increase in death attributable to ozone by 2030. Infant mortality from respiratory disorders such as asthma is on the rise. Millions are dying from polluted water. Some countries and populations are affected more than others. Deaths from unsafe water are 40.5 times higher in countries such as Brazil, Russia, India, China and South Africa than in OECD (Organisation for Economic Co-operation and Development) countries (OECD, 2008).

Vulnerability at the individual level is influenced by biological aspects (such as age and sex), social aspects (such as social connectedness and social policy) and environmental aspects (such as geographical location). Children, older people and people who are poor are most vulnerable. Children, for example, have a higher metabolic rate than adults and are exposed to different risks because of their activities. Asthma resulting from air pollution, leukemia

Figure 20.7
Degradation of Global Environment and its Effects on Human Health

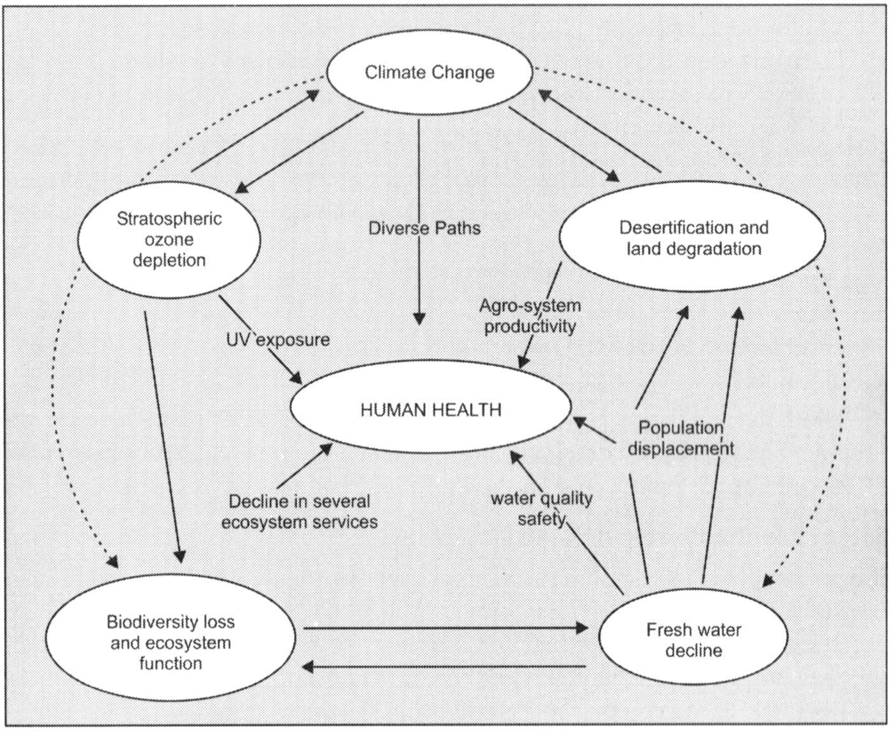

resulting from exposure to pesticides *in utero*, neurodevelopmental disorders resulting from lead poisoning and birth defects resulting from drinking contaminated water still occur because children's vulnerability is often no factored into policies (OECD, 2008).

Climate change is arguably the single biggest global health threat today (WHO, 2003). Pollution of air, water and land, and water stress will reduce agricultural production in many regions and exacerbate food security problems. Famines due to climatic influences are well recognized. Temperature changes and rising sea levels will give rise to changing patterns of disease. Traditional ways of life are already threatened in every nation. Displacement, loss of means to work and psychological stress due to climate change are all part of the new fabric of the social determinants of health. All regions are affected.

Biodiversity's relevance to human health is becoming an international political issue, as scientific evidence builds on the global health implications of biodiversity loss. This issue is closely linked with the issue of climate change, as many of the anticipated health risks of climate change are associated with changes in biodiversity (e.g. changes in populations and distribution of disease vectors, scarcity of fresh water, impacts on agricultural biodiversity and food resources etc). This is because the species most likely to disappear are those that buffer against infectious disease transmission, while surviving species tend to be the ones that increase disease transmission, such as that of West Nile Virus, Lyme disease and Hantavirus.

The growing demand and lack of drinkable water on the planet present an additional challenge to the future of human health. Partly, the problem lies in the success of water suppliers to increase supplies, and failure of groups promoting preservation of water resources. While the distribution of clean water increases, in some parts of the world it remains unequal. According to 2008 World Population Data Sheet, only 62 per cent of least developed countries are able to access clean water.

Some of the health issues influenced by biodiversity include dietary health and nutrition security, infectious disease, medical science and medicinal resources, social and psychological health. Biodiversity is also known to have an important role in reducing disaster risk, and in post-disaster relief and recovery efforts.

In the coming years, vulnerable populations such as children, those who are poor, those who are already frail and indigenous communities will bear the brunt of further environmental degradation. It is fair to say that those who already suffer social and environmental injustice will continue to do so unless policy decisions are made to address inequality in general, and poverty, discrimination, pollution and natural resource depletion in particular.

The health sector is already responsible for many protective measures such as health education, preventive programmes such as mosquito control and food hygiene, surveillance of disease, disaster preparedness and forecasting future health risks. These responsibilities and policy development extend beyond the sector and include early warning systems of extreme weather events, neighbourhood watch for vulnerable people, involvement in urban planning for green spaces and cool spaces, climate-proof housing and work spaces. As we have said, human health is determined by many factors and will be promoted and protected in many sectors. This introduces us to the idea of 'health in all policies', an idea that was first developed by the WHO and refined in 1988 at the Second International Conference on Health Promotion.

Economic inequality and environmental changes are closely connected to each other. Poor countries are unable to meet required emission standards to slow down climate change. The depletion of ozone in the stratosphere (middle atmosphere) also has an important impact on global climate and, in turn, human health, increasing the amount of harmful UV radiation that reaches the earth's surface. This results in diseases such as skin cancer.

Development strategies that do not incorporate ecological safeguards often lead to ill health, while strategies that can promote health invariably also protect the environment. Thus, environmental health and human health are closely interlinked. An improvement in health is central to sound environmental management. However this is rarely given sufficient importance in planning development strategies.

Examples of the Links between Environmental Health and Human Health

— Millions of children die every year due to diarrhoea from contaminated water or food. An estimated 2000 million people are affected by these diseases and more than 3 million children die each year from water-borne diseases across the world. In India, it is estimated that every fifth child under the age of 5 dies due to diarrhea. This is a result of inadequate environmental management and is mainly due to inadequate purification of drinking water. Wastewater and/or sewage entering water sources without being treated leads to continuous gastrointestinal diseases in the community and even sporadic large epidemics. Large numbers of people in tropical countries die of malaria every year and millions are infected. An inadequate environmental management of stagnant water, the breeding site of the Anopheles mosquito, is the most important factor in the spread of malaria. The resurgence of malaria in India is leading to cerebral malaria that affects the brain and has a high mortality.

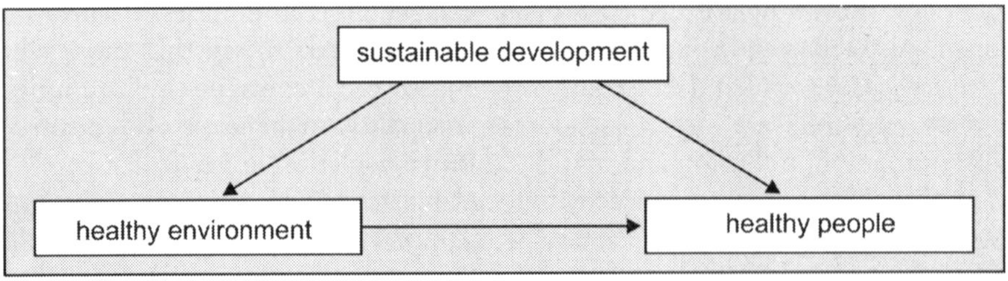

— Millions of people, mainly children have poor health due to parasitic infections, such as amoebiasis and worms. This occurs from eating infected food, or using poor quality water for cooking food. It is estimated that 36 per cent of children in low-income countries and 12 per cent in middle income countries are malnourished. In India, about half the children under the age of four are malnourished and 30 per cent of newborns are significantly underweight.

— Hundreds of millions of people suffer from serious respiratory diseases, including lung cancer and tuberculosis, from ill-ventilated homes and public places. Motor vehicle exhaust fumes, industrial fumes, tobacco smoke and cooking food on improper chulas, contribute to respiratory diseases.

— Millions of people are exposed to hazardous chemicals in their workplace or homes that lead to poor health due to industrial products where controls are not adhered to.

— Tens of thousands of people in the world die due to traffic accidents owing to inadequate management of traffic conditions. Ineffective first aid at the accident site and the frequent inability to reach a hospital within an hour causes a large number of deaths, especially from head injuries.

— Basic Environmental needs - such as clean water, clean air and adequate nutrition - which are all related to environmental goods and services do not reach over 1,000 million people living in abject poverty.

— Several million people live in inadequate shelters or have no roof over their heads especially in urban settings. This is related to high inequalities in the distribution of wealth and living space.
— Population growth and the way the resources are being exploited and wasted, threaten environmental integrity and directly affect the health of every individual.
— Health is an outcome of the interactions between people and their environment. Better health can only come from more sustainable environmental management.

Important Strategic Concerns

— The world must address people's health care needs and the sustainable use of natural resources, which are closely linked to each other.
— Strategies to provide clean potable water and nutrition to all people are an important part of a healthy living environment.
— Providing clean energy sources that do not affect health is a key to reducing respiratory diseases.
— Reducing the environmental consequences of industrial and other pollutants, such as transport emissions, can improve public health.
— Changing agricultural patterns away from harmful pesticides, herbicides and insecticides which are injurious to the health of farmers and consumers and using alternatives, such as IPM and non-toxic biopesticides, can improve the health of agricultural communities as well as food consumers.
— Changing industrial systems into those that do not use or release toxic chemicals that affect the health of workers and people living in the vicinity of industries can improve health and environment.
— There is a need to change from using conventional energy to cleaner and safer sources like solar, wind and ocean power that do not affect human health. Providing clean energy will lead to better health.
— The key factors are to control human population and consume less environmental goods and services which could lead to 'health for all'.
— Poverty is closely related to health and is itself a consequence of improper environmental management. An inequitable sharing of natural resources and environmental goods and services, is linked to poor health.

Environmental Warfare

Using weather as a weapon and manipulating the environment for military purpose is common and has come to focus in the Gulf War. The United States was the first to use weather modification as a weapon in Indo-China region from 1967–72. The intention was to increase normal monsoon rainfall and disrupt Vietnamese logistics, marking the emergence of the use of environment as a weapon in warfare – artificial triggering of earthquakes and tidal waves, modifying hurricanes or hailstorms to change their course, creating artificial fog to impede enemy traffic, etc.

The use of chemical weapons are not only harmful to people and other organisms but to the ecosystem also. The emergence of cruise and ballistic missiles and other types of bombs are responsible for a lot of pollution in the atmosphere. The Gulf Crisis (1990) has brought it back to the forefront. Dumping millions of barrels of crude oil in the sea to hamper naval operations and foul up desalination plants supplying drinking water, marked the beginning of environmental warfare. Military experts are of the opinion that every state has its environmental or geophysical vulnerabilities that can be exploited by an enemy in the event of war. All this can be checked through an international awareness along with international collaboration.

In brief, global environmental issues discussed in this chapter and many others highlight that now it is high time to protect the biosphere from environmental catastrophes. These environmental issues have become a part of discussion not only at environmental conferences but also in every international political meet. Several documents and agreements have been signed between nations. It will not be enough to have the agreement, but their implementation is necessary at all levels. The growing globalization might make the task even more difficult. With globalization, the rich have greater freedom to exploit the poor. With its attendant economic asymmetry, the globalization of the economy has clear consequences for environmental health and sustainable development. There is a need to evolve a global strategy and it should be supported by action at all levels, especially at the community level.

21

Environmental Problems and Planning in India

A country's environmental problems are affected by the level of its economic development, the availability of natural resources and the lifestyle of its population. In India, rapid growth of population, poverty, urbanization, industrialization and several related factors are responsible for the rapid degradation of the environment. Environmental problems have become serious in many parts of the country, and can no longer be neglected. The main environmental problems in India relate to air and water pollution particularly in the metropolitan and industrial zones, degradation of common property resources which affect the poor adversely due to a degeneration of their life support system, threat to biodiversity and inadequate system of solid waste disposal and sanitation with consequent adverse impact on health, infant mortality and birth rate. The degradation of natural resources imposes several ecological problems. In earlier chapters of this book, several problems of water, air, noise and soil pollution as well as their impact of human activities on environment in India have been discussed in brief. But their overall analysis is necessary not only for their understanding but also for the environmental planning in India.

Air Pollution

The main sources of air pollution in India are automobiles, thermal power plants, industries, agriculture, quarrying and also the natural processes. During the process of combustion of fossil fuels, mainly oxides of carbon and particulates are generated. Apart from this, various industries also generate other pollutants such as SO_2, H_2S, ammonia, oxides of nitrogen, hydrocarbons, and other toxic substances including ozone, lead and fluorides. The nature of urban and rural air pollution is quite different in India, therefore, their respective causes, effects and remedies are also different.

Urban Air Pollution

Urban air quality has generally deteriorated throughout the Indian cities. The reasons are rising number of motor vehicles, especially poorly maintained vehicles and growing industrialization without any priority for pollution

Figure 21.1

Vehicular Population and Percentage of Categories of Different Types of Vehicles in Major Metropolitan Cities

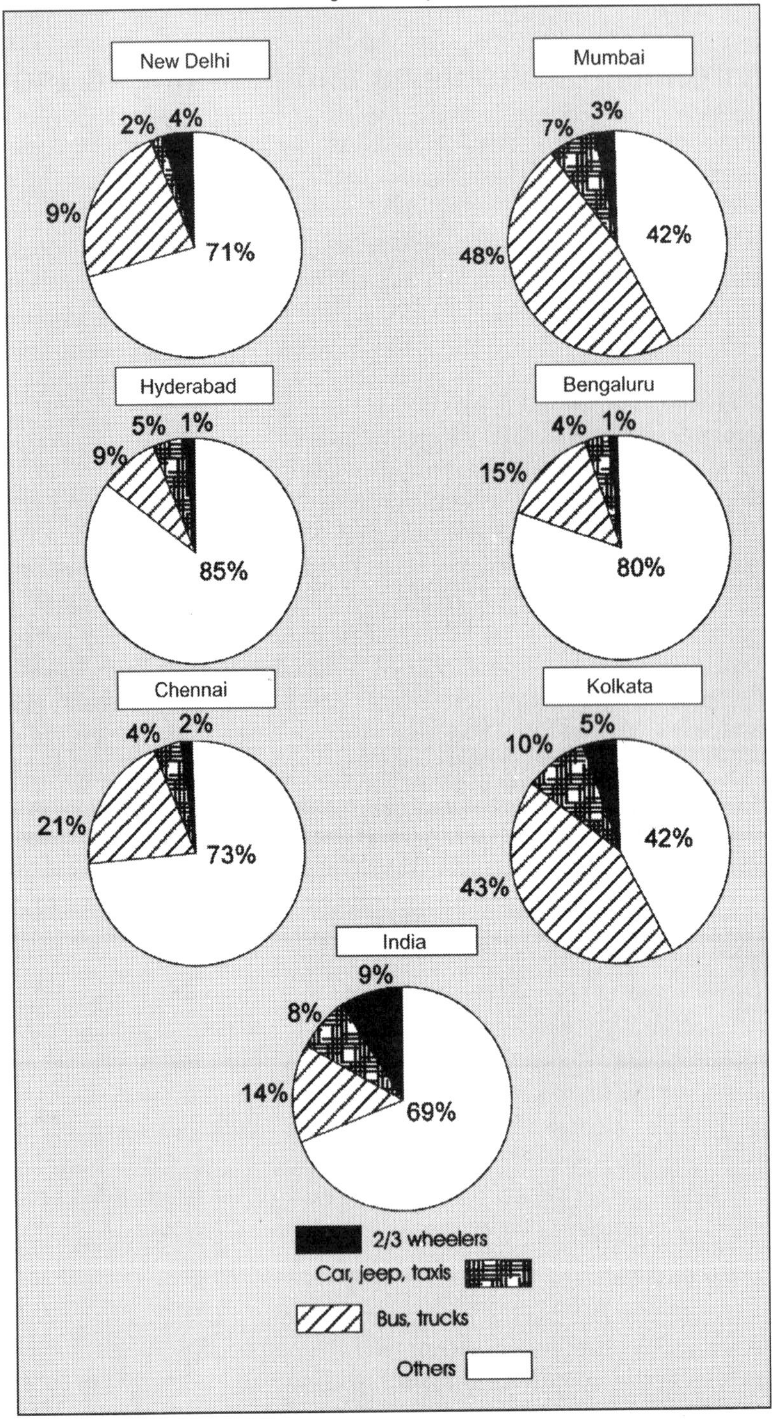

abatement. Indian cities qualify as some of the most polluted cities in the world and heading the list is the national capital, New Delhi. The single most factor responsible for this deterioration of air quality in cities is the exponential increase in petrol and diesel fuelled vehicles. Between 1986 and 1991, the number of vehicles on Indian roads has increased from 11 million to 21 million. Of these, 70 per cent are two- and three-wheelers with inefficient and highly polluting two-stroke engines. The number of two-wheelers has been growing at a rate of about 20 per cent a year and it is estimated that these vehicles will be around 36 million by the turn of the century. It is expected that the number of passenger cars and diesel vehicles will also be doubled by that time. The percentage of vehicular population in India in the six metros has been depicted in Figure 21.1.

It becomes clear from Table 21.1 that Delhi has the maximum vehicular pollution with total emission of 872 tons per day followed by Mumbai having 549 tons of total emission per day. The concentration of pollution is much higher at the highways, road signals and traffic junctions.

Table 21.1

Estimated Vehicular Emission Load in Metropolitan Cities

(vehicular pollution load – tonnes per day)

Name of City	*Suspended Particulate Matter*	*Sulphur Dioxide*	*Oxides of Nitrogen*	*Hydro-Carbons*	*Carbon Monoxide*	*Total*
Delhi	8.58	7.47	105.38	207.98	542.51	872
Mumbai	4.66	3.36	59.02	90.17	391.6	549
Bangalore	2.18	1.47	21.85	65.42	162.80	254
Kolkata	2.71	3.04	45.58	36.57	156.87	245
Ahmedabad	2.46	2.41	33.33	56.46	149.28	244
Pune	1.99	1.07	13.50	61.00	135.2	213
Chennai	1.95	1.68	23.51	42.05	119.35	188
Hyderabad	1.62	1.30	14.03	46.94	105.14	169
Jaipur	0.98	1.04	12.74	17.49	42.73	75
Kanpur	0.88	0.90	11.14	18.53	40.35	72
Lucknow	0.95	0.79	8.07	18.75	41.02	70
Nagpur	0.46	0.34	4.24	13.60	29.16	48
Grand Total	29.42	24.87	352.39	674.96	1916.01	2998

Industries are also responsible for the air pollution in India. Most of the industries are located near cities, irrespective of any environmental consideration. Most of these industries have not adopted normal pollution control measures, therefore, have become a main cause of air pollution. For example, Mumbai has various industries, mainly chemical (523), textile (531), and mechanical workshops (3,348). Besides, there are 32 factories processing rubber or plastic products. In addition, there is a large concentration of chemical units within 30 km of Mumbai. Similar is the story of the other cities of India.

Coal-based thermal plants have been established for power generation and are responsible for air pollution through gaseous emissions of sulphur dioxide, nitrogen oxide, etc. Air pollution can also cause acid rain, which not only adversely affects man but is also responsible for damage of soil, vegetation and aquatic life of the region and also produces huge amount of solid wastes, fly ash and bottom ash.

Due to air pollution, life in big cities of India has become intolerable due to health problems. Air pollution is the main cause of respiratory illness and death, physical damage and loss of vision.

Recent Trends in Air Quality

The National Air Quality Monitoring Programme, Government of India in 2011 has claimed that with the last 15 years of economic development and regulatory reforms, India has made progress in improving air quality. The table 21.2 present emissions sampled at many locations over time and data analysed by scientific methods by multiple agencies, including the World Bank. Over 1996–2008, average nationwide levels of major air pollutants have dropped by 25–45 per cent in India.

Table 21.2

Levels of Major Air Pollutants in India

	1996	*2005*	*2008*
Pollutant PMIO (Micrograms per cubic metre)	109	67	59
Pollutant CO_2 emissions (Kg per 2006) PPS of GDP	0.7	0.6	0.5
Health mortality rate (under 5 per 1000)	100	73	67
Pollutant methane (agricultural emissions % of total)	68.8	64.3	n. a
Pollutant nitrous oxide (agricultural emissions % of total)	75.2	73.4	n.a

For 2010, the key findings of India's Central Pollution Control Board are:

- Most Indian cities continue to violate India's and world air quality PM 10 targets. Respirable particulate matter pollution remains a key challenge for India.
- A decreasing trend has been observed in sulphur dioxide levels in residential areas of many cities such as Delhi, Mumbai, Lucknow, Bhopal during the last few years.
- A decreasing trend has been observed in nitrogen dioxide levels in residential areas of some cities such as Bhopal and Solapur during the last few years.
- Most Indian cities greatly exceed acceptable levels of suspended particular matter. This may be because of refuse and biomass burning, vehicles, power plants' emissions and industrial sources.
- The average annual SOx and NOx emission levels and periodic violations in industrial areas of India are significantly lower than the emissions in the residential areas of India.

- Of the four major India cities, air pollution was the worst in Delhi; Kolkata was close to the second followed by Mumbai and Chennai.

In India, there are several laws to control air pollution, but their implementation is not effective. Pollution by factories is perhaps the easiest to control by imposing standards of emission. Recently, the Supreme Court has also directed factories of Delhi and its adjoining areas to adopt either proper measures to check pollution or close the factory. Similar type of action and imposition of heavy penalties is necessary.

On the other hand, reducing emission by vehicles is much more complex. A few measures which can reduce the air pollution are as follows:

1. Strengthening and introduction of mass transit system as these use significantly less fuel per person per kilometre.
2. Vehicles with old technology should be banned from big cities.
3. Two/three-wheelers should be upgraded to four-stroke engines.
4. Heavy duty vehicles should not be allowed to ply on the inner city roads at peak hours (8 to 12 noon and 4 to 8 pm) so as to reduce traffic stagnation.
5. Catalytic converters should be made compulsory for all petrol driven vehicles and dieselized vehicles must have a particle trap.
6. Better urban design should be planned to avoid congestion and reduce transport needs.
7. Better and efficient traffic control plan.
8. Introduction of modern technology should be encouraged in vehicles and, in the long run, the use of electric vehicles or battery operated vehicles should be promoted.

Rural Air Pollution

In India, rural population uses substantial quantities of non-commercial fuel, i.e. crop residues, animal dung or wood. This still provides 80 per cent of rural energy for cooking. Air pollutants due to biofuels are largely released directly where people are, i.e. inside or near households at the mealtimes everyday. Although the emissions are relatively modest, the actual exposure to people is significant in many millions of households around the country. Several health effects are suspected to arise due to smoke such as respiratory infection in children, chronic lung diseases and lung cancer in adults, and adverse pregnancy outcomes, such as low birthweight and still birth, in women exposed to it, during pregnancy.

The rural air pollution can be reduced by: (i) providing an affordable clean fuel as substitute to biofuels, (ii) installation of efficient smokeless *chullahs* (much effort has been done in this direction by the government), (iii) good ventilation in houses (iv) providing bio-gas facilities, and (v) above all, the habits of the rural people be changed, therefore, awareness regarding ill-effects of the use of biofuels should be developed among the rural people.

Water Pollution

Once admired for its beautiful water resources, India has over the years, progressively depleted its supply of water due to various factors, including increasing population, thoughtless urbanization, deforestation and the resultant soil erosion and flawed water management. The available water is also polluted in some way or the other. Surface water sources mainly include rivers, springs, lakes and ponds.

The problem of water pollution in India is increasing day by day. The major sources of water pollution in the country are: (i) domestic effluents, (ii) sewage, (iii) industrial effluents, (iv) agricultural effluents, (v) thermal pollution, (vi) oil pollution, and (vii) pollution created by radioactive elements. Water pollution from domestic and human waste water causes many severe water-borne diseases. Sewage effluents from cities dotting the river's bank contribute to nearly three-fourths of the water pollution.

A study of World Health Organization claimed that out of India's 3,119 towns and cities, just 209 have partial sewage treatment facilities and only 8 have full waste water treatment facilities. A 1995 report claimed that 114 Indian cities were dumping untreated sewage and partially cremated bodies directly into the Ganges River. Open defecation is widespread even in the urban areas of India. According to another 2005 report, sewage discharged from cities and towns is the predominant cause of water pollution in India. A large number of Indian rivers are severely polluted as a result of discharge of domestic sewage.

River Water Pollution

Water pollution in India has now reached a crisis point. Almost every river system in India is now polluted to a great extent. As assessed by the scientists of the National Environmental Engineering Research Institute (NEERI), Nagpur, nearly 70 per cent of water in India is polluted. India has five major river systems, namely, the Ganga, the Brahmaputra and Indus river systems in the North, and the peninsular, East coast and the West coast river systems in the South.

North Indian Rivers

Many large rivers are closely associated with Indian culture and heritage. The pollution situation in our country is worse than that of some of the industrialized countries of Western Europe and America. *The Ganga*, the most sacred and important river of India, is regarded as the cradle of Indian civilization. The 2,525 km long river starts from Gangotri in the Himalayas and joins the Bay of Bengal, at Ganga Sagar. According to a report of the Central Pollution Control Board, despite river Ganga's considerable resilience as a self-purifying and fast-flowing river, its organic pollution load is significantly high. At Kanpur, 45 tanneries and 10 textile mills are the major sources of liquid wastes discharged into the river Ganga. The wastes contain heavy organic load and putrescible material. It is estimated that 1,400 million litres of sewage and 200 million litres of industrial effluents are being discharged everyday into the river Ganga. The BOD (106 tonnes/day), total solids (2,308 tonnes/day) and suspended solids (1,251 tonnes/day) are not only very high but also exceed the prescribed limits of Indian Standards Institution. Recently, discharges from the Barauni Oil Refinery caused gross pollution along a long stretch of the main Ganga. Preliminary observations were made on the pollution of the river Kali and a limnological survey was made of the river with reference to fish mortality. The main pollutants came from chemical and soap factories and large quantities of municipal sewage.

According to an estimate, about 19,65,000 kilograms of polluted matters are discharged into Ganga every year. The study undertaken by the Central Board for Preservation and control of water pollution has revealed the fact that long stretches of the Ganga have become unsuitable for any beneficial human use near the cities of Kanpur and Varanasi. The status of Ganga river pollution at selected locations has been reported as given in table 21.3:

Table 21.3

Status of Ganga River Pollution

Locations	*MPN (Coliform) Bacteria PL=500/100 ml*	*BOD PL=3g/l*	*DO PL=5g/l*
Rishikesh	about 500	3.00	9.0
Haridwar	—	3.50	7.0
Kannauj	10,000,000	7.80	—
Kanpur	—	—	—
Allahabad	7,10,000	—	—
Varanasi	—	15.25	3.0
Patna	2,500	—	—
Kolkata	10,000	—	—

Source: Report of Japanese Scientist (2005) BOD=Biological Oxygen Demand, DO=Dissolved Oxygen, PL=Permissible Limit.

A major step to control and clean the river Ganga has been taken in the form of Ganga Action Plan. The Ganga Action Plan Phase I was started in 1985 in order to reduce the pollution load on the river. But the efforts to decrease the pollution level in the river become more after spending ₹ 901.71 crore. Therefore, this plan was withdrawn on 31 March 2000. The steering committee of the National River Conservation Authority reviewed the progress of the GAP and Phase II was approved in stages. The GAP is still under implementation. The National River Ganga Basin Authority (NRGBA) was established by the Central Government of India on 20th February, 2009. It also declared Ganges as the *National River* of India. A project under the NRGBA with World Bank assistance for abatement of pollution of river Ganga at an estimated cost of ₹ 7,000 crore has been approved in April, 2011. The principal objective of the project is to fund creation of pollution abatement infrastructure for conservation and restoration of water quality of the river.

Also, Yamuna river has now become a polluted river mainly because of sewage and industrial flows. The upper segment of 224 km long stretch between Tejawala and Wazirabad receives enormous pollution load from agro-chemicals (fertilizers and pesticides) brought in by the surface run off from the agricultural fields of Haryana State. The Delhi segment of the river between Wazirabad barrage to Okhla is highly polluted mainly from sewage drains of the city. After Okhla to its confluence with the Chambal, the content of micro-pollutants has become so high that BOD has become maximum. The residents of Agra and Mathura have no option than to use the polluted water. Delhi metropolitan area is the biggest polluter of the Yamuna. About 1900 million litre of sewage from Delhi is discharged into the Yamuna per day, where as the total treatment capacity is only 1270 million litres per day, Yamuna Action Plan (YAP) was launched in 1993 to study and control pollution in the Yamuna river.

A survey was carried out in 1965 along 21 km stretch of the river Gomti in the vicinity of Lucknow, receiving 19.84 million gallons of wastes per day from pulp and paper factory, distillery and sewage. Heavy fish mortality was reported in Rihand reservoir due to high free chlorine content (62 ppm) discharged from Kanoria Chemical Industries. The organic wastes from a sugar factory and a distilleries plant cause year round pollution of the small river Dhaha. A case of heavy fish mortality was recorded in 1962 and 1966. Wastes from different factories such as those engaged in the manufacture of paper, chemicals, sugar, cement, etc. are the major sources, discharging over 4 million gallons of wastes per day into the river Sone.

The river Hooghly at Kolkata receives wastes from various types of factories dealing with pulp and paper, distillery, tannery, textile, heavy chemicals, paints and varnishes, shellac, hydrogenated oil, matches, cycle rim, petroleum oil, tar pigment, insecticides and fungicides. Of these, wastes from paper and pulp, distillery, chemicals, textiles, shellacs and a number of domestic outfalls contribute substantially to the pollution complex. River Damodar, which flows through the coal belt area in Bihar is also a seriously polluted river which is the receptacle for wastes released from large number of industries such as the Sindri unit of the Fertilizer Corporation of India, the Bihar government's superphosphates factory and the associated cement company. The entire Asansol-Durgapur industrial belt on lower Damodar valley suffers from severe pollution caused by the discharge of wastes containing high phenol, cyanide and ammonical nitrogen.

The indiscriminate discharge of large volumes of highly putrescible liquid wastes of the Orient Paper Mills creates serious pollution problems in the river Ib (Orissa) of the Mahanadi river system. The river Bhadra (Krishna river system) receives effluents from pulp and paper and steel industries.

Industries generate a significant quality of waste water which ultimately finds its way to a stream/river. Industrial discharges containing toxic and hazardous substances contribute to the severe kind of pollution in the aquatic systems. Industrial development is largely because of the production of chemicals resulting in the generation of toxic and hazardous substances which have been continuously on the increase during the last three decades.

Industrial effluents, though comparatively lesser in volume, cause serious menace to aquatic environment and biotic communities including fish and ultimately affect man, through the food chain.

South Indian Rivers

River Godavari at Rajahmundry (Andhra Pradesh) is polluted by the effluents of Andhra paper mill at Rajahmundry. The river Kalu in Mumbai receives highly acidic and untreated wastes from Amar Dye and Chemical Company, India Dyes, Century Rayon, National Rayon, Central Chemicals, etc. Effluents of the Gwalior Rayon factory at Mavoor, about 21 km from Beypore, have created a pollutional hazard in the river Chaliyar at Calicut, Kerala. A large scale fish mortality in 1966 was attributed to highly putrescent organic matter creating almost anaerobic conditions in the river with very low or nil oxygen.

In Tamil Nadu, river Cooum at Chennai gets polluted by the washings from a large number of slums, cattle yards, overflow from sewage pumping station, wastes from automobile workshops and many factories. Nearly 300 tanneries are spread along the banks of river Palar over a stretch of 120 km from Vaniambadi to Ranipet. The wastewater discharge from these tanneries affects the ground water quality because of sodium and chlorides present in the tannery waste. River Cauvery is polluted by Mettur Chemical and Industrial Corporation Ltd. River Vaigai receives effluents from many chemical and soap factories and large quantities of municipal sewage.

The rate of pollution in South Indian rivers seems to be higher than those of the North Indian rivers. The main reason for this is the summer season during which the icebergs get melted and the melted water is drained into the major North Indian rivers like the Ganga, the Brahmaputra, etc. During the rainy season, rivers get a good current from the natural showers itself. The only possibility of pollution for such rivers is by the catchment ot pollutants when they pass through the urban and the industrial belts, whereas the South Indian rivers do depend upon the north-east and the south-west monsoons. Deforestation also affects rainfall and ultimately changes the river flow in South Indian rivers. Already these rivers are under stress due to lack of rainfall caused by deforestation and pollution is an additional/secondary stress.

Access to safe drinking water remains an urgent need as only 72 per cent of the population in urban area receives well treated water supply. The situation in rural areas is much worse. In India, almost all surface water sources are contaminated and unfit for human consumption. The diseases commonly caused due to contaminated water are diarrhoea, trachoma intestinal worms, hepatitis, etc. Recent WHO data show that about 21 per cent of all communicable diseases in India are water borne diseases. Epidemiological studies have shown that diarrhoea and intestinal worm infections account for an estimated 10 per cent of the total burden of disease.

An uncontrolled disposal of urban waste into water bodies, open dumps and poorly designed landfills, cause surface water and ground water contamination. Industrial wastes containing heavy metals such as mercury, chromium, lead and arsenic can threaten or destroy marine life. Health hazards such as malaria, filariasis, hepatitis, cholera can also spread. The epidemic of jaundice is also a result of contaminated water supply. In 1955–56, in Delhi, about 29,300 persons developed jaundice, following contamination of the drinking water supply from contents of a severed drain from Najafgarh in West Delhi. Similarly, epidemic of hepatitis-E has been caused due to drinking of polluted water. In India, major incidence of hepatitis epidemics has occurred in various urban centres in years as shown in Figure 21.2.

A few remedies to control the water pollution are:

— industries should enforce standards for water effluents;
— the Pollution Control Boards must enforce the laws;
— adequate sewage disposal facilities should be developed, so that sewage is not directly released into the water stream;
— the sewage and water treatment plant should be established by every municipal body;
— for small and medium sized industries which cannot afford effluent plants, combined treatment plants be established;

Figure 21.2

Hepatitis Epidemics in India

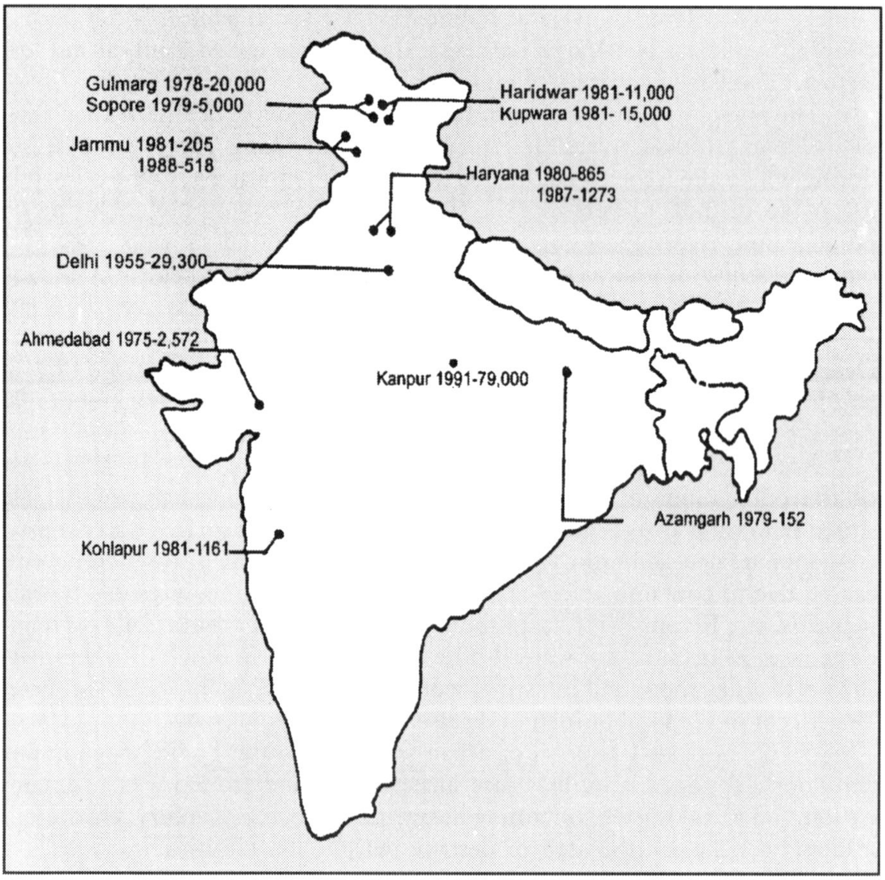

— a pollution tax be imposed, especially on industries responsible for water pollution;
— more and more *Sulabh* type of toilets be built;
— action plans concerning water pollution should be completed within specified period; and
— above all, awareness among people about sanitation and its related impact on their health should be promoted.

Groundwater Pollution in India

An example of groundwater pollution caused by excessive extraction is that of fluoride contamination or fluorosis. Fluorosis is not a localized problem. It has spread across 19 states and across a variety of ecological regions ranging from the Thar Desert, the Gangetic plains and the Deccan plateau. Each of these regions is distinct in terms of rainfall, soil type, groundwater recharge regime, climatic conditions and hydrology. High fluoride concentration in groundwater is a natural phenomenon in several countries like China, Sri Lanka, West Indies, Spain, Holland, Italy and Mexico. Experts claim that a fluoride belt stretches across the Middle East across Pakistan and India and then into South-east Asia and the South of China. According to a report of the Rajiv Gandhi National Drinking Water mission, the bedrock of the Indian peninsula consists of a number of fluoride-bearing minerals. When the bedrock weathers, the fluoride leaches into water and the soil. Although the Indian peninsular bedrock has always been the same, this problem has only surfaced during the last three decades. It is related to the over-extraction of groundwater, resulting in the tapping of aquifers with high fluoride concentrations.

The beginnings of this phenomenon can be traced back to the 1970s and the 1980s, when there was a massive state investment in rural water development for irrigation as well as for drinking. Encouraged by state subsidies on diesel and electricity, people invested in diesel and submersible pumps in a bid to extract groundwater through bore-wells. This policy aggravated the fluoride problem.

Fluoride mainly enters the human body through drinking water where 96–99 per cent of it combines with the bones, as it has an affinity for calcium phosphate in the bones. The excess intake of fluoride can lead to dental fluorosis, skeletal fluorosis or non-skeletal fluorosis. Dental fluorosis is characterized by discoloured, blackened, mottled or chalky-white teeth. Skeletal fluorosis leads to severe and permanent bone and joint deformities. Non-skeletal fluorosis leads to gastrointestinal problems and neurological disorders. Fluoride can damage the fotus and adversely affect the IQ of children.

Once fluoride is detected in water, the only solution is to deflouridate it. Various technologies are available for this process. However, the type of technology to to be selected depends upon the fluoride levels in the water and the volume of water to be deflouridated. None of the Indian technologies are foolproof. Deflouridation plants and household water treatment kits are only stop-gap solutions.

Solid Wastes Disposal

The solid wastes generation is increasing day by day not only in big cities of India but also in small towns and has become a threat to the environment. The composition and quantity of waste changes with increase in per capita income. The main change in solid waste material has been with the increased use of plastic and other synthetic materials in place of biodegradable organic materials. The solid wastes include domestic and municipal waste, industrial and agricultural waste, waste created by mining and also radioactive waste. The composition of solid wastes in five big cities of India has been given in the following Table 21.4.

Table 21.4

Composition of Solid Wastes from Cities

Cities	*Characteristics (%)*					
	Non-degradable					*Degradable*
	Paper	*Plastics*	*Metal*	*Glass*	*Ash and Earth*	
Kolkata	3.18	0.65	0.66	0.38	34.00	47.00
New Delhi	6.29	0.85	1.21	0.57	36.00	35.00
Nagpur	1.88	1.35	1.33	1.34	41.42	34.81
Bengaluru	4.00	2.00	—	1.00	15.00	78.00
Mumbai	10.00	2.00	3.6	0.2	44.20	40.00

Table 21.5

Collection of Solid Wastes in Cities and Annual Per Capita Solid Wastes Generated

Cities	*Solid Wastes*	
	Generated (Mt)	*Collected (%)*
Mumbai	0.20	90
New Delhi	0.44	77
Chennai	0.29	90
Bengaluru	0.16	96
Lucknow	0.31	74
Varanasi	0.14	87
Hubli-Dharwad	0.15	89
Mysore	0.26	60
Gulbarga	0.10	40
Tumkur	0.13	50

When this solid waste is not collected and disposed of efficiently and effectively, it attracts rodents and flies which then spread diseases and also pollute water and cause land degradation. Urban India generates 1,88,500 tonnes per day (68.8 million tonnes per year) of municipal solid wastes at per capita waste generation rate of 500 grams per person. The status of solid wastes collection in selected cities is given in Table 21.5.

The quantity of solid waste generation varies in cities of India according to their size and nature of industrialization. Figure 21.3 shows quantity of waste generated per day:

Figure 21.3

Municipal Solid Waste Generation Quantity in Indian Cities

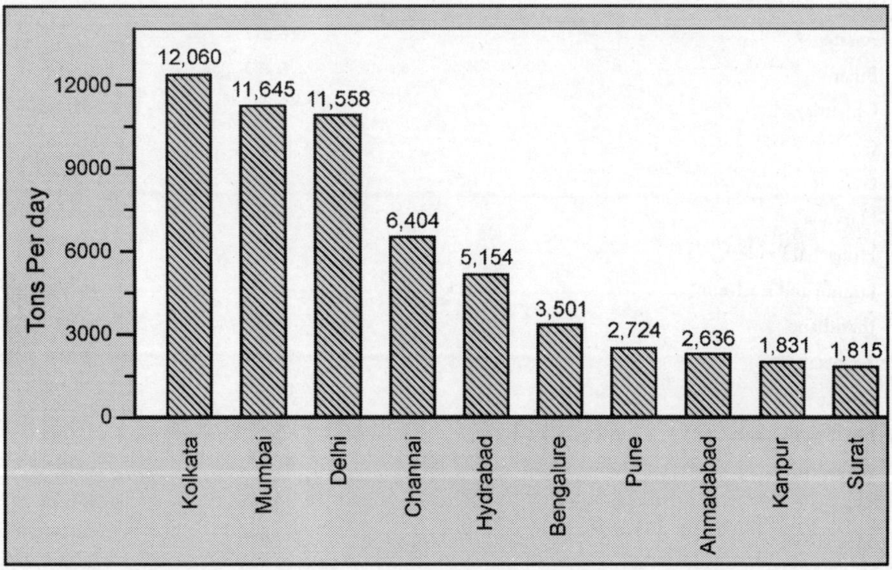

These wastes can be utilized after sorting for generating energy and useful resources by way of composing, vermicompost and anaerobic digestion. The potential of power generation from these wastes is tremendous. It is estimated that urban municipal wastes have a potential to generate upto 100 MW of electricity. Industrial wastes coming from dairies, distilleries, perfumes, tanneries, pulp and paper and food processing industries have the potential to generate about 700 MW of power.

These wastes are left untreated; they ferment slowly and produce biogas that contains 65–75 per cent methane gas, which is a greenhouse gas. Therefore, development of suitable technologies for utilization of wastes is essential to minimize adverse health and environmental consequences.

Deforestation

Forests have a many-faced ecological role to play which affects human life in a variety of ways. They thwart the dangers of cloud drifting, soil erosion, floods, wind erosion and ground water evaporation. They also protect a wide variety of flora and fauna, provide recreation and can effectively control air pollution of moderate magnitudes. But, due to industrialization, commercialization, growth of population habitat and expansion of agriculture in India, forest cover is fast disappearing. Table 21.6 indicates the state-wise forest cover in India.

Table 21.6
States of Forests in India, 2009

State/UTs	*Forest Area in sq km*	*Percentage*
Andhra Pradesh	63,814	23.20
Arunachal Pradesh	51,540	61.55
Assam	26,832	34.21
Bihar	6,473	6.87
Chhattisgarh	59,772	44.21
Goa	1,224	33.06
Gujarat	18,927	9.66
Haryana	1,559	3.53
Himachal Pradesh	37,033	66.52
Jammu and Kashmir	20,230	9.10
Jharkhand	23,605	29.61
Karnataka	38,284	19.96
Kerala	11,265	28.99
Madhya Pradesh	94,689	30.72
Maharashtra	61,939	20.13
Manipur	17,418	78.01
Meghalaya	9,496	42.34
Mizoram	16,717	79.30
Nagaland	9,222	55.62
Orissa	58,136	37.34
Punjab	3,050	6.07
Rajasthan	32,639	9.54
Sikkim	5,841	82.31
Tamil Nadu	22,877	17.59
Tripura	6,294	60.02
Uttarakhand	34,651	64.79
Uttar Pradesh	16,583	6.88
West Bengal	11,879	13.38
A&N Islands	7,171	86.93
Chandigarh	34	29.82
Dadra and Nagar Haveli	204	41.55
Daman & Diu	8	7.14
Lakshadweep*	–	–
Delhi	85	5.73
Pondicherry	13	2.71
Total	7,69,512	23.41

* No discernible forest cover.

Source: The State of Forest Report 2009; *Statistical Year Book*, India, 2012.

India has a forest area of 7,69,512 sq km, which constitutes only 23.41 per cent of the total land area as against 33 per cent of National Forest Policy of 1988. A large part of these forests is degraded and productivity is very poor. Even in protected forests, unrecorded changes are going on. Actually the closed type of forest having forest cover density of 40 per cent or more is approximately only 11 per cent of the country's total land area.

A wide range of flora and fauna are fast disappearing as their natural habitats are destroyed. There is a need for massive reforestation programmes. An effective afforestation, reforestation and regeneration activities should be carried out. Agro-forestry should also be encouraged. For rural population, cheap and clean fuel should be provided to save the forest from denudation.

Soil Degradation

In India, about 130 million hectares of land (45% of total geographical area) is affected by serious soil erosion through ravine and gully, shifting cultivation, cultivated wastelands, sandy areas, deserts and waterlogging (Government of India, 1989). The soil formation is a very long process and under favourable conditions, in India, it takes almost a thousand years to form only 2.3 cm of soil layer. It is reported that the loss of one mm of cultivated soil could cost 10 kg of nitrogen and 2 kg of phosphorus. Soil erosion by rain and river that takes place in hilly areas causes landslides and floods, while cutting trees for firewood and timber, grazing by livestock, traditional agricultural practices, construction of roads, indiscriminate quarrying and other activities all lead to the soil erosion. In desert regions of Rajasthan, wind erosion causes heavy soil erosion and drifting of sand.

The main causes of soil erosion are depletion of forests, wrong use of land such as cultivation on steep slopes, cattle rearing, etc. Apart from this India is losing 5,334 million tonnes of soil annually due to erosion because of indiscreet and excess use of fertilisers and insecticides over years. About one millimetre of top soil is being lost each year. The rate of loss is 16.4 tonnes per hectare every year. Experiments conducted by ICAR indicated that non-judicious and imbalanced use of inorganic fertilizers (NPK) over years may result in deterioration of soil fertility/nutrient deficiencies.

The government has launched 'National Project on Management of Soil Health and Fertility' during 2008–09 to promote soil test based judicious use of fertilizers for improving soil health and productivity. An area of about 8000 hectares is under ravines. Such areas are distributed in Uttar Pradesh, Madhya Pradesh, Bihar, Rajasthan, Gujarat, Punjab, West Bengal, Tamil Nadu and Maharashtra.

In India, wastelands are spread in more than 1,295 lakh hectares of land, out of which 937 lakh hectares is non-forest degraded land and 358 lakh hectares is forest degraded area. There is a need to intensify the efforts to check soil erosion, to retain and increase the productivity of agricultural land and to control the expansion of the desert areas and landslides. The Government of India has already established a Wasteland Development Board in order to check further degradation of soil and to make useful the existing wastelands of the country. By adopting soil conservation technologies, regeneration of wastelands should be done.

Biodiversity

India has a rich heritage of species and genetic strains of flora and fauna. Overall, 8 per cent of world species are found in India. It is estimated that India is tenth among the plant rich countries of the world. India has diversity in its ecosystem consisting of forests, wetlands, grassland, deserts, marine areas, coral reefs, etc. and each system supports thousands of species. The total number of living species identified in India so far is 2,00,000 (many more not yet discovered). The pattern of biodiversity in India is given in Table 21.7.

Table 21.7

Biodiversity in India

Species	*Number*
Total Plants	45,000
Mammals	350
Reptiles	453
Amphibians	182
Fishes	1693
Insects	67,000
Molluscs	4,000
Birds	1,200
Invertebrates	6,800

But India's rich biological diversity is rapidly eroding. More than half of its forests are gone, two-thirds of its wetlands drained out or polluted, almost all of its dry scrub and grasslands habitat destroyed. Habitat destruction, hunting and overexploitation, pollution and poisoning and introduction of exotics have caused a decline in many species. Many species have gone extinct, of what remains, over 10 per cent of flowering plants, 21 per cent mammals and five per cent of birds are believed to be under threat.

At the Convention on Biological Diversity (CBD), signed at the Earth Summit in 1992, India was the first country to sign the CBD. Not only this, but several efforts have been made to conserve the rich heritage of biodiversity. In order to preserve it, 358 wildlife sanctuaries and 63 national parks have been developed in India. There is a need to implement the laws of the protection of forest and wildlife. What is needed more is the people's participation. Two measures which might greatly help are:

1. Local communities must be given much more secured and meaningful rights over public lands and waters in their own localities, following the lead to joint forest management programmes.
2. Local communities should be given substantial incentives to continue conserving biodiversity resources on private and public lands in their own localities by linking financial inputs.

It is hoped that India would, in coming years, rise to this important challenge. (for details see Chapter 15).

Natural Hazards and their Impact on Environment

In every part of the world, several natural phenomena such as volcanic eruptions, earthquakes, cyclones, landslides, floods, droughts and famines have had their impact on environment. Sometimes their impact is causal and normal but often their impact is adverse not only on man but also on environment, therefore, termed as natural hazards, environmental hazards, environmental stresses, environmental disasters, extreme events, etc. These natural hazards are common in various parts of India.

Since India is not located in the volcanic zone, volcanic eruption is not common. But, in November-December 1991, at Andaman Island, a dormant volcano became active and caused damage to vegetation, marine life and also polluted the atmosphere. The incidence of landslide is common in hilly areas of Jammu and Kashmir, Himachal Pradesh and Uttar Pradesh. Similarly, coastal regions are susceptible to cyclonic storms, which cause massive destruction to life and property as well as to vegetation. The coastal regions of Orissa, West Bengal, Andhra Pradesh and Tamil Nadu often face the problem of cyclones.

In India, earthquake-affected areas include Jammu and Kashmir, Himachal Pradesh, Uttar Pradesh, West Bengal and the north-eastern states. These areas are located in the foothills of the Himalayas, while there are other areas, which include Punjab, Haryana, Delhi, Rajasthan and Maharashtra states. The impact of earthquakes on environment is much more in the form of destruction of flora and fauna, occurrence of rifts and cracks in land, changes in course of rivers, cracks in dams and a great loss of man and property, depending upon the intensity of the earthquake. On 11 October, 1937, an earthquake claimed the lives of over three lakh people. Other important earthquakes have occurred on 30 May 1885 in Kashmir; 12 June 1887 at Shillong; 15 June 1870 in Kacch; 4 April 1905 in Kangra Valley; 5 January 1934 in Bihar; 15 August 1995 in Assam; 11 December 1967 at Koyana (Maharashtra); 21 August 1988 at Darbhanga (Bihar); 20 October 1991 in Uttar Kashi; 30 September 1993 in Maharashtra, 26 January 2001 in Bhuj (Gujarat) etc.

The Indian subcontinent is very vulnerable to droughts, floods, cyclones, earthquakes, landslides, avalanches and forest fires. Among the 36 states and Union territories in the country, 22 are prone to disasters.

Among all the disasters that occur in the country, floods are the most frequently occurring natural disasters, due to the irregularities of the Indian monsoon. About 75 per cent of the annual rainfall in India is concentrated in three to four months of the monsoon season. As a result there is a very heavy discharge from the rivers during this period causing widespread floods. Approximately 40 m ha of land in the country has been identified as being prone to floods. The major floods are mainly caused in the Ganga-Brahmaputra-Meghna basin, which carries 60 per cent of the total river flow in our country.

India has a long coastline of 5,700 km, which is exposed to tropical cyclones arising in the Bay of Bengal and the Arabian Sea. The Indian Ocean is one of the six major cyclone-prone regions of the world. In India, cyclones occur usually between April and May and also between October and December. The eastern coastline is more prone to cyclones as it is hit by about 80 per cent of the total cyclones generated in the region.

The term 'Tsunami' comes from the Japanese language, meaning harbor (tsu) and wave (nami). A tsunami can be generated by any disturbance that rapidly displaces a large mass of water, like an undersea earthquake, volcanic eruption or submarine landslide. The wave travels across the ocean at speeds of 500–1000 km/h. As the wave approaches the land, it 'compresses' – sometimes up to a height of 30 metres – and the sheer weight of the water is enough to crush objects in its path, often reducing buildings to their foundations and scouring exposed ground to the bedrock. The tsunami on 26 December 2004, killed 3,10,000 people, making it the deadliest tsunami in the recorded history.

Droughts are a perennial feature in some states of India; 16 per cent of the country's total area is drought prone. Drought is a significant environmental problem as it is caused by a lower-than average rainfall over a long period of time. Most of the drought-prone areas identified by the Government lie in the arid and semi-arid areas of the country. While a good monsoon becomes a cause of floods in India, the failure of monsoon is responsible for drought resulting in famines. Similarly, droughts become the cause of environmental degradation. The famine that occurred in year 1899, known as *Chappan-ka-Akkal*, is well known in the history of India, because of its widespread impact and death of several lakhs of people. A similar famine occurred in 1917. In 1986–87, about two-thirds area of India was under the grip of famine. The worst hit area of famine is the Rajasthan state, where famine often occurs due to desertic conditions and failure of monsoon. During the last 20 years, Rajasthan has had famine in 16 years. The development of irrigation facilities and proper water management can help to solve this problem.

Environmental Planning in India

In earlier chapters as well as in this chapter, references have been made regarding various efforts done in India for the protection of the environment. In fact, India is a country where all the components of environment have been considered as life support systems right from the ancient times and their protection has been of paramount importance. The air, water, land, vegetation, soil, rivers, mountains, trees, animals, etc. have been considered as sacred in one form or the other. There was a general public opinion to protect all these as a part of religious activity. Although at that time there was no problem of environment because of limited population and of simple pattern of life, but, gradually, especially during the present century, with the growth of population, industrialization, urbanization as well as growth of other economic activities, environmental problems have developed, and therefore, a planned strategy is necessary.

For the protection of environment and to control pollution, several efforts have been undertaken, both by the government and NGOs. Even before independence, some laws have been enacted for the protection of the environment. In Indian Penal Code of 1860, Sections 268, 290, 291, 426, 430, 431 and 432 are related with the environment. Similarly, Section 277 is related with water pollution and Section 278 with air pollution. In Motor Vehicle Act, 1938, there was a provision to control pollution and Indian Forest Act was passed in 1927. After independence, serious efforts have been made in the form of legislation for the conservation and protection of the environment; the Act, and the laws passed during last fifty years, is a major step for the protection of environment in India. The following are the important environmental legislations:

1. Damodar Valley Corporation (Prevention of Pollution of Water) Regulation Act, 1948.
2. River Board Act, 1956.
3. Water Preservation and Control of Pollution Act, 1974 and 1977.
4. Atomic Energy Act, 1972.
5. Radiation Protection Rules, 1971.
6. Wild Life Protection Act, 1972.
7. Factories Amendment Act, 1987.
8. Environmental Protection Act, 1986.
9. Central Motor Vehicles Rules, 1989.
10. Public Liability Insurance Act, 1991.

Above all, in 1977, a major step in this direction has been taken in the form of 42nd Constitution Amendment. Accordingly, Article 48-A imposes a duty on the state 'to protect and improve the environment and to safeguard the forests and wildlife of the country'. Article 51A(g) imposes a duty, on citizens of India 'to protect and improve the natural environment including forests, lakes, rivers and wildlife and to have compassion for living creatures'.

The establishment of the Ministry of Environment and Forest in November 1980 is another step towards environment protection and also for policy making and planning. Similarly, every state government has established a ministry of environment. The Central Pollution Control Board has been established to take measures to control water and air pollution and also prepare plans for countrywide awareness about pollution. In the country, national laboratories, IITs, university departments, NEERI and several other autonomous bodies are engaged in research in the field of environment. The role of non-governmental organizations (NGOs) is also important and many NGOs are doing good work in this field.

The National Forest Policy has been framed for the protection and development of forest in the country. A strategy has also been evolved to develop forests resources. The programmes like social forestry and agro-forestry are quite successful for the conservation of the biodiversity of the country.

Similarly, national parks and game sanctuaries have been established throughout the country to protect biodiversity.

Efforts are going on for the development of non-conventional sources of power like solar energy, wind energy, biogas, tidal energy, etc. This will not only solve the demand of energy but also improve safety from the point of view of environmental pollution.

In India, Ministry of Environment and Forests is regulating the policies and programmes relating to conservation of country's natural resources including lakes and rivers, its biodiversity, forests and wildlife, ensuring the welfare of animals and prevention and abatement of pollution. While implementing these policies and programmes , the Ministry is also the nodal agency for the United Nations Environment Programme (UNEP), South Asia Co-operative Environment Programme (SACEP), International Centre for Integrated Mountain Development (ICIMOD) and United Nations Conference on Environment and Development (UNCED). The Ministry also coordinates with multilateral bodies such as the Commission on Sustainable Development (CSD), Global Environment Facility (GEF) and regional bodies such as Economic and Social Council for Asia and Pacific (ESCAP) and South Asian Association for Regional Co-operation (SAARC) on matters pertaining to environment.

The main objective of the environmental planning in India is to protect the environment and to help the government of India in policy frameworking and its proper implementation. In this direction, substantial work has been done (which has been referred in this book at several places, legislative measures are described in the next chapter). Some of the recent measures, policies etc. taken in the direction of environmental planning are as follows:

- For biodiversity conservation, Govt. of India had enacted Biodiversity Act, 2002 (BDA, 2002). A national level National Biodiversity Authority (NBA) was established in 2003 at Chennai and also State Biodiversity Boards (SBB) have been established by State Government and Biodiversity Management Committees (BMC) constituted by the local bodies.
- Wildlife conservation is under the wildlife wing in the Ministry; has two divisions, namely Project Elephant division and Wildlife division. In addition, there are three autonomous bodies, i.e. National Tiger Conservation Authority (NTCA), Central Zoo Authority (CZA) and Wildlife Institute of India (WII). Wildlife Crime Control Bureau (WCCB) came into existence on 6th June, 2007.
- The Ministry of Environment and Forests has used Environment Impact Assessment Notification, 2006 as a tool to regulate rapid industrial development of the country and minimizing and reversing the trends which may lead to climate change in the long run. The EIA notification was amended in December, 2009.
- The Central Pollution Control Board (CPCB) is responsible for planning and executing comprehensive nationwide programmes for the prevention

and control of water and air pollution. This Board is also providing scientific information, rendering technical inputs for formulation of national policies and programme training and development of manpower and organizing activities for promoting awareness at different levels of the government and public at large.

- Management strategies include:
  - Development and promotion of clean technology
  - Industrial pollution abatement through preventive strategies (waste minimization)
  - Hazardous waste management
  - Bio-Medical waste management
  - E-waste management
  - Municipal solid waste management
- Conservation of water bodies:
- National River Conservation Directorate (NRCD) launched Ganga Action Plan (GAP). Phase-I scheme in 1985 was the beginning of the river cleaning programme. GAP Phase-I was extended to GAP Phase-II during 1993–96
  - National River Conservation Plan (NRCP) was initiated in 1996. The NRCP presently covers 38 rivers spread in 20 states.
  - National Lake Conservation Plan (NLCP) was approved in May, 2001. Out of five lakes targeted for completion during 2010–11, rejuvenation/conservation works in two lakes have been completed.
  - National Ganga River Basin Authority (NGRBA) was constituted on 20th February 2009 and *Mission Clean Ganga* will be ensured by 2020.
- Under National Wetlands Conservation Programme (NWCP), wetlands (Conservation and Management) Rules 2010 were notified under Environmental Protection Act, 1986.
- The 'Convention of Wetlands' was signed in Ramsar, Iran in 1971.
- National Afforestation and Eco-Development Board (NAEB) in involved in:
  - National Afforestation Programme (NAP) scheme
  - NAEB scheme
  - Eco-Development Forces (EDF) scheme
- National Natural Resource Management System (NNRMS) involves utilization of remote sensing technology for accurate inventory of resources and to utilize this information for monitoring changes in ecological system.

In India, there is a need for sustainable growth, i.e. growth which is not harmful to environment. The eco-friendly development can solve the problem of environmental degradation. Efforts are going on for the environmental management. At all levels of education, provisions have been made for the knowledge of environment and its conservation. In the country, many centres are providing special training for environmental management. The programmes of environmental awareness have been launched through media. India is an active member of international organizations concerning environment. Several programmes are going on under UNEP. The government has recently started emphasizing the combined use of regulatory and economic instruments for improving environmental quality. This will be more effective if enforced systematically. The environmental audit introduced in 1993 is a measure to deal with this problem through improved self-regulation. The use of GIS (Geographical Information Systems) will be useful for effective planning. There is a need for coordination between government agencies, NGOs and the public for the proper management of environmental quality in the country.

22

Environmental Ethics and Legislations in India

'Environmental ethics concerns the question of whether or not our interactions with natural environment, or our interactions between each other with respect to the natural environment, are subject to moral constrains or moral requirements' (Carter, 2011: 111). It is a moral duty of all the people not to harm the environment. If we poison our environment through our daily activities, then we will harm each other. Alan Carter has further elaborated the environmental ethic frame and suggested five points which should be taken into consideration, these are:

- future generations
- people in other countries
- non-human animals
- non-sentient living beings
- species and ecosystems

All this means that the policies we require are those that respect the interest of all persons, whether geographically or temporally near or far, that respect the interest of sentient non-human and that respect ecosystem integrity.

Environmental ethics deals with issues related to the rights of individuals that are fundamental to life and well being. These concern not only the needs of each person today, but also those who will come after us. It also deals with the rights of other living creatures that inhabit Earth.

Resource Consumption and Need for Equitable Utilization

Environmental ethics deals with issues that are related to how we utilize and distribute resources. Can individuals justifiably use resources so differently that one individual uses resources many times more lavishly than other individuals who have barely enough to survive? In a just world, there has to be a more equitable sharing of resources than exists at present. The even distribution of resources has global, national and local concerns that we need to address. There are rich and poor nations, rich and poor communities and families in every country. In this era of modern economic development, the disparity between the haves and have-nots is widening. Our human environment in the urban, rural and wilderness sectors use natural resources that shift from the wilderness (forests, grasslands, wetlands, etc.) to the rural sector and from there to the urban sector. Wealth also shifts in the same direction. This unequal distribution of wealth and access to land and its resources is a serious environmental concern. An equitable sharing of resources forms the basis of sustainable development for urban, rural and wilderness-dwelling communities. As the political power base is in the urban centres, this itself leads to inequalities and a subsequent loss of sustainability in resource management in the rural and forest sectors.

The first report on the *Status of India's Environment* (1985) published by Anil Agarwal has brought forth 8 propositons which are of relevance to the ethical issues related to environmental concerns. These are:

1. Environmental destruction is largely caused due to consumption patterns of the rich.
2. The worst sufferers of environmental destruction are the poor.
3. Even where nature is being 'recreated', as in afforestation, it is being transformed away from the needs of the poor and towards those of the rich.
4. Even among the poor, the worst sufferes are the marginalized cultures and occupations and, mostly, women.
5. There cannot be proper economic and social development without a holistic understanding of society and nature.
6. If we care for the poor, we cannot allow the Gross Nature Product to be destroyed any further. Conserving and recreating nature should be our highest priority.
7. The Gross Nature Product will be enhanced only if we can arrest and reverse the growing alienation between the people and the common property resources. For this, we have to learn a lot from our traditional cultures.
8. It is totally inadequate to talk only of sustainable rural development, as the world conservation strategy does. We cannot save the rural environment or rural people dependent on it, unless we bring about sustainable urban development.

Disparity in Resource Ownership

Environmental ethics are concerns with who owns resources and how they are distributed. This can be looked at from different levels. At the global level, it deals with the great North-South divide between the rich industrialized nations of North America and Europe, as against the needs of developing countries of the South such as in South-east Asia and South America. People living in the economically advanced nations use greater amounts of resources and energy per individual and also waste more resources. This is at the cost of the poor people who are resource-dependent and live in developing nations.

The economically advanced west has exploited their natural resources to such an extent that they have exhausted them nearly everywhere else. They now buy their resources from resource-rich but economically-deprived nations at a low cost. This depletes the developing nations of natural resources on which the poor depend for their livelihood.

Changing this unfair economic practice to a more just and fair way in managing trade would require a new thinking on the part of people who live in the super-rich countries.

Indian Traditions of Environmental Ethics

Although many countries of the world are having traditional environmental ethics and value system but, India's conservation ethics and traditional value system is of immense importance and needs special mention.

India has inherited a culture of tolerance, non-violence, equity and compassion for animate objects. In the olden times, animals were a part of daily life and synthesized with the religion. Religions teachings, social and political norms, and economic policies treated man as a part of nature, not as a moulder or superior to it. Air, water, land, animals and human beings are the creations of one superior power – God. Therefore, the fundamental ethic of behavior was to live in harmony with each other.

In ancient Indian tradition, people have always valued mountains, rivers, forests, trees and several animals. Thus, much of the nature was venerated and protected. Forests have been associated with the names of forest gods and goddesses, both in the Hindu religion as well as in tribal cultures. 'Tree' goddesses have been associated with specific plant species. *Ficus religiosa*, the Peepal tree, is venerated and is not to be cut down. The Banyan tree in some regions, such as Maharashtra, is venerated once a year by tying a thread around it as a symbol of respect. The Tulsi plant is grown in every home.

Patches of forests, now called 'sacred groves', have been dedicated to a deity in many Indian cultures especially in tribal areas. These traditionally protected forest patches depict the true nature of undisturbed vegetation and have a large number of indigenous plant species as their exploitation has been controlled through local sentiments.

Certain species of trees have been protected as they are valued for their fruits or flowers. The mango tree is protected for its fruit around most farms even when wood becomes scarce. The Mahua tree (*Madhuca indica*) is protected by tribal people as it provides edible flowers, oil from its seeds and is used to make a potent alcohol. Many plants, shrubs and herbs are used in Indian medicines, which were once available in the wild in plenty. These are now rapidly vanishing. Many species of animals are venerated as being the 'vahan' or vehicle of different gods on which they are said to travel through the cosmos.

In Indian mythology, the elephant is associated with Ganesha. The elephant-headed Ganesha is also linked to the rat. Vishnu is associated with the eagle. Rama is linked to monkeys. In mythology, Hanuman, the monkey god, rendered invaluable help to Rama during his travels to Lanka. The Sun god, Surya, rides a horse and has a superb chariot on which he moves through the sky. The lion is linked to Durga and the blackbuck to the moon . The cow is associated with Krishna, the snake with Shiva and Vishnu, the swan with Saraswathi. Vishnu's incarnations have been represented as taking various animal forms which serially include a fish, a tortoise, a boar and a *dwarf*, and a half-man, half-lion form.

The associations to various plants that have been given a religious significance include Tulsi, which is linked to Lakshmi and Krishna. The Tulsi plant is also linked to the worship of one's own ancestors. The peepal tree is said to be the tree under which Buddha attained enlightenment. It is also associated with Vishnu and Krishna. Several trees are associated with the goddess Lakshmi, including Amalaki, Mango and the Tulsi shrub.

Traditions also held that these species, which were considered as an important aspect of nature, were the basis of local life-support systems and were integral to bring about a harmonious life. In traditional societies of the past, these examples were all a part of ethical values that protected nature. As modern science based on the exploitation on nature spread into India, many of these traditions began to lose their effectiveness as measures that led to conserving nature.

Environmental Ethics in Indian Constitution

The above mentioned philosophy of peaceful co-existence with nature also got mentioned, in one way or another, in the Indian Constitution. Two provisions relating to environment were incorporated in the Indian Constitution in the form of Articles 48A and 51A(g). Article 48 A is a constitutional pointer to the state to protect and improve the environment and Article 51A(g) confers a fundamental duty of the citizens of India to protect and improve the environment and have compassion for living creatures. This clearly shows that the Indian Parliament fell in line with old traditional values. The language used in these Articles clearly indicates that the principle of equity, co-existence, reverence for nature and non-violence have been given legal recognition. The use of the term 'protect and improve' implies the improvement of the natural environment as well as quality of life. Further, protection of the environment implicitly directs us not to cut trees, keep the water of rivers, lakes, etc. clean and wholesome. This is reminiscent of centuries old 'Chipko movement' of village Kherjarilli of Rajasthan where Amrita Bai, her four family members and other 359 persons sacrificed their lives to save trees of the village (popularly known as 'Green Khejris'). Thus, we have a culture where trees are regarded more precious and revered than our lives. The government has also declared the villages of Bishnois in Punjab and Rajasthan as 'reserved areas' and cutting of trees and killing of animals have been declared an offence in those territories. If one happens to visit these villages, he can see deer and other innocuous animals roaming freely and fearlessly.

The term 'to have compassion for living creatures' used in Article 51A(g) impliedly recognizes the principle that all creatures are made equal and that animal killing should be prohibited as taught by the principle of non-violence; or as the moral code of conduct says that 'killing of animals and birds is a sin of highest order'. Abiding by this philosophy of olden times, the Indian Parliament has also passed two statutes, viz. The Wild Life (Protection) Act, 1972 and the Cruelty Against Certain Animals Act, 1962. It is also in consonance with another ethical principle that unarmed enemy and mute birds and animals should not be harmed or attacked. This reminds us of an old story of Prince Siddhartha who saved a swan, who was shot at with an arrow by his brother. The same Prince Siddhartha later became Lord Buddha – a great preacher of non-violence (*ahimsa*) and harmonious relationship with nature. The object behind the theory of 'to have compassion for living creatures', scientifically speaking, is that animals and birds are the second tropic level of the 'food chain' or foodwebs and help in energy flow. The utility of the animals and their products need no special description here. It should not also be forgotten that wildlife is the guardian and protector of our forests. Decay of wildlife implies decay of forests. Thus, forests and wildlife are interdependent and complementary to each other.

Indian social and moral ethics, relating to environment protection and conservation, have been accepted by Indian judiciary time and again. Some of the judicial pronouncements upholding old Indian traditions concerning environment are quoted below:

> Water and rivers have dominated the destiny and fortune of man. Plentiful rivers have brought prosperity to those who lived on their banks. If Bhagirath brought salvation, Ganga sustains life ... ages have rolled up by it and it has remained eternal (*F.K. Husian* vs. *Union of India*, [AIR 1990] Ker 321).

> In order to arouse amongst the people the consciousness of cleanliness of environment, the Government of India and the governments of the states and union territories may consider the desirability of organizing "keep the city clean week", "keep the town clean week" and "keep the village clean week" ... at least once a year. During that week, the entire city, town and village should be kept as far as possible clean, tidy and free from pollution of land, water and air (*M.C. Mehta vs. Union of India*, [AIR 1988] SC 1115).

> Environmental law is based on the realization of mankind of the dire physical necessity to preserve these invaluable and none too easily replenishable gift of Mother Nature to man and his progeny from the reckless wastage and raparious appropriation that common law permits (*T. Damondhar Rao vs. Municipal Corporation, Hyderabad*, [AIR 1987] SC 171).

The essence of their teachings in the form of Vedas, Upanishads, Smritis and Dharmashastras was to protect nature and its elements, i.e. air, water, soil, land, vegetation and animals. In Indian culture, a worshipful attitude has been developed towards trees, earth, sky (*aakash*), air (*vayu*), water (*jal*), rivers, sun, mountains, etc. It was regarded a sacred duty of every person to protect them. The Hindu religion enshrined a respect for nature, environmental harmony and conservation. Many verses in the *Rigveda* and *Artharaveda* have been devoted to the praise of *Surya Devta* (Sun), *Vayu Devta* (Lord Air), *Agni Devta* (God of Fire), *Varuna Devta* (God of Water), *Prithvi Mata* (Mother Earth),

Vanya Devi (Goddess of Forests), etc. Therefore, cutting of trees and polluting air, water and land were regarded as sins. Protection of their purity was considered to be the duty of everyone. *Rigveda, Manusmriti,* and *Charak Sanhita* have emphasized on purity of water and its healing and medicinal value. The society, as a whole, has developed a code of conduct (*maryada*) to protect nature and live in co-existence with local and regional environment.

Environmental Legislations in India

Before independence, in India, some laws have been enacted for the protection of the environment. In Indian Penal Code of 1960, Articles 268, 290, 291, 426, 430, 431 and 432 are related to the environment. Similarly, Article 277 was related to water pollution and 278 to air pollution. In Motor Vehicle Act, 1938, there was a provision to control pollution and Indian Forest Act was passed in 1927. After independence, serious efforts have been made in the form of legislations for the conservation and protection of the environment. Originally, the Indian Constitution of 1950 did not have explicit reference to environment protection, so there was no independent and separate provision dealing with the protection or improvement of the environment. But, taking note of Stockholm Conference and growing awareness for environmental pollution and eco-imbalances, the Indian Parliament passed a historic amendment - 42nd Constitution Amendment Act, 1976. This Act incorporated two significant Articles - Articles 48A and 51A(g) - to protect and improve the environment. Further, it introduced certain changes in the seventh Schedule of the Constitution. These changes are as follows:

> *Article 48A*: 'The State shall endeavour to protect and improve the environment and to safeguard the forests and wildlife of the country.'
>
> *Article 511(g)*: 'It shall be a duty of every citizen of India ... to protect and improve the natural environment including forests, lakes and wildlife, and to have compassion for living creatures.'

The environmental laws, which have been passed by central and state legislatures, are based on the recognition of clean environment as a human right or fundamental right. As it has been recognized that clean environment is the basic need for the survival of humanity and it cannot be ensured without ecological balance, thus, this right belongs to all as survival of mankind depends on clean, healthful or pollution-free environment. Any attempt to defile, damage the natural environment would amount to violation of human right to clean environment. The Stockholm Conference of 1972 also declared that man has the fundamental right to freedom of equality and adequate conditions of life, in an environment of a quality that permits a life of dignity and well-being.

To meet these challenges to mankind, various measures have been adopted in India including legal measures. Many laws have been passed by the Indian Parliament and the state legislatures to contain the problem of air, water, land, radiation pollution and eco-imbalances. These laws include the Wildlife (Protection) Act, 1972, the Water (Prevention and Control of

Pollution) Act, 1974, the Water Prevention and Control of Pollution) Cess Act, 1977, the Forest (Conservation) Act, 1980, the Air (Prevention and Control of Pollution) Act, 1981 and the Environment (Protection) Act, 1986. A brief note on these laws provides a deep insight into the legislative measures taken in India regarding protection of environment.

Environment (Protection) Act, 1986

The Environment (Protection) Act, 1986 was passed by Indian Parliament in order to fulfil constitutional obligation as provided under Article 48A as well as to show its strong commitment toward Stockholm Declaration of 1972, in which it was suggested that all governments must evolve necessary laws to protect and improve the flora and fauna, non-renewable resources, wildlife and human health.

The objects and reasons as provided by the Act are as follows: Concern over the environment has grown the world over since the sixties. The decline in environmental quality has been evidenced by increasing pollution, loss of vegetal cover and biological diversity, excessive concentrations of harmful chemicals in the ambient atmosphere and in food chains, growing risks of environmental accidents and threats to life-support systems. The world community's resolve to protect and enhance the environmental quality, found expression in the decision taken at the United Nations Conference on Human Environment held in Stockholm in June 1972. The Government of India, participating in the conference, strongly voiced the environmental concerns. While several measures have been taken for environmental protection both before and after the conference, the need for a general legislation to further implement the decisions of the conference has become increasingly evident. Besides, existing laws generally focus on specific types of pollution or on specific categories of hazardous substances. Some major areas of environmental hazards are not covered. There also exists uncovered gaps in areas of major environmental hazards. There are inadequate linkages in handling matters of industrial and environmental safety. Control mechanisms to guard against slow, insidious build up of hazardous substances, especially new chemicals, in the environment are weak. Because of a multiplicity of regulatory agencies, there is need for an authority which can assume the lead role in studying, planning and implementing long-term requirements of environmental safety and to give direction to, and coordinate a system of speedy and adequate response to emergency situations threatening the environment.

The Environment (Protection) Act, 1986 is a special law that extends to the whole nation. It was passed in March 1986 and came into force on 19 November, 1986. The Act consists of 26 sections divided into four chapters. Seven schedules dealing with emission standards of air, noise, effluents, etc. have also been appended to it. Various rules, including the Environment (Protection) Rules, 1986, have also been provided with it. They are as follows:

1. Hazardous Wastes (Management and Handling) Rules, 1989.
2. Hazardous Microorganism Rules, 1989.

3. Manufacture, Storage and Import of Hazardous Chemicals Rules, 1989.
4. Chemical Accidents (Emergency Planning, Preparedness and Response), Rules 1996.
5. Bio-Medical Water (Management and Handling) Rules, 1998.
6. Municipal Solid Wastes (Management and Handling) Rules, 2000.
7. Recycled Plastics Manufacture and Usages Rules, 1999.
8. Noise Pollution (Regulation and Control) Rules, 2000.

According to Section 2(a) of the Act, environment includes: (a) water, air and land, and (b) the interrelationship which exists among and between (i) water, (ii) air, (iii) land, (iv) human beings, (v) living creatures, (vi) plants, (vii) microorganisms, and (viii) property. Thus, it includes animate and inanimate objects and their interrelationship.

Section 2(b): Environmental pollutant means 'any solid, liquid or gaseous substance present in such concentration as may be, or tend to be, injurious to environment'.

Section 2(c): 'Environmental pollution' means 'the presence in the environment of any environmental pollutant'.

Section 2(d): 'Handling', in relation to any substance, means the manufacture, processing, treatment, package, storage, transportation, use, collection, destruction, conversion, offering for sale, transfers or the like of such substance.

Section 3: Power of Central Government to take measures to protect and improve environment.

Section 4: It authorizes the Central Government to appoint offices with proper designations.

Section 5: It empowers the Central Government to issue directions to any person/officer/authority appointed under Section 4 of the Act.

Sections 7 to 17: These are related to prevention, control and abatement of environmental pollution.

The Act also provides power to Central Government to constitute a Board of Environment Protection. In fact, this Act is a very comprehensive document on environment protection in India. The other Acts on environment have also interconnection with 1986 Act.

Air (Prevention and Control of Pollution) Act, 1981

This Act was passed by the Indian Parliament in the exercise of its powers conferred under Article 253 of the Constitution. The aims and objects of the Act provide:

> Whereas decisions were taken at the United Nations Conference on Human Environment held at Stockholm in June 1972, in which India participated to take appropriate steps for the preservation of the natural resources of the earth which, among other things, include the preservation of the quality of air and control of air pollution.

And, whereas it is considered necessary to implement the decisions aforesaid in so far as they relate to the preservation of the quality of air and control of air pollution.

The Act aims to achieve the following goals:

1. to provide for the prevention, control and abatement of air pollution;
2. for the establishment of Board with a view to carry out the above mentioned purpose;
3. to confer on and assign to such Board powers and functions relating to prevention, control and abatement of air pollution and other matters connected thereto; and
4. to lay down the standards to maintain the quality of air.

This Act is applicable to the whole of India.

The term 'air pollution' means the presence in the atmosphere of any pollutant and 'air pollutant' means any solid, liquid or gaseous substance (including noise) present in the atmosphere in such concentration as may be or tend to be injurious to human beings or other living creatures or plants or property or environment. Thus, air pollutants include smoke, smoot, heat, fly ash, suspended particulate matter(SPM), noise, radioactive substances, etc.

The above definition emphasizes two things: (a) high concentration of solid, liquid or gaseous substances, which (b) is injurious or likely to be injurious to human beings, flora and fauna, property or environment. Some small quantity of gases, SPM or vapour is tolerable or cannot be designated as pollutants, unless and until the concentration of such things reaches such a volume that it becomes injurious. For example, sound at low pitch is acceptable and liked, but sound at a very high pitch (known as noise) is not acceptable and becomes a health hazard. The same is true of other gases, liquids and solid substances. A small quantity of pollutants usually does not affect human health adversely. Such quantity or volume may be described as permissible/tolerable limit as the nature also has its self-purification mechanism. But, if the volume or the quantity of the pollutants is such which is deleterious/injurious to the health of human beings, flora, fauna, etc. it becomes environmental pollution. In Taj Trapezium case, the apex court observed that emission of sulphur dioxide from coke/coal-using industries was causing acid rain (the sulphur dioxide when combined with moisture forms sulphuric acid called acid rain) which had a corroding effect on the gleaming white marble of the Taj Mahal. Therefore, 292 industries were ordered either to close down or to switch to using gas.

Sections 3 to 15 provide for the constitution of a board, qualifications of its members, terms and conditions of their service, meetings, vacation of seats and temporary association of persons with the board, etc.

The Act envisages two types of boards – one at the central level and the other in the respective states.

Water (Prevention and Control of Pollution) Act, 1974

The main aim and object of the Act of 1974 is to maintain or restore the wholesomeness of water and to prevent, control and abate water pollution. To achieve these objectives, the Act has provided various provisions, which are very comprehensive. In view of sub-section 2(e) read with Sections 17 and 18 of this Act, the fundamental objective of the statute is to provide clean water to citizens.

Section 2 of the Act has defined certain basic terms used in the Act. While defining water pollution it provides that:

> Pollution means such contamination of water or such alteration of physical, chemical or biological properties of water or such discharge of any sewage or trade effluent or of any other liquid, gaseous or solid substance into water (whether directly or indirectly) as may, or is likely to, create a nuisance or render such water harmful or injurious to public health or safety, or to domestic, commercial, industrial, agricultural or other legitimate uses, or to the life and health of animals or plants or of aquatic organism [Section 2(e)].

Thus, pollution means:

1. contamination of water, or
2. alteration of physical, chemical or biological properties of water, or
3. discharge of sewage or trade effluent, or
4. any other solid, liquid or gaseous substance, which may or is likely to create
   (i) nuisance, or
   (ii) render such water harmful or injurious to, public health or safety, or to domestic, commercial, industrial, agricultural or other legitimate uses.

Thus, it is a very comprehensive definition and covers all changes in physical, chemical or biological properties of water. The definition also covers the rise in the temperature of water and discharge of radioactive substances in the water.

Further, the Act has used two terms in relation to water pollution: stream and well. The 'stream' here includes (a) river, (b) watercourses (whether flowing or for the time being dry), (c) inland water (whether natural or artificial), (d) subterranean water (underground water), and (e) sea or tidal water.

Section 3 provides that a Board shall be appointed and constituted by the Central Government to perform the functions under the Act.

Environment (Protection) Act, 1986

The Indian Parliament enacted the Environment (Protection) Act, 1986 to prevent, control and abate environmental pollution which came into force on 19 November 1986 – the birth anniversary of the late Prime Minister, Indira Gandhi. Section 2 of the Act has defined various terms including 'environment', 'environmental pollutants' and 'environmental pollution' comprehensively, but it dose not specifically refer to noise. Only Section

6(2)(b) mentions the word 'noise' and provides that the government may make rules for allowable limits of environmental pollutants including noise for different areas. Rule 5 of the Environment (Protection) Rules, 1986 under clause (1)(ii) empowers the Central Government to make rules regarding the maximum allowable limits of various environmental pollutants (including noise) for an area. Thus, in the exercise of Sections 3, 6 and 25, read with Rule 5 of the Environment (Protection) Rules, 1986, the Central Government has notified Noise Pollution (Regulation and Control) Rules, 2000 on February 19 2000 for preventing adverse impact on human health, including harmful psychological effects.

Noise Pollution (Regulation and Control) Rules, 2000

Noise Pollution (Regulation and Control) Rules, 2000, framed by the Central Government under provisions of the Environment (Protection) Act, 1986 and read with Rule 5 of the Environment (Protection) Rules, 1986. Rule 3 provides for ambient air quality standards in respect of noise for different areas/zones as specified in the schedule annexed to the Rules, which is as under:

Schedule

Ambient Air Quality Standards in Respect of Noise

Area Code	*Category of Area/Zone*	*Limits in dB (A) Leq.*	
		Day-time	*Night-time*
(A)	Industrial Area	75	70
(B)	Commercial Area	65	55
(C)	Residential Area	55	45
(D)	Silence Zone	50	40

Notes: (1) Day-time shall mean from 6.00 a.m. to 10.00 p.m.
(2) Night-time shall mean from 10.00 p.m. to 6.00 a.m.
(3) Silence zone is defined as an area comprising not less than 100 metres around hospitals, educational institutions and courts. The silence zones are zones which are declared as such by the competent authority.
(4) Mixed categories of areas may be declared as one of the four above mentioned categories by the competent authority.

It is for the State Government to categorize the area into industrial, commercial, residential or silence zone and to take measures for abatement of noise including noise emanating from vehicular movements. Clause 4 provides that all development authorities, local bodies and other authorities concerned, while planning developmental activity or carrying out functions relating to town and country planning, shall take into consideration all aspects of noise pollution as a parameter of quality of life to avoid noise menace and to achieve the objective of maintaining the ambient air quality standards in respect of noise.

Rule 4: Responsibility as to enforcement of noise pollution control measures:

1. The noise level in any area/zone shall not exceed the ambient air quality standards in respect of noise as specified in the schedule.
2. The authority shall be responsible for the enforcement of noise pollution, control measures and the due compliance with the ambient air quality standards in respect of noise.

Rule 5: Restrictions on the use of loudspeakers/public address systems:

1. A loudspeaker or a public address system shall not be used except after obtaining written permission from the authority.
2. A loudspeaker or a public address system shall not be used at night (between 10.00 p.m. and 6.00 a.m.) except in closed premises for communication within, e.g. auditorium, conference rooms, community halls and banquet halls.

Rule 6: Consequences of any violation in silence zone/area:

Whoever, in any place covered under the silence zone/area, commits any of the following offences, shall be liable for penalty under the provisions of the Act:

1. whoever plays any music or uses any sound amplifiers,
2. whoever beats a drum or tom-tom or blows a horn either musical or pressure, or
3. whoever exhibits any mimetic, musical or other performances of a nature to attract crowds.

Rule 7: Complaints to be made to the authority:

1. A person may, if the noise level exceeds the ambient noise standards by 10dB (A) or more given in the corresponding columns against any area/zone, make a complaint to the authority.
2. The authority shall act on the complaint and take action against the violator in accordance with the provision of these rules and any other law in force.

Rule 8: Power to prohibit, etc. continuance of music, sound or noise:

1. If the authority is satisfied from the report of an officer in charge of a police station or other information received by him that it is necessary to do so in order to prevent annoyance, disturbance, discomfort or injury or risk of annoyance, disturbance, discomfort or injury to the public or to any person who dwell or occupy property in the vicinity, he may, by a written order, issue such directions as he may consider necessary to any person for preventing, prohibiting, controlling or regulating:
   (i) the incidence or continuance in or upon any premises of
      (a) any vocal or instrumental music,

(b) sounds caused by playing, beating, clashing, blowing or use in any manner whatsoever of any instrument including loudspeaker, public address systems, appliance or apparatus or contrivance which capable of producing or reproducing sound, or

(ii) the carrying on in or upon, any premises of any trade, avocation or operation or process resulting in or attended with noise.

2. The authority empowered under sub-rule (1) may, either on its own motion, or on the application of any person aggrieved by an order made under sub-rule (1), either rescind, modify or alter any such order.

Wildlife (Protection) Act, 1972

In 1972, the Indian Parliament passed a comprehensive national law – The Wildlife (Protection) Act, 1972 – with the sole aim of protecting wildlife. It not only prohibits hunting but also created protected areas and controls trade in wildlife products. To achieve these objectives, it has created a separate and independent authority to protect and improve wildlife. The Act has been accepted and adopted by all the states except Jammu and Kashmir.

The first and foremost purpose of this Act is to protect the habitats of wild animals. As a sequel to it, various national parks and game sanctuaries have been established to ensure greater protection to wildlife. Some special provisions also aim to preserve endangered species like Project Tiger, Gir Lion Sanctuary, Himalayan Musk Deer Project, etc. The Act consists of 60 Sections and VI Schedule – divided into Seven Chapters.

Wildlife, according to the 1972 Act, includes any animals, bees, butterflies, crustaceans, fish and moths, and aquatic or land vegetation which forms part of any habitat. This definition is wider in its connotation. The term 'wild animal' has also been defined as any animal found wild in nature and includes any animal specified in Schedule I, II, III, IV or V. Schedule I includes list of mammals (1 to 41-B), amphibians and reptiles (1 to 17-A), birds (1 to 18), crustacea and insects. It has been clarified that 'animal' includes amphibians, birds, mammals, and reptiles and their young, and in the case of birds and reptiles, their eggs.

Hunting of wild animals has been prohibited altogether. But, the chief wildlife warden has been authorized to grant permission to hunt animals under certain circumstances. These are as follows:

1. if he is satisfied that any wild animal specified in Schedule I has become (i) dangerous to human life; or (ii) is disabled; or (iii) diseased as to be beyond recovery; or
2. when the chief wildlife warden, or the authorized officer, is satisfied that any wild animal specified in Schedule II, III, or IV, has become dangerous to human life or to property, or is so disabled or diseased so as to be beyond recovery.

The killing in good faith of any wild animal 'in defence' of oneself or of any other person is not an offence.

The Act has also specified provisions to protect specified plants of sanctuaries, national parks, forests or areas specified by notification. All the specified plants, or part or derivation thereof, have been declared to be the property of the State Governments, and if they are a part of a sanctuary or national park, they shall be the property of the Central Government. Therefore, a person must have a licence from the Chief Wildlife Warden, or other authorized officer, to commence or carry on business or occupation as a dealer in specified plants.

The Act also prohibits to (a) wilfully pick, uproot, damage, destroy, acquire or collect, or (b) possess, sell, offer for sale or transfer by way of gift or otherwise, or transport any specified plant, whether dead or alive or part or derivative. But, this provision is not applicable to the member of a Scheduled Tribe.

Similarly, cultivation, dealing in, possession of specified plants without a licence has also been declared to be a punishable act. Further, no person shall purchase, receive or acquire any specified plant, or part or derivative otherwise than from a licensed dealer.

Laws Relating to Forest Protection

Forest or natural vegetation is an integral element of natural environment. It not only affects but controls the environment and its degradation. As stated earlier, due importance has been given to forests during ancient as well as medieval periods but during British rule, overexploitation of forests for commercial purposes had begun. But, soon, its importance has been realized and several acts/laws have been introduced. In Forest Act, 1865, the forest has been considered as a state government property. The Indian Forest Act, 1927 was enacted with the objective to protect the forests. Under this, forests have been classified into four categories, viz. reserved forests, village forests, protected forests and non-government private forests. This Act was a very elaborate one which deals not only with the use and conservation of forests but also empowers to punish those who are responsible for the unauthorized cutting of forests.

The Forest (Protection) Act, 1980 has been passed in the spirit of the agreement of 1972 Stockholm Conference on environment. This was further modified in 1988. This Act has given elaborate powers to Central Government for the protection and conservation of forests including biodiversity as well as also for the establishment of national parks in consultation with State Governments. In order to make this Act more effective, six regional centres at Bangalore, Bhopal, Bhubaneswar, Lucknow, Shillong and Chandigarh have been established. At present, governments (both at centre and in states) are very particular about the forest conservation and National Forest Policy was also declared in 1988.

Laws Relating to Hazardous Wastes Management

There were various laws, which directly or indirectly dealt with hazardous wastes and toxic substances, but they were not effective under the changing modern conditions. Therefore, the Central Government has passed important rules in exercise of powers conferred by the Environmental (Protection) Act, 1986. These rules are as follows:

1. Hazardous Wastes (Management and Handling) Rules, 1989.
2. Manufacture, Storage and Import of Hazardous Chemical Rules, 1989.
3. Hazardous Microorganisms Rules, 1989.
4. National Environment Tribunal Act, 1995.
5. National Environment Appellate Authority Act, 1997.
6. Bio-Medical Wastes (Management and Handling) Rules, 1998.
7. Recycled Plastics Manufacture and Uses Rules, 1999.
8. Municipal Solid Wastes (Management and Handling) Rules, 2000.

A brief review of the above rules is presented as under:

Hazardous Wastes (Management and Handling) Rules, 1989

These rules aim to deal with the problem of hazardous wastes comprehensively.

Although, the term 'hazardous wastes' has not been defined in the rules, but Rule 3(i) provides that 'hazardous wastes' means – waste substances which are generated in the processes as indicated in the schedule.

Rule 4 fixes the responsibility of the occupier and operator of the facility for proper collection, reception, treatment, storage and disposal of hazardous wastes without any adverse effect on the environment.

Rule 4-A describes the duties of the occupier and operator of a facility as follows:

1. contain contaminants and prevent accidents and limit their consequences on humans and the environment; and
2. provide persons working on the site with information, training and equipment necessary to ensure their safety.

'Proper authorization' by the State Pollution Control Board is really the control mechanism. This authorization letter or permit is issued to a man who is well versed in this technique and possesses proper and adequate facilities, technical capabilities and equipment to handle hazardous wastes safely (Rule 5).

Import and export of hazardous wastes have been completely prohibited for dumping and disposal. However, as an exception, such wastes may be permitted to be imported only as raw material for recycling or reuse (Rule 12). Such wastes can be used for processing and reuse as raw material only after procuring a proper permit from the State Pollution Control Board. In such a case, the exporting country should also inform and seek permission from the Central Government, which may be granted or refused by it. The occupier exporting or importing hazardous wastes from or to India shall comply with the Articles of the Basel Convention to which India is a signatory. Further, the Ministry of Environment and Forests shall be the nodal agency to deal with transboundary movement of hazardous wastes.

Manufacture, Storage and Import of Hazardous Chemical Rules, 1989

Hazardous chemicals have been dealt with separately under these rules of 1989. The following are the requirements which a chemical industry shall be required to fulfil before it starts working or in case of an existing industry within a period of ninety days of coming into operation of these rules, i.e. 27 November 1989. Following are the main provisions of these rules:

1. *Notification of sites*: No one can undertake any industrial activity which would produce a threshold quantity or more of an additional hazardous chemical unless he submits a written detailed report about the industry in accordance with Schedule 7 of the Rules at least three months before commencing the activity (Rule 7). Rules of 1989 have provided a list of 684 such hazardous chemicals. Therefore, an industry before its commencement is required to give the above notice.
2. *Safety report*: Before a chemical industry starts functioning, it has to prepare a safety report containing information specified in Schedule 8 at least ninety days before commencing that activity and in case of an existing industry within a period of six months after coming into operation of these rules (Rule 10). The report should be prepared with the help of experts not associated with such industrial activity.
3. *Preparation of on-site emergency plan by the occupier*: Under Rules 13, the occupier is required to prepare and keep up-to-date on-site emergency plan furnishing the details of how major accidents will be dealt with. This plan shall include the name of person responsible for safety and who can take action according to the plan. A mock drill based on the on-site emergency plan shall be conducted every six months.
4. *Preparation of off-site emergency plans*: Rule 14 makes it obligatory on the authorities concerned to prepare and keep up-to- date an adequate off-site emergency plan detailing how emergencies relating to a possible major accident on that site will be dealt with. Further, such plans should be prepared by the authority before a new industry starts its activity or in case of an existing industrial activity within six months of coming into operation of these rules (Rule 14). The authority concerned shall ensure the rehearsal of the off-site emergency plan at least once in a calendar year.
5. *Information to the persons liable to be affected*: One of the important features of the rules is to impose a mandatory duty on the occupier of the industry to take appropriate steps to inform persons outside the site who are likely to be affected by a major accident (Rule 15).

National Environment Tribunal Act, 1995

The Act has been passed with a view to provide for 'strict liability' for damages arising out of any accident occurring while handling any hazardous substances, and to dispose of such cases expediously. It has also recognized the principle of 'no-fault liability' in accidents occurring due to hazardous substances. It also provides heads under which compensation for damage may be claimed in such accidents.

Section 3 enunciates the principle of 'no-fault' of the owner to pay compensation for death, injury or damage to human beings and environment. Further, the claimant shall not be required to plead and establish that the death, injury or damage was due to any wrongful act, neglect or default of any person.

The Tribunal will hear the applicant and would be guided by the principles of natural justice and shall also have the power of the civil court as provided under the Civil Procedure Code of 1908 (Section 5). Both the parties shall be heard before an injunction is issued by the Tribunal.

The Central Government shall establish the National Environment Tribunal consisting of one chairperson, vice-chairperson, judicial members and technical members; and the bench shall consist of one judicial member and one technical member each (Section 9). The chairperson either must be or has been a judge of the Supreme Court or a High Court; or has for at least two years held the office of vice-chairperson, who will also be a judicial person (Section 10).

National Environment Appellate Authority Act, 1997

Recently, the Government of India has established a National Environment Appellate Authority which is chaired by a retired Supreme Court judge. It is aimed to hear appeals with respect to restriction of areas in which any industry process/operation of industry shall not be carried out, subject to safeguards provided by the Environment (Protection) Act, 1986. It is a type of check on the establishment of the industries and their likely fall-outs.

Bio-Medical Wastes (Management and Handling) Rules, 1989

These rules are applicable to all those persons who generate, collect, receive, store, transport, treat, dispose, or handle bio-medical wastes in any form. Bio-medical waste has been defined as 'any waste, which is generated during the diagnosis, treatment or immunization of human beings or animals or in research activities pertaining thereto or in the production or testing of biologicals including categories mentioned in Schedule I'. Schedule I has mentioned 10 categories of bio-medical wastes, which include human anatomical waste, animal waste, microbiology and biotechnology waste, waste sharps, discarded medicines and cytotoxic drugs, solid waste, chemical waste, incineration ash, etc.

Municipal Solid Wastes (Management and Handling) Rules, 2000

It consists of nine rules and four schedules. Rule 3 (xv) defines 'municipal solid waste' as follows:

> '"Municipal solid waste" includes commercial and residential wastes generated in a municipal or notified area in either solid or semi-solid form including industrial hazardous wastes *but* excluding treated bio-medical wastes.'

Rule 3 has also defined various terms used in the rules such as anaerobic digestion, bio-degradable substance, bio-methanation, disposal, land-filling, recycling, leachate segregation, vadose water, vermicomposing, etc.

Rule 4 declares that every municipal authority shall be responsible for the implementation of the rules and for any infrastructural development for collection, storage, segregation, transportation, processing and disposal of municipal solid wastes. Overall responsibility has to be given to the Secretary of the Department of Urban Development of a State/Union Territory and the District Collector/Deputy Commissioner in particular to implement the provisions of the rules.

The Central Pollution Control Board and the State Pollution Control Boards have the responsibility of monitoring compliance with standards regarding groundwater, ambient air quality, leachates and compost quality including incineration standards as provided in Schedules II, III and IV of the rules of 2000. Further, the municipal authority will apply for the setting up of waste processing and disposal facility including landfills, to the State Pollution Control Board, which shall issue the authorization within forty-five days for a given period. Such authorization shall be valid for one year only.

The municipal solid wastes shall be managed and handled with the compliance criteria and procedures laid down in Schedule II. The Schedule provides parameters of (i) collection of municipal solid wastes, (ii) segregation, (iii) storage, (iv) transportation, (v) processing, and (vi) disposal of municipal solid wastes, and their compliance criteria. Schedule III provides for specifications for landfill sites and Schedule IV lays down the standards for composting, treated leachates and incineration.

Other Acts

Apart from the above mentioned Acts directly related with environment, there are other Acts and Rules also, which are applicable in certain cases. These are:

1. Atomic Energy Act, 1962.
2. Radiation (Protection) Rules, 1971.
3. Insecticides Act, 1968.
4. Indian Fisheries Act, 1897.
5. Prevention of Food Adulteration Act, 1954.
6. Mines Act, 1952.
7. Mines and Minerals (Regulation and Development) Act, 1957.
8. Motor Vehicles Act, 1988.

In fact, proper legislative measures are existing in India for the protection and conservation of environment. There is a need for proper monitoring and implementation of rules and regulations so that our environment becomes healthy and the quality of life of the people of India improves.

Reference

Carter, Alan (2011), Environmental Ethics, in Tony Fitzpatric (ed) *Understanding the Environment and Social Policy,* Policy Press, Bristol, pp. 111–130.

Glossary

acid rain A condition in which natural precipitation becomes acidic after reacting chemically with pollutants in the air.

adaptation Measures to ensure that organisms not limited to humans can adopt to the global warming that is already unavoidable.

aerosol A substance consisting of small particles, typically having diameters that range from 1/100 micrometer to 1 micrometre.

air pollution The deterioration of the quality of air that results from the addition of impurities.

albedo A measure of the reflectivity of a surface, measured as the ratio of light reflected to light received. A mirror or bright snowy surface has a high albedo, whereas a rough, flat road surface has a low albedo.

arithmetic growth In population studies, growth characterized by the addition of a constant number of individuals during a unit interval of time. For instance, if there are x individuals in year O and x + a in year 1, arithmetic growth implies x + na in year n.

atmosphere The predominantly gaseous envelope that surrounds the earth.

atom The fundamental unit of an element.

atomic nucleus The small positive central portion of the atom that contains its protons and neutrons.

atomic number The number of protons in an atomic nucleus.

autotroph An organism that obtains its energy from the sun, as opposed to a heterotroph, which is an organism that obtains its energy from the tissue of other organisms. Most plants are autotrophs.

background radiation The level of radiation on earth from natural sources.

benthic organism A plant or animal that lives at or near the bottom of a body of water.

biodegradable Refers to substances that can readily be decomposed by living organisms.

biogas Methane produced by fermentation of organic waste. Biogas is chemically identical to methane produced in any other manner.

biogeochemical cycles The transport of elements between rock, water, air, and living organism in a circular fashion that periodically returns them to each stage.

biomass The total weight of all the living organisms in a given system.

biome A group of ecosystems characterized by similar vegetation and climate that are collectively recognizable as a single large community unit. Examples include the Arctic Tundra, the North American Prairie, and the tropical rain forest.

biosphere That part of the earth and its atmosphere that can support life.

birth cohort A group of individuals born in a given period, such as in a particular year.

birth rate The number of individuals born during some period, usually a year, divided by an appropriate population. For example, the crude birth rate in human populations is the number of live children born during a given year divided by the mid-year population of that year.

black lung disease A series of debilitating and often fatal diseases that affect the lungs of miners who work in underground coal mines.

bloom A rapid and often unpredictable growth of a single species in an ecosystem.

branching chain reaction A chain reaction in which each step produces more than one succeeding step.

breeder reactor A nuclear reactor that produces more fissionable material than it consumes.

cancer An abnormal, unorganized, and unregulated growth of cells in an organism.

carnivore An animal that eats the flesh of other animals.

carrying capacity The maximum number of individuals of a given species that can be supported by a particular environment.

chain reaction A reaction that proceeds in a series of steps, each step being made possible by the preceding one. See also branching chain reaction.

chemical bond A linkage that holds atoms together to form molecules.

chemical energy The energy that is absorbed when chemical bonds are broken or released when they are formed.

clear-cutting The practice of cutting all the trees in a designated area. Clear-cutting leaves behind barren, open regions but is sometimes the cheapest way to harvest timber if only short-term economic evaluation is used. See selective cutting.

climate The composite pattern of weather conditions that can be expected in a given region. Climate refers to yearly cycles of temperature, wind, rainfall, and so on and not to daily variations.

climax community A natural system that represents the end, or apex, of an ecological succession.

common mode failure An accident in which a single event causes multiple failures and thereby knocks out redundant systems. See redundancy.

competition An interaction in which two or more organisms try to gain control of a limited resource.

composting The controlled, accelerated biodegradation of moist organic matter to form a humus-like product that can be used as a fertilizer or soil conditioner.

cooling pond A lake or pond used to cool water from an electric generating station or any other industrial facility.

cooling tower A large tower-like structure used to cool water from an electric generating station or any other industrial facility.

critical condition In nuclear science, a critical condition is said to exist when a chain reaction continues at a steady rate, neither accelerating nor slowing down.

critical minimum size In general, a critical condition relates to a point at which some property changes very abruptly in response to a small change in some other property of the system. In ecology, a population is said to be reduced to its critical minimum size when its numbers are so few that it is in acute danger of extinction.

crude rate A vital rate with the entire population of some area as the denominator.

death rate The number of individuals dying during some period, usually a year, divided by an appropriate populations. For example, the crude death rate in human populations is the number of deaths during a given year divided by the mid-year population of that year.

decibel (*dB*) A unit of sound intensity equal to 1/10 of a bel. The decibel scale is a logarithmic scale used in measuring sound intensities relative to the intensity of the faintest audible sound.

demographic transition The pattern of change in vital rates typical of a developing society. The process can be outlined briefly as follows: Birth and death rates in a pre-industrial society are typically very high; consequently, population growth is very slow. Introduction or development of modern medicine causes a decline in death rates and hence a rapid increase in population growth. Finally, birth rates fall, and the population grows slowly once more.

demography The branch of sociology or anthropology that deals with the statistical characteristics of human populations, with reference to total size, density, number of deaths, births, migrations, marriages, prevalence of disease, and so forth.

desert A climate system in which rainfall is less than 25 centimetres per year. These systems are barren and support relatively little plant or animal life.

desertification The diminution or destruction of the biological potential of the land, which can ultimately lead to desert-like conditions.

doubling time The time a population takes to double in size, or the time it would take to double if its annual growth rate were to remain constant.

ecological niche The description of the unique functions and habitats of an organism in an ecosystem.

ecology The study of the interrelationships among plants and animals and the interaction between living organisms and their physical environment.

economic externality That portion of the cost of a product that is not accounted for by the manufacturer but is borne by some other sector of society.

ecology the science of the relations of all organisms and their environment.

ecosystem A group of plants and animals occurring together plus that part of the physical environment with which they interact. An ecosystem is defined to be nearly self-contained, so that the matter that flows into and out of it is small compared with the quantities that are internally recycled in a continuous exchange of the essentials of life.

ecosystem homeostasis The control mechanisms within an ecosystem that act to maintain constancy by opposing externals.

entropy A thermodynamic measure of disorder. It has been observed that the entropy of an undisturbed system always increases during any spontaneous process; that is, the degree of disorder always increases.

environment The sum total of all conditions, agencies and influences which affect the development, growth, life and death of an organism, species or race.

environmental resistance The sum of various pressures, such as prediction, competition, adverse weather, and so on, that collectively inhibit the potential growth of every species.

epidemiology The study of the distribution and determination of health and its disorders.

estuary A partially enclosed shallow body of water with access to the open sea and usually a supply of fresh water from the land. Estuaries are less salty than the open ocean but are affected by tides and, to a lesser extent, by wave action of the sea.

euphotic zone The surface volume of water in the ocean or a deep lake that receives sufficient light to support photosynthesis.

eutrophication The enrichment of a body of water with nutrients, with the consequent deterioration of its quality for human purposes.

expectation of life The number of years an infant can be expected to live under a specified schedule of death rates.

extrapolation The prediction of points on a graph outside the range of observation.

evolution, theory of A theory that states that species are not unchangeable but arise by descent and modification from pre-existing species.

fire climax A condition in which the continuance of a given ecosystem is maintained by fire.

fission (*of atomic nuclei*) The splitting of atomic nuclei into approximately equal fragments.

food chain An idealized pattern of flow of energy in a natural ecosystem. In the classic food chain, plants are eaten only by primary consumers; primary consumers are eaten only by secondary consumers; secondary consumers only by tertiary consumers, and so forth.

food web The actual pattern of food consumption in a natural ecosystem. A given organism may obtain nourishment from many different trophic levels and thus give rise to a complex, interwoven series of energy transfers.

fusion (*of atomic nuclei*) The combination of nuclei of light elements (particularly hydrogen) to form heavier nuclei.

Gaia The ancient Greek goddess of the earth. This word has recently been used to describe the biosphere and to emphasize the interdependence of the earth's ecosystems by likening the entire biosphere to a single living organism.

gas A state of matter that consists of molecules that are moving independently of each other in random patterns.

gasohol A motor fuel consisting of approximately 90 per cent gasoline and 10 per cent ethyl alcohol.

genetic engineering The process of artificially removing specific genes from one organism and replacing them with genetic information from another.

geo-metric growth In population studies, growth such that in each unit of time, the population increases by a constant proportion. Also called exponential growth.

geothermal energy Energy derived from the heat of the earth's interior.

Gloger's rule A rule, proposed by C.W.L. Gloger, stating that animals living in warm climates are more darkly coloured than closely related animals living in cold climates.

green revolution The realization of increased crop yields in many areas due to the development of new high-yielding strains of wheat, rice, and other gains in the 1960s.

green-house effect The effect produced by certain gases, such as carbon dioxide or water vapour, that cause a warming of the earth's atmosphere by absorption of infra-red radiations. The term is an inappropriate analogy to green-houses, which were once thought to keep themselves warm by admitting sunlight but retaining infra-red radiations. However, it has been shown that most of the heat retention in green-houses results from the conservation of warm air. The atmosphere, of course, is open, so the mechanism is not the same.

gross national product (*GNP*) The total value of all goods and services produced by the economy in a given year.

half-life (*of a radioactive substance*) The time required for half of a sample of radioactive matter to decompose.

hard energy system An energy system based on large centralized electric-generating facilities. See also soft energy system.

hazardous waste The material that poses a threat to human health and the environment.

heat A form of energy. Every object contains heat energy in an amount that depends on its mass, its temperature, and the specific heat of the materials of which it consists. Heat can also be generated by chemical reactions such as combustion.

heat engine A mechanical device that converts heat to work.

heavy oil Natural petroleum deposits that are too thick to be extracted by conventional pumping techniques. Heavy oils are found in 'dry' conventional wells and also in oil shales and tar sands.

hectare A metric measure of surface area. One hectare is equal to 10,000 square metres or 2.47 acres.

herbicide A chemical used to control unwanted plants.

herbivore An animal that eats only plant matter.

heterotroph An organism that obtains its energy by consuming the tissues of other organisms. See also autotroph.

home range The area in which an animal generally travels and gathers its foods.

hormones These compounds can be used as insecticides, for if sprayed properly, they will disrupt the life cycle of specific insects' larvae and kill them.

humus The complex mixture of decayed organic matter that is an integral part of healthy soil.

hydroelectric power Power derived from the energy of falling water.

hydrological cycle (*water cycle*) The cycling of water, in all its forms, on the earth.

injunction A legal order requiring a defendant to stop doing a wrongful act.

innate capacity for increase The maximum rate of growth of a population as determined by the reproductive capacity of the organism.

inversion A meteorological condition in which the lower layers of air are cooler than those at higher altitudes. This cool air remains relatively stagnant and causes a concentration of air pollutants and unhealthy conditions in congested urban regions.

isotopes Atoms of the same element that have different mass numbers.

joule A fundamental unit of energy, 4.184 joules = 1 calorie.

kinetic energy Energy of motion.

krill Small, shrimp-like crustacean that grows in large numbers in the cold waters of the southern oceans. Krill represents a primary food supply for many species of whales.

legume Any plant of the family Leguminosae, such as peas, beans, or alfalfa. Bacteria living on the roots of legumes change atmospheric nitrogen, N_2, to nitrogen-containing salts that can be readily assimilated by most plants.

mass number The sum of the number of protons and neutrons in an atomic nucleus.

megalopolis An agglomeration of several smaller cities and towns into a single expanded urban area.

meteorology The science of the earth's atmosphere.

mine spoil The earth and rock removed from a mine and discarded because the mineral or fuel content is too low to warrant extraction.

mineral reserves The estimated supply of ore in the ground.

mitigation To slow the rate at which global warming increases so that the rate of growth can eventually be haulted and reversed.

molecule The fundamental particle that characterizes a compound. It consists of a group of atoms held together by chemical bonds.

mutagen A substance that causes mutations. An inheritable change in the genetic material of an individual.

mutation A change in the structure or amount of gerietic material.

mutualism An interaction beneficial and necessary to both interacting species.

natural selection A series of events occurring in natural ecosystems that eliminate some members of a population and spares those individuals endowed with certain characteristics that are favourable for reproduction.

neutron A fundamental particle of the atom, that is, electrically neutral.

nodules Small pieces of rocks, typically having a mass of about 1 kg, that are found on the floor of the ocean. These nodules are rich in manganese and also contain several other minerals such as copper, iron, aluminum and cobalt.

noise Unwanted sound.

nuclear reactor A device that utilizes nuclear reactions to produce useful energy.

omnivore An organism that eats both plant and animal tissues. Common omnivores include bears, pigs, rats, chickens and people.

open dump A site where solid waste is deposited on a land surface with little or no treatment.

ore A rock mixture that contains enough valuable minerals to be mined profitably with currently available technology.

organic farming A set of farming practices that uses no synthetic fertilizers or chemicals.

organochlorides A class of organic chemicals that contain chlorine bonded within the molecule. Some organochlorides, such as DDT, are effective pesticides. They are generally broad-spectrum and long-lived in the environment.

organophosphates A class of organic compounds that contain phosphorus and oxygen bonded within the molecule. Some organophosphates, used as pesticides, are broad-spectrum and extremely poisonous, although they are not long-lasting in the environment.

Ozone Triatomic oxygen, O_3.

Ozone layer A region of the stratosphere, at 20–25 km, where the concentration of Ozone (O_3) is higher than elsewhere.

parasitism A special case of predation in which the predator is much smaller than the victim and obtains its nourishment by consuming the tissue or food supply of a larger living organism.

photosynthesis The process by which chlorophyll bearing plants use energy from the sun to convert carbon dioxide and water to sugars.

phytoplankton Any microscopic, or nearly microscopic, free-floating autotrophic plant in a body of water. There are a great many different species that exist in a community of phytoplanktons; these plants occur in large numbers and account for most of the primary production in deep bodies of water.

plankton Any small, free-floating organism living in a body of water.

pollution The impairment of the quality of some portion of the environment by the addition of harmful impurities.

polychlorinated biphenyl (PCB) A class of organochloride chemicals that are structurally related to DDT. They were used widely in the plastics and electrical industries until they were found to be potent environmental poisons.

population The breeding group to which an organism belongs in practice. A population is generally very much smaller than an entire species, for all the members of a species are seldom in close proximity to each other.

population distribution The composition of a population categorized by several variables, often age and sex.

population ecology The branch of ecology dealing with the size, growth, and distribution of populations of organisms.

power The amount of energy delivered in a given time interval. Power = energy/time.

prairie An extensive area of fairly level, predominantly treeless land. Prairies are characterized by an abundance of various types of grasses.

predation An interaction in which some individuals eat others.

predator An animal that attacks, kills and eats other animals.

primary consumer An animal that eats plants.

radioactivity The emission of radiation by atomic nuclei.

radioisotope A radioactive isotope.

radiolysis A chemical change initiated by radiation.

rate of natural increase The difference between the crude birth and crude death rates. Also called the crude reproductive rate.

rate of population growth The increase in population over a given time divided by the initial population.

recycling The process whereby waste materials are re-used for the manufacture of new materials and goods.

redundancy Superabundance. In the context of safety systems, redundancy refers to the provision of a series of devices that duplicate each other's functions and that are programmed to go into operation in sequence if a preceding device in the series fails.

rendering The cooking of animal wastes such as fat, bones, feathers, and blood to yield tallow (used in the manufacture of soap) and high-protein animal feed.

replacement level The level of the total fertility rate that, if continued unchanged for at least a generation, would result in an eventual population growth of zero.

resource Any source of wealth or means of producing wealth.

respiration The process by which plants and animals release energy by the step-wise oxidation of organic molecules.

saltwater intrusion The movement of salt water from the ocean to terrestrial groundwater supplies that occurs when the water table in coastal areas is reduced.

sanitary landfill A site where solid waste is deposited on a land surface, compacted, and covered with dirt to reduce odours and prevent disease and fire.

savannah A type of tropical or sub-tropical prairie that is subject to seasonal patterns of rainfall. Savannahs are common in central Africa.

secondary consumer A predator that eats an animal that eats plants.

selective cutting The process of cutting selected trees in a stand of timber. See clear-cutting.

smog Smoky fog. The word is used loosely to describe visible air pollution.

soft energy system An energy system based on the use of small, home-sized, decentralized devices using renewable resources.

solar cell A semi-conductor device that converts sunlight directly into electrical energy.

solar collector A device designed to concentrate solar energy for a useful purpose.

solar design active systems: Heating systems that use a controlled flow of some substance; usually air or water, to collect, store, and transmit heat from the sun. *passive systems*: Construction systems that use the structure of a building itself to collect, store, and transmit heat from the sun.

solar energy Energy derived from the sun.

species A group of organisms that interbreed with other members of the group, but not with individuals outside the group.

species diversity The number of species present in a particular area or community.

strip mining Any mining operation that operates by removing the surface layers of soil and rock, thereby exposing the deposits of ore to be removed.

subsidence The settling of the surface of the ground as deep ground water or minerals are removed.

subsidies Economic incentives to induce polluters not to despoil the environment.

succession The sequence of changes through which an ecosystem passes during the course of time. *Primary succession* is a sequence that occurs when the terrain is initially lifeless, or almost so. *Secondary succession* is the series of community changes that takes place in disturbed areas previously occupied by a biological community synfuels An abbreviation for synthetic fuels, liquid fuels produced from oil shales, tar sands, or coal.

sustainable development The development that meets the needs of the present without compromising the ability of future generations to meet their own needs.

synthetic chemicals Compounds produced by chemical reactions performed in a laboratory, as opposed to those formed naturally.

technological fix An approach to solving a social problem, such as environmental degradation, by technical or engineering methods.

thermal pollution A change in the quality of an environment (usually an aquatic environment) caused by raising its temperature.

thermodynamics The science concerned with heat and work and the relationships between them. *First Law of Thermodynamics*: Energy cannot be created or destroyed. *Second Law of Thermodynamics*: It is impossible to derive mechanical work from any portion of matter by cooling it below the temperature of the coldest surrounding object.

threshold level The minimum dose of a toxic substance that causes harmful effects.

tidal energy Energy derived from the movement of the tides.

total fertility rate (*TFR*) The total number of infants a woman can be expected to bear during the course of her life if birth rates remain constant for at least one generation.

toxic substance Any substance whose physiological action is harmful to health.

trespass An illegal intrusion or invasion of other people's property.

trophic Pertaining to nourishment.

tropical rain forest A biome that occurs when high, fairly constant rainfall and temperature permit plants to grow rapidly throughout the year.

tumour An abnormal growth of unregulated cells.

tundra Arctic or mountainous areas that are too cold to support trees and are characterized by low mosses and grasses.

urban sprawl The growth of suburban areas and the expansion of cities to pre-empt increasingly large areas of land.

urbanization A demographic process characterized by movement of people from rural to urban settlements, from small towns to large cities, and from large cities to their suburbs.

vital event In demography, a birth, a death, a marriage, a termination of marriage, or a migration.

water pollution The deterioration of the quality of water that results from the addition of impurities.

water table The upper level of water in the zone of saturated subsurface soil and rock.

weathering The breaking down of rocks into smaller particles by physical and chemical processes.

Xerophyte A plant adopted to arid conditions.

zero population growth A condition in which the birth and death rates in a population are equal therefore, the size of the population remains constant.

Bibliography

Arvill, R. (1967) *Man and Environment-Crisis and the Strategy of Choice*, Penguin.

Bach, Wilfried (1972) *Atmospheric Pollution*, McGraw Hill, New York.

Bahuguna, S. (1990) Tehri-The Dam of Discontent, *Yojana*, Vol. 34 (10), pp. 9–12.

Billihgs, W.D. (1971) *Plant, Man and the Ecosystem*, Macmillan.

Blair, A. and D. Hitchcock (1995) *Environment and Business*, Routledge, London.

Brij Gopal and N. Bhardwaj (1979) *Elements of Ecology*, Vikas Publications, New Delhi.

Brown, E.H. (1970) Man Shapes the Earth, *Geographical Journal*, Vol. 136, pp. 74–85.

Burton, Robert, W.K. and Gilbert, F.W. (1978) *The Environment as Hazard*, Oxford University Press, New York.

Clapham, W.B. (1973) *Natural Ecosystems*, Macmillan, London.

Colinvaux, Paul. A. (1973) *Introduction to Ecology*, John Wiley and Sons, New York.

Darlingten, P.J., (1957) *Zoogeography*, John Wiley, New York.

Dasseman, R.F. (1976) *Environmental Conservation*, John Wiley, New York.

Detwyler, T.R. (1971) *Man's Impact on Environment*, McGraw-Hill, New York.

Dikshit, R.D. (1984) *Geography and the Teaching of the Environment*, Department of Geography, Poona, University, pp. 68–83.

Dogra, Bharat (1980) *Forest and People*, Himalaya Darshan Prakashan Samiti, Rishikesh.

Dowdeswell, W.H. (1984) *Ecology: Principles and Practice*, Heinemann, Oxford.

Dutt, N. and Dutt, P.K. (2005) *The Environment & its Problems*, Sarat Book Distributors, Kolkata.

Fitzpatrick, T. (ed.) (2011) *Understanding the Environment and Social Policy*, Policy press, University of Bristol, Bristol (UK).

Flanagan, D. (1970) *The Biosphere*, W.H. Freeman and (eds) Co., San Francisco.

Furley, P.A., and Newey, W.W. (1983) *Man and the Biosphere*, Butterworths, London.

Futehally, I. (1988) *Our Environment*, NBT, New Delhi.

Gerasimov, I.P. (1983) *Geography and Ecology*, Progress Publishers Moscow.

Glan, L.A. and Rhoda, L.M. (eds.) (1970) *Ecological Crisis Readings for Survival*, Harcourt Brace Jovanovich, New York.

Goudie, Andrew (1986) *The Human Impact on the Natural Environment*, Basil Blackwell, Oxford.

Haggett, P. (1972) *Geography: A Modern Synthesis*, Harper and Row, London.

Hewitt, K. and Burton, I. (1971) *The Hazardousness of a Place: A Regional Ecology of Demaging Events*, Research Paper No. 6, University of Toronto Department of Geography.

Hohenemser, C., Kates, R.W. and Slovle, P. (1983) 'The Nature of Technological Hazard', *Science*, 220, pp. 378–384.

Irving, R.M. and Priddle, G.B. (1971) *Crisis-Reading in Environmental Issues and Strategies*, Macmillan, London.

Jackas, G.V. and White, R.O. (1939) *The Rape of the Earth*, Fabev, London.

Kates, R.W. (1978) Risk Assesment of Environmental Hazard, ICSU/SCOPE, Report No. 8, Wiley.

Keith, Smith (1996) *Environmental Hazards*, Routledge, New York.

Kerski, J. and Ross, Simon (2005) *The Environment* , Hodder Arnold, London.

Khemani, L.T. et al. (1989), 'Spread of Acid Rain Over India', *Atomos Environ*, 23, pp. 757–762.

Kumar, V.K. (1982) *Kanpur City: A Study in Environmental Pollution*, Tara Book Agency, Varanasi.

Maddox, J. (1972) *The Doomsday Syndrome*, McGraw-Hill, New York.

Manivasakam, N. (1992) *Environmental Pollution*, N.B.T., New Delhi.

Marquita, K.H. (2007) *Understanding Environmental Pollution*, Cambridge University Press.

Mash, G.P. (1964) *Man and Nature (Physical Geography as Modified by Human Action)*, Charles Scriber, New York.

Maxwell, K.E. (1973) *Environment of Life*, Dickenson Publishers, Co., California.

Nicholson and Max (1970) *The Environmental Revolution*, McGraw-Hill, New York.

Odum, Eugene P. (1971) *Fundamentals of Ecology*, W. B. Saunders Co., London.

Park, C.C. (1980) *Ecology and Environmental Management*, Butterworths, London.

Petak, W.J. and Atkinson, A.A. (1982) *Natural Hazard Risk Assessment and Public Policy: Anticipating the Unexpected.* Springer-Verlag, New York.

Peter, C. and Susan, C. (1998) *The South, The North and the Environment*, Pinter, London.

Pfajjlin, J.R. and Ziegler, E.N. (eds) (1976) *Encyclopedia of Environmental Science and Engineering*, Vol. I and II Gordon and Breach Science Publishers, New York.

Rao, K.L. (1975) *India's Water Wealth*, Orient Longman, N. Delhi.

Raven, P.H., Berg, L.R. and Johnson, G.B. (1993) *Environment*, Sounders College Pub. Orlando.

Saraf, R.K. (eds) (1973) *Proceedings of Symposium on Environmental Pollution.* CEPHERI, Nagpur.

Seshadri, B. (1969) *The Twilight of India's Wild Life*, Oxford, London.

Singh, H.H. et al. (eds) (1986) *Geography and Environment*, Concept, New Delhi.

Singh, L.R. et al. (eds) (1983) *Environmental Management*, Allahabad Geog. Society, Allahabad.

Sharma, H.S. (eds) (1988) *Environmental Design and Development*, Scientific Pub., Jodhpur.

Sharma, P.D. (1988) *Elements of Ecology*, Rastogi, Meerut.

Singh, S.P. (1990) Tehri Dam-An Assessment and Vision, *Yojana*, Vol. 34 (10), pp. 5–8.

Singh, Savindra and Tiwari, R.C. (eds) (1989) *Geomorphology and Environment*, Allahabad Geog. Society, Allahabad.

Smith, G.H. (ed.) (1965) *Conservation of Natural Resources*, New York.

Stephen, T. (1996) *Environmental Management Plans Demystified*, Spon Press, London.

Stoddard, D.R. (1965) *Geography and the Ecological Approach*, Geography, Vol. 50, pp. 242–51.

Strahler, A.N. and Strahler, A.H. (1976) *Geography and Man's Environment*, John Wiley, New York.

Sutton, P.W. (2007) *The Environment*, Policy Press, Cambridge

Taylor, G. (1967) *Geography in the 20th Century*, Methuen, London.

Thomas, W.L. (1965) *Man's Role in Changing the Face of the Earth*, University of Chicago Press.

Turk, J. (1985) *Introduction to Environmental Studies*, C.B.S. College Publishers, Chicago.

UNESCO (1981) 'Biosphere Reserves', MAB Information System, Publication No. SC-81/WS/61.

UNESCO (1970) *Use and Conservation of Biosphere*, Natural Resources Research 10, Paris.

World Commission on Environment and Development (1987) *Our Common Future*, Oxford University Press, Oxford.

Zimmermann, E.W. (1951) *World Resources and Industries*, Harper and Row, London.

Index